Digital Logic and State Machine Design

DIGITAL LOGIC and STATE MACHINE DESIGN

THIRD EDITION · · · · · · · · · · · · · · · · · · ·

David J. Comer

Brigham Young University

OXFORD

UNIVERSITY PRESS

Oxford University Press

Oxford   New York
Athens Auckland Bangkok Bombay
Calcutta Cape Town Dar es Salaam
Delhi Florence Hong Kong Istanbul
Karachi Kuala Lumpur Madras Madrid
Melbourne Mexico City Nairobi Paris
Singapore Taipei Tokyo Toronto

and associated companies in
Berlin Ibadan

Copyright © 1995, 1990, 1984 by Oxford University Press, Inc.

Reprinted 2002

Published by Oxford University Press, Inc.,
198 Madison Avenue, New York, New York 10016

Visit our website at http://www.oup-usa.org

Oxford is a registered trademark of Oxford University Press.

ISBN 0-19-510723-3
CIP data available upon request.

Cover design ©John Wilkes/Photonica

9 8 7 6 5 4 3 2

Printed in India by Gopsons Papers Ltd.

Preface to Third Edition

This book is written to fill a need for a concise, practical book in the important area of digital system design at the undergraduate level. The primary goal of the textbook is to demonstrate that sequential circuits can be designed using state machine techniques. These methods apply to sequential circuit design as efficiently as Boolean algebra and Karnaugh mapping methods apply to combinatorial design. Once the techniques are presented, the more important task of designing digital systems is pursued. This task consists of producing the schematic or block diagram of the system based on nothing more than a given set of specifications. The design serves as the basis for the construction of the actual hardware system.

Three significant changes have been made in this third edition. The first is that state machines are introduced earlier than in the second edition. The second is the addition of an entire chapter on programmable logic devices. Older state machine architectures have been deleted to make room for this topic. The third change is the inclusion of a chapter on computer organization.

Since an understanding of binary number systems and combinatorial logic circuits is prerequisite to a discussion of sequential circuits, these topics are treated in the first three chapters. Chapter 4 introduces the basic element in state machine design, the bistable flip-flop. Chapters 5 through 7 are dedicated to sequential circuit design. Some minor topics have been removed or relegated to the Appendix so that the fundamentals of state machine design can be emphasized earlier in the text.

Chapter 8 discusses programmable logic along with some rather recent developments in this area. Chapter 9 has been added to discuss the basic organization of computer systems. Earlier chapters consider sequential circuits in noncomputing systems. Chapter 9 allows the reader to see how these controllers might apply to the design of simple computers. Chapter 10 is essentially identical to previous editions in its coverage of asynchronous sequential circuits.

I am grateful to the reviewers that offered suggestions and pointed out errors in the manuscript. These reviewers include James J. Carroll, Clarkson University; C.W. Caldwell, University of Arkansas; Dwight Gordon, Kansas State University; Paul Gray, University of Wisconsin; David Mundie, Memphis State University; Carol Qiao Tong, University of Colorado. I was also greatly assisted by Professors Thomas Jannett of the University of Alabama and Gene Ware of Brigham Young University and wish to thank them for their considerable effort.

I am always grateful for the many students that find the field of electrical or computer engineering as interesting as I have. Without them there would be no reason to write such a book.

David J. Comer
Provo, Utah
November 1994

Contents

Introduction to Digital Systems

1.1
OVERVIEW

The digital systems area is an exciting and important branch of the electronics field and is becoming more important every year. Few other areas affect the lives of so many people as the products of this area. A list of fields that rely heavily on digital systems includes computing, communications, entertainment, automatic control of mechanisms, and instrumentation.

One characteristic of the digital systems area is its rapid change. New developments take place yearly to advance the capability of several fields. While this area can lead to a challenging career, a digital systems engineer must continue to learn new concepts throughout his or her professional career to remain productive. The following paragraphs expand on some of the fields that have changed dramatically as a result of developments in the digital systems area.

Computing: The dominant digital system is the computer. The obvious significance of this remarkable piece of electronic hardware often overshadows many other noteworthy digital systems. Two products that typify the rapid technological advancement in computing systems are the personal computer and the handheld calculator. These devices did not even exist in 1970. In the early 1960s, the IBM 7090 represented the classic mainframe computer [1]. This transistorized computer had a speed comparable to the IBM-AT personal computer and offered about 150 kilobytes of main memory. The purchase price was $1,600,000 with a monthly rental fee of $30,000. Less than 30 years later personal computers were available for 1/1000 of the 7090 purchase price. These computers may store 500 times more data in main memory, perform operations 20 times faster, and occupy one-tenth the space required for the 7090. Advancements in digital systems, primarily the development of integrated circuits, carried the computer from the expensive scientific or business installation to the home.

In 1965, a computer company assigned an engineer to evaluate the feasibility of building an electronic desktop calculator to compete with the leading desktop electromechanical calculator that sold for $1400. After a thorough study of this task, the engineer concluded that a desktop calculator using the first generation of integrated circuits would sell for about $5000. The company decided for obvious reasons not to pursue this project. Just six years later, Hewlett-Packard introduced the first handheld electronic calculator. This device was much quicker and performed far more calculations than the electromechanical calculator. It sold for about $300 and occupied the palm of a hand rather than a desk top. Again, the integrated circuit made such an advancement possible. Improvement in the fabrication of integrated circuits has led to the continual decrease in prices for computers and calculators. While very few fields create products that sell for less as time passes, products based on digital electronics generally decrease in cost each year.

In addition to computers and calculators, digital systems are also used in equipment related to computing applications. These devices are called computer peripherals. Devices such as CRT monitors and keyboards interface with computers by digital methods. Dot matrix and laser printers are driven by digital circuits. Disk drives are controlled and data is exchanged between computer and disk using digital schemes.

Communications: This is also an exceptionally significant field that applies digital methods to many areas. In earlier days, analog methods were used exclusively to transmit signals over phone lines. In the 1960s, AT&T began converting analog systems to digital to improve the quality of transmission and to operate more efficiently. Present-day telephone systems transmit several simultaneous long-distance conversations over the same pair of wires or fiberoptic link using digital pulse-code modulation techniques.

Facsimile (FAX) communications are based on digital transmission methods as are transmissions of information between modem and computer. The cellular telephone industry was originally based on analog signal transmission, but now must apply digital methods to handle the demand. Cellular phones and FAXs now have a substantial influence on modern business procedures.

Scientific data from outer space explorations are transmitted back to earth using digital transmission methods to minimize the adverse effects of noise. Visual information is transmitted from one point to another using digital methods to pack more information into fewer data bits using data compression methods.

Entertainment: The compact disk or CD player offers the highest quality sound available in commercial sound reproduction equipment. This device, introduced in 1983 by both SONY and Phillips, is based on digital representation of sound. It is more immune to noise than any analog recording method.

Recovery of music from old records and tapes is done by digital methods that eliminate much of the noise and interference. Digital techniques will be very prominent in the new High Definition Television (HDTV) systems that promise to greatly improve present TV quality.

Television commercials and programs are now edited with digital techniques under computer control. Studio programming is largely controlled by digital control circuits.

Automatic Control: The field of automatic control of mechanisms is also an important field. Digital automatic control schemes are now replacing older analog methods. Such systems are used in a wide variety of applications ranging from controlling the precise movement of positioning equipment involved in integrated circuit fabrication to controlling the path of the Apollo or Challenger space vehicles.

The guidance systems of the SCUD and Patriot missiles used in the Gulf War were implemented by digital circuits. The missiles launched from an airplane use digital automatic control theory to hone in on targets. Most rapid transit systems are now controlled by digital methods.

Instrumentation: This important field has been dominated by analog methods for many years. Recently, digital instruments have been replacing the older analog instruments. The digital oscilloscope used on the electronic workbench outperforms the older analog counterpart in many tasks. The digital voltmeter replaced analog meters in the 1970s. The logic analyzer that assists in the development of computer and other digital systems is itself a digital instrument. Medical equipment has applied digital methods in instruments such as MRI and CAT scan devices. Medical instruments are also following the trend of conversion to digital methods to offer more accurate and visible readouts of measurements.

These few important areas to which digital methods are applied have been cited to demonstrate the significance and the diversity of digital systems. For those who will work in these exciting areas, it is important to understand the fundamental concepts of digital design. The goal of this textbook is to cover these concepts at a level that would allow entry into any digitally based field. The coverage is not directed specifically toward any particular field such as computer design, but provides the necessary framework to allow the reader to progress in any digital systems area.

1.2

THE INTEGRATED CIRCUIT

More than any other electronic device, the integrated circuit is responsible for the rapid development of the digital systems area. The minimization of size and cost of circuit components allows quite complex systems to be implemented inexpensively and in a small space.

The early computers of the 1940s used relays or vacuum tubes and were very large, inefficient, and unreliable. An example of this is the ENIAC (electronic numerical integrator and computer) developed at the University of Pennsylvania between the years 1943 and 1946. This pioneering computer contained over 18,000 vacuum tubes, weighed over 30 tons, and occupied a 30-by-50-foot room [2]. Not only was this system large and expensive, but also the failure rate of the vacuum tubes was very high with 19,000 replacement tubes required in 1952. The mean time between failures was measured in minutes for this system.

The bipolar transistor was invented in 1948 and led to the second generation of digital computers. The transistor was much smaller, required less power to operate, was less expensive, and was more reliable than the vacuum tube. The IBM 7090 discussed in the preceding section was representative of the large computers based on the transistor. Large and expensive, the second generation of computers primarily served big business organizations and large scientific laboratories. A small business or individual scientist or engineer could not afford access to these mainframe computers.

The effect of the bipolar transistor on the electronics field was considered to be significant enough to win the inventors, Shockley, Bardeen, and Brattain, a Nobel prize. Perhaps a more important result of the transistor is that it laid the groundwork for the invention of the integrated circuit.

The integrated circuit consists of a silicon chip that can contain many transistors, resistors, and capacitors within the chip with thin metal interconnections deposited on top of the chip. Entire circuits are fabricated on a small silicon chip that is encased in a package with metal pins to provide connections from the internal chip to external circuits. The first integrated circuits began to appear commercially in the mid-1960s. While the decreased space requirements over discrete transistor circuits were obvious, no one fully realized how great an advantage this would become in future integrated circuits.

Over the next 30 years, the integrated circuit capability increased from small-scale integration to medium-scale integration to large-scale integration to very-large-scale integration and finally to ultra-large-scale integration. In the mid-1990s, silicon computer or memory integrated circuits were capable of containing over 10,000,000 transistors on a single chip. These high-density circuits are based on the MOSFET, a transistor that is quite different from the bipolar device, but requires less space to fabricate. As fabrication processes improved from the capability of constructing 20 transistors on a chip to the capability of constructing several million transistors on a chip, the price of digital systems dropped by several orders of magnitude. In addition, the capability of digital systems increased very dramatically.

The history of the computer and especially that of the personal computer is intertwined with the history of the integrated circuit. With the advent of the large-scale integrated circuit, the possibility of a small personal computer became a reality. The decade of the 1970s brought the personal computer from conception to a rather sophisticated digital system. The 1980s brought continual improvement in capability of the personal computer and the more advanced workstation to the point that large mainframe computers were being challenged in performance.

This textbook will discuss the use of small-scale and medium-scale integrated circuits to demonstrate the concepts involved in digital logic design. While these concepts could be divorced from a consideration of the type of circuits used for implementation, some practical considerations must be incorporated into design procedures. This will be done by occasionally referring to a particular type of integrated circuit. Although several logic families are described in the Appendix, we have chosen to use TTL (transistor–transistor logic) circuits to demonstrate logic design. This family of circuits is one of the more popular and versatile, representing a reasonable choice for implementing digital system examples.

1.3
. . . .

DESIGN AND ANALYSIS

In recent years, more emphasis has been placed on teaching the design of circuits and systems as opposed to teaching the analysis of these components. Very few engineering or computer companies generate income by solving analysis problems. Most high-tech companies develop and sell products that are either entirely new or that represent an improvement over existing products. The improvement may be increased performance, easier operation, lower cost, or a combination of these factors. Each of these new products must be created to achieve the desired result. A large number of engineers will be involved in the development or design of these new products. Consequently, there is an important need to teach design methods.

Since we intend to discuss the design of digital systems, we will first consider the meaning of the word "design."

 ● A design is an underlying scheme or plan that governs the development of some entity.

This definition can be expanded to apply to digital system design.

 ● A digital design is a plan, often expressed in schematic or block diagram form, that governs the development of a digital system.

Digital system design consists of developing a plan for interconnecting a number of simple elements called logic circuits in such a manner to perform useful, specified functions. In a digital computer, the logic blocks are used to create systems that do arithmetic operations. In a traffic control system, the circuits determine the sequence and timing of several traffic lights. In a digital watch or clock, the logic circuits determine the numbers that should be displayed at any particular time of day.

In the context of this textbook, digital system design consists of choosing components such as gates, flip-flops, or registers and determining the interconnection of those components to produce a system that satisfies a set of specifications. The design itself is typically expressed on paper or is stored in a computer as a schematic diagram, block diagram, or wiring connection list.

We emphasize the fact that this plan or design must be produced before the actual system can be built. Typically, the designer is given a set of specifications that are to be met. As the design is created, it is directed toward the goal of satisfying the given specifications. After construction, the system will be tested to ensure that the specifications are met.

In engineering practice, it may take weeks or months to actually construct a digital system. If the system has not been designed properly on paper, the finished product may not perform acceptably. Those weeks or months of construction time might be entirely wasted in such a case. On the other hand, if the paper design is done properly, construction will lead to a reliable digital system that performs as expected with little or no modification. The success of a company may depend on the capabilities of its designers.

One important requirement for any design method is the ability to analyze. It is imperative, however, to understand the difference between design and analysis. The next few paragraphs will consider this difference.

1.3.1 ANALYSIS VS. DESIGN

In elementary school we are given analysis problems such as the following:

> *A parent buys 3 cans of corn at 47¢ per can, 2 loaves of bread at $1.28 per loaf, a gallon of milk at $2.50, and 2 pounds of steak at $2.65 per pound. How much money did this person spend on groceries?*

This type of problem serves a useful purpose in teaching a child how math can be used. It has a single answer that immediately measures a person's ability to apply the required mathematics. Furthermore, design methods are generally based on analysis; thus, it is logical to teach analysis prior to design.

A design problem based on this same example might read:

> *A parent has $80 with which to purchase groceries. Enough food must be purchased to serve 3 people for 3 days. The parent desires to include items from the four basic food groups. Given a price list for food, as posted in the store, construct an appropriate shopping list.*

Obviously, this problem has many answers, depending on the length of the price list, the rate of sales tax, the appetites of the people involved, and other factors. This design problem is more closely related to a real-world problem than is the analysis

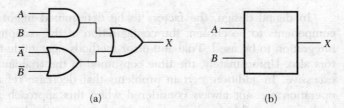

FIGURE 1.1   (a) A circuit to be analyzed. (b) A circuit to be designed.

problem. It is a situation that a parent may face daily, and its solution requires an ability to analyze. Generally, it is solved by a trial-and-error approach and may require several trials to reach an acceptable answer.

Analysis can be considered to be an important tool in solving a design problem. A designer must fully understand analysis before he or she can effectively apply design methods.

A simple example of the difference between analysis and design in the digital area can be demonstrated from Fig. 1.1. In part (a) of the figure, certain logic gates have been interconnected in a specific way. If inputs A and B are known, the calculation of output X is an analysis problem. There is a single correct answer for this system. Part (b) of the figure shows an empty block fed by inputs A and B. A design problem might specify all possible values that these inputs can have and require that the output X take on a certain value for each combination of input values. A set of gates and the interconnections between these gates must be designed to achieve the specified output values. This is a much more difficult problem than is the analysis problem.

An analysis problem specifies all factors involved and requires that an answer be found. Typically, there is a single conclusive answer to an analysis problem. A design problem specifies an answer that must be obtained and requires that the factors influencing that answer be determined. Generally, there are numerous combinations of factors that will produce the specified answer; thus there are multiple acceptable solutions to the design problem.

1.3.2   TRIAL-AND-ERROR METHODS IN DESIGN

As mentioned previously, a trial-and-error approach might be used to construct a desired shopping list. A trial solution (trial shopping list) is proposed, and then the proposal is analyzed. By comparing the results for analysis to the specifications, an error can be evaluated. If this error is too large, the trial solution is modified in a way that decreases the error. This procedure continues until acceptable results are achieved. If the list of food in each group is extensive, there may be thousands of possible combinations that satisfy the constraints imposed. There may also be thousands of unacceptable combinations.

In digital design, the factors to be determined might consist of the type of components to be chosen, the configuration of the components, and the level of integration to be used. Trial-and-error methods may guide the choice of these factors also. Unfortunately, the time consumed in the trial-and-error method is often excessive. In addition, certain problems that decrease reliability of digital system operation are not always considered when this approach is used. Although it is perhaps impossible to eliminate trial-and-error methods, a good design procedure should minimize the time required for such methods while also producing a reliable system.

1.3.3 TOP-DOWN DESIGN

The engineer may be considered an inventor in many situations. In order to design new products, new ideas or new concepts will often be applied. Thus, it is essential that the engineer be concerned with learning to design. As in many creative endeavors, the process of design is not well defined. It is more open ended and may have several acceptable solutions to a given problem. For this reason, it is useful to develop structure for the design process.

The method applied to the creation of complex digital systems in this text is called "top-down" design. Later chapters, especially Chapter 7, will emphasize this method for the design of state machines. The earlier chapters cover analysis methods and provide background information on the circuits necessary to proceed to digital design methods.

This text is intended to introduce the reader to the fundamental concepts of logic circuits and extend these concepts to the design of relatively complex systems. In so doing, the reader should acquire the capability of understanding the concepts required in the design of computers, communication circuits, digital control systems, and other types of digital systems.

REFERENCES AND SUGGESTED READING

1. Paul Ceruzzi, "A few words about this picture," *American Heritage of Invention and Technology*. Forbes. Spring 1994.

2. A. Ralston, ed., *Encyclopedia of Computer Science*. New York: Van Nostrand Reinhold, 1976.

Chapter 1

Binary Systems and Logic Circuits

Before we begin our study of digital circuits and systems, we should become aware of the reasons for using the two-level logic scheme that is so prevalent in the digital field. This scheme is closely related to the binary number system, thus the first section of this chapter will justify the use of binary numbers by considering the advantages of binary number systems over decimal number systems when used in logic circuits. Having established this justification, the second section will introduce several important binary-related numbering systems and codes. Succeeding sections will discuss the physical relationships among different binary representations and the actual logic circuits. The basic concepts needed to use integrated circuits or IC chips effectively will then be considered.

Although several logic families are described in the Appendix, we have chosen to use TTL (transistor–transistor logic) circuits to demonstrate logic design. This family of circuits is the most popular and versatile, representing a reasonable choice for implementing digital system examples. While CMOS (complementary metal-oxide-semiconductor) circuits are also popular, several circuits within this family are designed to be TTL compatible. This means that the CMOS circuit can be used in a system consisting primarily of TTL circuits.

Another point that should be discussed is the emphasis on gates and small-scale integrated circuits of this textbook. Many very effective programmable chips allow the implementation of large state machines or other logic circuits on a single chip. Several companies use these devices in developing digital systems. While such chips are considered in this textbook, they are not emphasized. Once the fundamental concepts are understood at the gate level, it is easy to extend these concepts to more complex programmable chips. It is not the intent of this textbook to provide the detailed technology required to use programmable chips. This can be obtained from the manufacturers' handbooks. Rather, the intent is to provide a solid basis in fundamental concepts of logic design, demonstrating these concepts with small- and medium-scale integrated circuits.

1.1

THE ADVANTAGES OF BINARY

SECTION OVERVIEW This section indicates that systems based on a binary numbering scheme can be more reliable than analog systems in certain numeric applications. It also considers several electronic circuits or components that can generate the necessary two-level output required to represent binary digits.

The binary number system forms the basis for all digital computer and most other digital system operations. That fact alone should underscore the tremendous importance of the binary number system. While it is true that many systems use hexadecimal, octal, or decimal numbering, these numbers must be coded or represented to the digital system in some sort of binary-type code. Hence, all computers use binary or a binary-related system to represent numbers and alphabetic characters.

Why is binary so universal in digital systems? There are at least two reasons for the popularity of this numbering scheme. The first is that yes–no or on–off situations are much easier to classify than are quantitative situations. It is easy to look at a light switch and determine if it is off or on. This is a binary situation having only two possible states. We can readily tell if a water faucet is off with no water flow or if it is on with a significant flow of water. It is more difficult to tell how much water is flowing unless we have some rather elaborate equipment. If we must determine the flow rate, we are concerned with an analog situation. The number of cubic feet per second could be a continuously changing quantity, meaning our measuring system would have to monitor and identify this changing flow rate. The reliability and accuracy in classifying a digital variable as on or off are easy to achieve. It is more difficult to determine an accurate value of some analog quantity such as flow rate. As we shall see, the binary system uses only two values, 1 and 0, which can correspond to an "on–off" or "yes–no" situation. Thus, a binary scheme is easy to use and has high reliability and accuracy.

A second reason for using binary is that many electronic components are capable of producing two easily identifiable states. An output transistor stage can be saturated, leading to a low output voltage, or cut off, leading to a high output voltage. A popular computer circuit called a flip-flop has an output that is stable only at a high voltage level or at a low voltage level. This binary nature of many electronic circuits or components makes the decision to use the binary system a very reasonable choice.

As an example of the accuracy problem, an analog circuit that is to represent numbers varying from 0 to 1000 by an analog voltage might be considered. If the maximum output voltage is 6 V, then each of 1000 different voltage ranges will represent the 1000 numbers. Each voltage range would equal

$$\Delta V = \frac{6 \text{ V}}{1000 \text{ ranges}} = 0.006 \text{ V/range}$$

The number 8 would correspond to an output voltage of 8×0.006 V $= 0.048$ V for this analog circuit. A 6-mV range of voltage to represent each number leads to accuracy problems. When the circuit output generates a particular voltage, a specific number is indicated. A temperature change can cause this output voltage to vary more than 6 mV. Thus, the analog circuit may reflect an incorrect output number.

This problem is eliminated in binary systems by using several binary circuits to represent each number by some precise digital code. The next two sections will consider this type of representation.

Although multilevel circuits having more than two output levels have been studied for years, binary circuits are essentially universal in computer systems.

1.2
NUMBER SYSTEMS

SECTION OVERVIEW This section studies the basis of the decimal, binary, octal, and hexadecimal number systems. Since digital circuits are binary in nature, all numbers represented by these circuits must use some binary-related code. Hence, binary-coded octal, binary-coded hexadecimal, and binary-coded decimal numbering methods are reviewed. Conversion of numbers from one base to another is emphasized. The section contains an introduction to alphanumeric codes used to represent numbers, characters, or control information.

The decimal number system is quite useful for human transactions and is used throughout the world. This system uses the positional method to assign values to the number. That is, the position of a digit relative to the decimal point determines the power of 10 by which the digit is to be multiplied. The number 312.4 means

$$3 \times 10^2 + 1 \times 10^1 + 2 \times 10^0 + 4 \times 10^{-1}$$

In the decimal system, the number 10 is the *base* or *radix*. The digit position to the left of the radix point (decimal point if the radix is 10) is multiplied by the radix to the power of zero. The next position to the left indicates that this digit should be multiplied by the radix to the first power. In general, a number of any radix can be expressed as

$$N_r = \cdots + a_4 r^4 + a_3 r^3 + a_2 r^2 + a_1 r^1 + a_0 r^0 + a_{-1} r^{-1} + a_{-2} r^{-2} + \cdots \quad (1.1)$$

where r is the radix and a_i is the digit value. To conserve space in writing a number, we can eliminate the plus signs and eliminate the radix multiplier if we allow the position of each digit to indicate this multiplier. When we do this, however, we must insert the radix point as a reference point and we must indicate the radix to be used. For example, the number

$$5 \times 8^3 + 2 \times 8^2 + 1 \times 8^1 + 3 \times 8^0 + 2 \times 8^{-1} + 6 \times 8^{-2}$$

can be expressed as

$$5213.26_8$$

using this shorthand method. The subscript indicates the radix used. We will use this method of radix identification whenever there is a possibility of confusion.

We note that this numbering system requires exactly r symbols. The decimal system uses the symbols 0 through 9. The highest symbol value is one less than the radix, since adding 1 to this value is accounted for by a 1 carried to the next column of higher significance.

Equation (1.1) is useful for several purposes. It can be used to express a number in base r form or to convert to a decimal number. As we have seen, a number can be expressed in base r simply by writing the coefficients while deleting the powers of r indicated in Eq. (1.1). To convert the number to decimal, the individual terms are added. The base 8 number 5213.26_8 is converted to decimal by forming the sum

$$5 \times 8^3 + 2 \times 8^2 + 1 \times 8^1 + 3 \times 8^0 + 2 \times 8^{-1} + 6 \times 8^{-2}$$

$$= 2560 + 128 + 8 + 3 + 0.25 + 0.09375$$

$$= 2699.34375_{10}$$

This equation also forms the basis of converting from a decimal number to a number expressed in some other base. We will develop this application in the succeeding paragraphs.

1.2.1   THE BINARY SYSTEM

The binary system uses the base 2; consequently only the two symbols 0 and 1 are necessary to express any binary number. The number

$$1101.101_2$$

is converted to the decimal system using Eq. (1.1):

$$1 \times 2^3 + 1 \times 2^2 + 0 \times 2^1 + 1 \times 2^0 + 1 \times 2^{-1} + 0 \times 2^{-2} + 1 \times 2^{-3} = 13.625_{10}$$

TABLE 1.1 Decimal equivalent of 2^n.

2^n	Decimal Equivalent	Binary Equivalent											← Col. no.
		11	10	9	8	7	6	5	4	3	2	1	
2^0	1											1	
2^1	2										1	0	
2^2	4									1	0	0	
2^3	8								1	0	0	0	
2^4	16							1	0	0	0	0	
2^5	32						1	0	0	0	0	0	
2^6	64					1	0	0	0	0	0	0	
2^7	128				1	0	0	0	0	0	0	0	
2^8	256		0	1	0	0	0	0	0	0	0	0	
2^9	512	1	0	0	0	0	0	0	0	0	0	0	
2^{10}	1024	1	0	0	0	0	0	0	0	0	0	0	

Conversion of Integers

There are two popular methods of converting from decimal to binary numbers. The first method merely applies a knowledge of the powers of 2 as shown in Table 1.1. We start this procedure by identifying the highest power of 2 contained in the decimal number, that is, the highest power of 2 that does not exceed the number. We then begin to construct the binary equivalent by placing a 1 in the column that accounts for this power of 2. Next we subtract from the original number the decimal equivalent of that power of 2 now included in the binary conversion. We then examine the difference and repeat this procedure. Continual repetition will lead to termination when the difference between the remaining decimal number and the power of 2 is zero.

EXAMPLE 1.1

Convert 789_{10} to a binary number.

SOLUTION

The highest power of 2 contained in 789_{10} is 9 since 2^9 is 512 and 2^{10} is 1024, which exceeds 789. We start construction of our binary number by placing a 1 in the tenth column. The complete procedure is demonstrated in Table 1.2. The final number is obtained by inserting zeros in all blank columns to give $789_{10} = 1100010101_2$. ●

TABLE 1.2 Conversion of decimal to binary numbers.

				10	9	8	7	6	5	4	3	2	1
	789_{10}												
Contains	512_{10}	=	2^9		1								
Difference	277_{10}												
Contains	256_{10}	=	2^8			1							
Difference	21_{10}												
Contains	16_{10}	=	2^4							1			
Difference	5_{10}												
Contains	4_{10}	=	2^2									1	
Difference	1_{10}												
Contains	1_{10}	=	2^0										1
Difference	0_{10}			1	1	0	0	0	1	0	1	0	1

Another version of this procedure identifies the highest power of 2 contained in the decimal number and places a 1 in the binary number to be constructed. The difference between the original number and this power of 2 is formed, then examined to see if it contains the next lower power of 2. If not, a 0 is placed to the right of the 1 in the binary number to be constructed, and the next lower power of 2 is checked. When the difference does contain a power of 2, a 1 is inserted in the binary number being formed and the decimal equivalent of this power of 2 is subtracted from the remaining partially converted decimal number.

Returning to Example 1.1, we see that 512_{10} is contained in 789_{10}. We write a 1 to start the construction of the binary equivalent. The entire procedure is demonstrated in Table 1.3.

A second method used to convert from a decimal to a binary number involves repeated division of the number by 2 and examination of the various remainders. We illustrate this method by converting 16_{10} to a binary number in Table 1.4.

The procedure is terminated when the result of zero is reached. The converted binary number is simply the remainders arranged such that the last remainder is the most significant bit, that is,

$$16_{10} = 10000_2$$

Each time a 2 is divided into the decimal number, it reflects the fact that a higher power of 2 is contained in the number. In a binary number, each digit position moving from right to left represents a higher power of 2. Thus, each time we divide

TABLE 1.3 A modification of the procedure in Table 1.2.

	789_{10}	
Contains	512_{10}	1
	—	
Difference	277_{10}	
Contains	256_{10}	1
	—	
Difference	21_{10}	
Does not contain	128_{10}	0
Does not contain	64_{10}	0
Does not contain	32_{10}	0
Contains	16_{10}	1
	—	
Difference	5_{10}	
Does not contain	8_{10}	0
Contains	4_{10}	1
	—	
Difference	1_{10}	
Does not contain	2_{10}	0
Contains	1_{10}	1
	—	
Difference	0_{10}	1 1 0 0 0 1 0 1 0 1

a 2 into the decimal number, the higher power of 2 is accounted for in the binary number by recording the remainder bit.

In the preceding example, the remainder was zero for all but the fifth division, indicating that 2^4 was contained in the original number. Now suppose we want to convert 17_{10} to a binary number. This is done in Table 1.5.

Each division by 2 indicates 2 bits of information about the number. If the nth division results in a nonzero number, we know that the decimal number is greater than or equal to 2^n. If the remainder is 1, it means that 2^{n-1} is contained in the number. If the remainder is zero, then 2^{n-1} is not contained in the number. These points are demonstrated in Table 1.6 in converting 23_{10} to a binary number. This

TABLE 1.4 Conversion of 16_{10} to binary.

Divide number by 2	$16 \div 2 = 8$	r0	Record remainder	0
Divide result by 2	$8 \div 2 = 4$	r0	Record remainder	0
Divide result by 2	$4 \div 2 = 2$	r0	Record remainder	0
Divide result by 2	$2 \div 2 = 1$	r0	Record remainder	0
Divide result by 2	$1 \div 2 = 0$	r1	Record remainder	1

TABLE 1.5 Conversion of 17_{10} to binary.

Divide number by 2	$17 \div 2 = 8$	r1	Record remainder	1
Divide result by 2	$8 \div 2 = 4$	r0	Record remainder	0
Divide result by 2	$4 \div 2 = 2$	r0	Record remainder	0
Divide result by 2	$2 \div 2 = 1$	r0	Record remainder	0
Divide result by 2	$1 \div 2 = 0$	r1	Record remainder	1

$17_{10} = 10001_2$

TABLE 1.6 Conversion of 23_{10} to binary.

1st division	$23 \div 2 = 11$ r1	Nonzero result indicates that number is $\geq 2^1$ Remainder indicates that 2^0 is contained in number	Indicated by recording a 1 in the 2^0 column
2nd division	$11 \div 2 = 5$ r1	Nonzero result indicates that number is $\geq 2^2$ Remainder indicates that 2^1 is contained in number	Indicated by recording a 1 in the 2^1 column
3rd division	$5 \div 2 = 2$ r1	Nonzero result indicates that number is $\geq 2^3$ Remainder indicates that 2^2 is contained in number	Indicated by recording a 1 in the 2^2 column
4th division	$2 \div 2 = 1$ r0	Nonzero result indicates that number is $\geq 2^4$ Zero remainder indicates that 2^3 is not contained in number	Indicated by recording a 0 in the 2^3 column
5th division	$1 \div 2 = 0$ r1	Zero result indicates that number is $< 2^5$ Remainder indicates that 2^4 is contained in number	Indicated by recording a 1 in the 2^4 column

$23_{10} = 10111_2$

method is particularly useful in converting decimal numbers greater than 2048_{10} to binary.

We can understand why this method works in converting integer numbers from base 10 to base 2 by considering the expression

$$N_{10} = a_4 2^4 + a_3 2^3 + a_2 2^2 + a_1 2^1 + a_0 2^0$$

This equation is based on Eq. (1.1), but assumes that N_{10} is small enough to be represented by a 5-bit binary number equal to $a_4 a_3 a_2 a_1 a_0$. We are given N_{10} and must determine the coefficients a_0 through a_4. When N_{10} is divided by 2, the result is

$$\frac{N_{10}}{2} = a_4 2^3 + a_3 2^2 + a_2 2^1 + a_1 2^0 + a_0 2^{-1}$$

We note that the first four terms will have integer values and their sum is an integer value while the fifth term will be either zero or 1/2. We could also express the result of this division as

$$\frac{N_{10}}{2} = a_4 2^3 + a_3 2^2 + a_2 2^1 + a_1 2^0 + \text{remainder } a_0$$

The value of the remainder is then equal to a_0 and will either be 0 or 1.

Now that the value of a_0 has been determined, we can drop this remainder and deal only with the integer part of $\frac{N_{10}}{2}$. If we designate this integer part N'_{10} and divide by 2 again, we get

$$\frac{N'_{10}}{2} = a_4 2^2 + a_3 2^1 + a_2 2^0 + \text{remainder } a_1$$

In this case, the remainder now identifies the value of a_1. We continue this procedure to get

$$\frac{N''_{10}}{2} = a_4 2^1 + a_3 2^0 + \text{remainder } a_2$$

$$\frac{N'''_{10}}{2} = a_4 2^0 + \text{remainder } a_3$$

$$\frac{N''''_{10}}{2} = \text{remainder } a_4$$

This procedure is seen to be implemented in the method outlined in Tables 1.4 and 1.5.

Conversion of Nonintegers

If a decimal number contains both an integer and noninteger part, the preceding method should be applied only to the integer part. A second method must be used to convert the noninteger part. This part of a binary number can be written as

$$N_{10}^d = a_{-1}2^{-1} + a_{-2}2^{-2} + a_{-3}2^{-3} + \cdots$$

Given the decimal number N_{10}^d, the binary number can be found by the following procedure.

The noninteger part is multiplied by 2 giving a result of

$$2N_{10}^d = a_{-1} + a_{-2}2^{-1} + a_{-3}2^{-2} + \cdots$$

At this point, we note that a_{-1} will be either 0 or 1 while the remaining terms represent a decimal number that is always less than unity. The first coefficient a_{-1} is identified as the integer part of $2N_{10}^d$. If we now drop this integer part of $2N_{10}^d$ and call the remaining noninteger part $N_{10}^{d'}$, we see that multiplying by 2 gives

$$2N_{10}^{d'} = a_{-2} + a_{-3}2^{-1} + \cdots$$

The coefficient a_{-2} is identified as the integer part of the result. This procedure is continued until all coefficients are identified.

EXAMPLE 1.2

. . . .

Convert the number 0.625_{10} to a binary number.

SOLUTION

Using the method outlined, we first multiply this noninteger number by 2 to get

$$2N_{10}^d = 1.250$$

The integer part identifies a_{-1} as 1. We now discard this integer leaving $N_{10}^{d'} = 0.250$. Multiplying by 2 gives

$$2N_{10}^{d'} = 0.500$$

Since the integer part of this term is zero, the coefficient a_{-2} is equal to zero. We again multiply by 2 to get

$$2N_{10}^{d''} = 1.000$$

This identifies a_{-3} as 1 and terminates the process since the noninteger part is zero. The resulting binary number is

$$0.625_{10} = 0.101_2$$

●

The methods of converting between decimal and binary can be easily extended to other base conversions as demonstrated by the following example.

EXAMPLE 1.3

Convert 863.203125_{10} to (a) binary and (b) base 8.

SOLUTION

(a) The integer part is first converted to binary using successive division by 2 resulting in

$$863_{10} = 1101011111_2$$

The decimal part is converted to binary using successive multiplication by 2 resulting in

$$0.203125_{10} = 0.001101_2$$

The resulting conversion is $863.203125_{10} = 1101011111.001101_2$
(b) The integer part is converted to base 8 by successive division by 8 as outlined below.

$$\frac{863}{8} = 107, \qquad r = 7 = a_0$$

$$\frac{107}{8} = 13, \qquad r = 3 = a_1$$

$$\frac{13}{8} = 1, \qquad r = 5 = a_2$$

$$\frac{1}{8} = 0, \qquad r = 1 = a_3$$

Thus,

$$863_{10} = 1537_8$$

The noninteger part is converted to base 8 by successive multiplication by 8.

$$8 \times 0.203125 = 1.625 \rightarrow a_{-1} = 1$$

$$8 \times 0.625 = 5.000 \rightarrow a_{-2} = 5$$

Thus,

$$0.203125_{10} = 0.15_8$$

The complete number is

$$863.203125_{10} = 1537.15_8$$ •

Accuracy of Binary Conversion of Nonintegers

When a conversion from a noninteger decimal number to binary is terminated after n binary bits, the maximum error, expressed in decimal, is 2^{-n}. A binary number that could be expressed accurately as $(0.1010ijk\cdots)_2$ where i, j, k are 0 or 1, can be terminated after four places and written as 0.1010_2. If the fifth and all succeeding bits were 0, this four-bit number would accurately represent the actual number. If the fifth and all succeeding bits were 1, the error between the four-bit number and the actual number would be maximum and would equal

$$E_2 = 0.00001111\cdots_2$$

In decimal, this error can be expressed as

$$E_{10} = 2^{-5} + 2^{-6} + 2^{-7} + 2^{-8} + \cdots$$

This error can also be written as

$$E_{10} = 2^{-5}(1 + 2^{-1} + 2^{-2} + 2^{-3} + \cdots)$$

The term in parentheses is a geometric series that approaches a value of

$$\frac{1}{1 - \frac{1}{2}} = 2$$

as the number of terms approaches infinity. The upper bound on the error is then

$$E_{10} = 2^{-5}(2) = 2^{-4}$$

In general, if n bits are used to represent the noninteger part of a binary number, the maximum error due to terminating the remaining bits is

$$E_{10} = 2^{-n} \tag{1.2}$$

Equation (1.2) is important in determining when to terminate a decimal to binary conversion. If a decimal number is to be converted to binary with a specified accuracy, this equation is used to calculate the number of bits to include in the binary number. With this number of bits, the error is guaranteed to be less than the specified value. Example 1.4 will demonstrate this method.

EXAMPLE 1.4

Convert the decimal number 0.252_{10} to binary with an error less than 1%.

SOLUTION

The absolute value of allowable error is found by calculating 1% of the number. This gives

$$E_{\text{allow}} = 0.01 \times 0.252 = 0.00252_{10}$$

The maximum error due to truncation of the binary number is set to be less than this allowable error by solving for n from Eq. 1.2. This equation is written as

$$2^{-n} < 0.00252$$

Inverting both sides of the inequality results in

$$2^n > 397$$

Taking the log of both sides and solving for n gives

$$n = \frac{\log 397}{\log 2} = 8.63 \approx 9 \text{ (next largest integer)}$$

This indicates that the use of nine bits in the binary number will guarantee an error less than one percent. The conversion, carried out to nine places, is

$$0.252_{10} = 0.010000001_2$$

The actual error can now be checked by converting the nine-bit binary number to decimal yielding 0.251953125_{10}. This number differs from the original number by less than one percent.   ●

1.2.2   OCTAL AND BINARY-CODED OCTAL (BCO)

If the base or radix chosen to express a number is 8, the number system is referred to as *octal*. The eight symbols 0 through 7 are required to express an octal number. Conversion from octal to decimal is accomplished with Eq. (1.1); thus

$$217_8 = 2 \times 8^2 + 1 \times 8^1 + 7 \times 8^0 = 143_{10}$$

The octal system is used in some computers and is closely related to the binary system. Of course, when the computer applies the octal system, the number must be represented by ones and zeros since the other six symbols are not available as outputs of binary circuits. Rather than applying octal code directly, the computer uses *binary-coded octal* (BCO). In this scheme, three binary positions are used to represent one digit of the octal number. The correspondence between the two systems is shown below.

2	1	7	Octal
010	001	111	Binary-coded octal

Each three-bit binary number can represent any number from zero to seven. One of the interesting features of BCO is the ease of conversion to and from binary. The octal number $217_8 = 143_{10}$. Converting to binary gives

$$143_{10} = 10001111_2 = 217_8$$

Now we will compare the binary equivalent of 217_8 to the BCO representation above. Except for the leading zero in the BCO system, the two numbers are equivalent. Conversion from BCO to binary consists of concatenation of the three-digit groups and dropping any leading zeros. In order to convert from binary to octal, we must subdivide the digits to the left of the octal point into groups of three, adding any necessary zeros to the leftmost group to yield three digits in that group. The following examples demonstrate these points:

$$6524_8 = 110\ 101\ 010\ 100\ (BCO) = 110101010100_2$$

$$2.51_8 = 010.101\ 001\ (BCO) = 10.101001_2$$

$$1011101110_2 = 001\ 011\ 101\ 110\ (BCO) = 1356_8$$

We can show why the conversion from binary-coded octal to binary is so straightforward by considering the following octal and binary-coded octal number:

$$a_2 \times 8^2 + a_1 \times 8^1 + a_0 \times 8^0$$

$$= b_8b_7b_6 \times 8^2 + b_5b_4b_3 \times 8^1 + b_2b_1b_0 \times 8^0$$

where $b_8b_7b_6$ is the binary equivalent of a_2, $b_5b_4b_3$ is the binary equivalent of a_1, and $b_2b_1b_0$ is the binary equivalent of a_0. We can convert the binary-coded octal number to straight binary by converting the radix raised to various powers to binary. Noting that

$$8^0 = 1_2, \qquad 8^1 = 1000_2, \qquad \text{and} \qquad 8^2 = 1000000_2$$

we can write the number as

$$b_8b_7b_6 \times 1000000 + b_5b_4b_3 \times 1000 + b_2b_1b_0 \times 1$$

Carrying out the indicated multiplications and additions gives the binary number:

	b_8	b_7	b_6	0	0	0	0	0	0
+				b_5	b_4	b_3	0	0	0
+							b_2	b_1	b_0
	b_8	b_7	b_6	b_5	b_4	b_3	b_2	b_1	b_0

From this result we see that the binary number is formed by concatenation of the binary-coded octal number.

A corollary of this result is that conversion between octal and binary can be accomplished easily by using binary-coded octal as an intermediate number. For example, binary to octal can be done as follows:

$$10110110_2 = 010\ 110\ 110\ \text{(BCO)}$$

$$= 266_8$$

The method of conversion indicated by Tables 1.4, 1.5, and 1.6 can be used for octal (or hexadecimal) as well as binary. It is left to the student to demonstrate the validity of this statement.

1.2.3 HEXADECIMAL CODE

A very important code in computer work is *hexadecimal* or *hex* code. This code uses 16 as the base; thus 16 symbols, 0 to 15, are required. Since positional notation is utilized, each column must have a single digit or symbol. Because of this requirement, the following symbols are used:

Decimal	Hex	Decimal	Hex
0	0	8	8
1	1	9	9
2	2	10	A
3	3	11	B
4	4	12	C
5	5	13	D
6	6	14	E
7	7	15	F

The number $A29_{16}$ is converted to decimal by writing

$$A29_{16} = 10 \times 16^2 + 2 \times 16^1 + 9 \times 16^0 = 2601_{10}$$

Binary-coded hex uses four columns to represent each digit of the hex number. The binary-coded hex equivalent of $A29_{16}$ is

$$1010 \ 0010 \ 1001 \ (\text{BCH})$$

Conversion between binary-coded hex and binary can be done by the method used to convert binary-coded octal. The groups of four digits are concatenated and any leading zeros dropped to form the binary number. Any base that is an integer power of 2 can be converted in this way. Again conversion between hex and binary can be facilitated by using binary-coded hex; thus

$$1E2_{16} = 0001 \ 1110 \ 0010 \ (\text{BCH}) = 111100010_2$$

1.2.4 BINARY-CODED DECIMAL CODE (BCD)

This code is used in applications involving the interface between a digital device and a human being. Since humans are trained to deal with the decimal system, the BCD code is useful for such applications. A typical situation utilizing this code is a digital voltmeter or DVM that displays decimal results. The measurement circuit outputs BCD codes which then drive BCD to 7-segment code converters to activate the light-emitting diode displays.

The BCD code uses four binary bits to represent each decimal position of the number. The number 843_{10} is represented by

$$843_{10} = 1000 \ 0100 \ 0011 \ (\text{BCD})$$

This code is rather inefficient because a 4-bit code that can contain 16 unique combinations is used for only 10 numbers. This inefficiency can be demonstrated by considering the number of bits required to represent the decimal numbers 999 and 999,999. In binary, 10 and 20 bits are necessary to express these two numbers. BCD code requires 12 and 24 bits, respectively.

1.2.5 ALPHANUMERIC CODES

Thus far in our discussion of binary we have considered representing only numbers with binary and binary-related codes. While numbers are important in many digital applications, alphabetic or other characters are necessary in several instances also.

There are many choices of code sets to represent *alphanumeric* (alphabetic and numeric) characters; however, one well-accepted code set has become a standard in microprocessor work. This code is called the ASCII code, which is an abbreviation for American Standard Code for Information Interchange. The ASCII code is a 7-bit code that contains alphanumeric characters along with several control characters used in exchanging information between two or more digital systems. This code is shown in Table 1.7.

The ASCII code is quite popular for personal and other computers. The keyboard on all PCs uses this code to communicate with the computer.

● ● ●   **DRILL PROBLEMS** Sec. 1.2

1.   Convert 683_{10} to binary, octal, hexadecimal, binary-coded octal, and binary-coded hex.

2.   Convert 79.156_{10} to a binary number using 6 bits to express the noninteger part.

3.   Convert $A2F_{16}$ to binary, octal, and decimal.

4.   Convert 712_8 to binary and binary-coded decimal.

5.   Convert 1000 0110 (BCD) to decimal and binary.

6.   Convert 21.36_{10} to binary with an error of less than 0.01_{10}.

1.3

THE USE OF BINARY IN DIGITAL SYSTEMS

SECTION OVERVIEW This section considers methods of representing the binary symbols 1 and 0 by voltage levels of an electronic circuit. Extension from single-bit numbers to multiple-bit numbers is also discussed.

We now understand how to represent binary or binary-related codes by a series of ones and zeros. We can easily write 1011 on paper as the binary code for the decimal number 11. The next matter we must consider is how this same information can be represented by electronic circuits.

Electronic circuits used for digital systems are designed to generate only two possible output voltage levels. For example, the higher level may be near 5 V, while the low level may be approximately 0 V. The circuits are designed to be quite tolerant to voltage level. The high level may vary from 2.4 to 5 V, while the low level may vary from 0 to 0.8 V. When a circuit output of 2.4 to 5 V is applied to a second digital circuit, this second circuit must interpret this input as the high voltage level. The circuits are also designed such that when proper inputs are applied, the output will never exist in the ambiguous region between 0.8 and 2.4 V. Of course, this

TABLE 1.7   7-bit ASCII code.

	BINARY	HEX		BINARY	HEX
A	100 0001	41	g	110 0111	67
B	100 0010	42	h	110 1000	68
C	100 0011	43	i	110 1001	69
D	100 0100	44	j	110 1010	6A
E	100 0101	45	k	110 1011	6B
F	100 0110	46	l	110 1100	6C
G	100 0111	47	m	110 1101	6D
H	100 1000	48	n	110 1110	6E
I	100 1001	49	o	110 1111	6F
J	100 1010	4A	p	111 0000	70
K	100 1011	4B	q	111 0001	71
L	100 1100	4C	r	111 0010	72
M	100 1101	4D	s	111 0011	73
N	100 1110	4E	t	111 0100	74
O	100 1111	4F	u	111 0101	75
P	101 0000	50	v	111 0110	76
Q	101 0001	51	w	111 0111	77
R	101 0010	52	x	111 1000	78
S	101 0011	53	y	111 1001	79
T	101 0100	54	z	111 1010	7A
U	101 0101	55	0	011 0000	30
V	101 0110	56	1	011 0001	31
W	101 0111	57	2	011 0010	32
X	101 1000	58	3	011 0011	33
Y	101 1001	59	4	011 0100	34
Z	101 1010	5A	5	011 0101	35
a	110 0001	61	6	011 0110	36
b	110 0010	62	7	011 0111	37
c	110 0011	63	8	011 1000	38
d	110 0100	64	9	011 1001	39
e	110 0101	65	SP	010 0000	20
f	110 0110	66	!	010 0001	21

region is crossed as an input or output voltage switches from one level to another. Different types of digital circuits may use different voltage levels. One digital system may use voltage levels from 0 to 2 V as the low level and 8 to 12 V as the high level. Another may use -2.1 to -1.7 V and -1.3 to -0.9 V as the two levels. Generally, system designers avoid mixing circuits with different levels, but it is sometimes unavoidable.

We have now established that digital circuits exist with two well-defined voltage levels. The binary number system requires two symbols; hence it is logical to identify

TABLE I.7 *(continued)*   7-bit ASCII code.

	BINARY	HEX		BINARY	HEX
"	010 0010	22	BEL	000 0111	07
#	010 0011	23	BS	000 1000	08
$	010 0100	24	CAN	001 1000	18
%	010 0101	25	CR	000 1101	0D
&	010 0110	26	DC1	001 0001	11
'	010 0111	27	DC2	001 0010	12
(	010 1000	28	DC3	001 0011	13
)	010 1001	29	DC4	001 0100	14
°	010 1010	2A	DEL	111 1111	7F
+	010 1011	2B	DLE	001 0000	10
,	010 1100	2C	EM	001 1001	19
-	010 1101	2D	ENQ	000 0101	05
.	010 1110	2E	EOT	000 0100	04
/	010 1111	2F	ESC	001 1011	1B
:	011 1010	3A	ETB	001 0111	17
;	011 1011	3B	ETX	000 0011	03
<	011 1100	3C	FF	000 1100	0C
=	011 1101	3D	FS	001 1100	1C
>	011 1110	3E	GS	001 1101	1D
?	011 1111	3F	HT	000 1001	09
@	100 0000	40	LF	000 1010	0A
[	101 1011	5B	NAK	001 0101	15
\	101 1100	5C	NUL	000 0000	00
]	101 1101	5D	RS	001 1110	1E
^	101 1110	5E	SI	000 1111	0F
_	101 1111	5F	SO	000 1110	0E
`	110 0000	60	SOH	000 0001	01
{	111 1011	7B	STX	000 0010	02
|	111 1100	7C	SUB	001 1010	1A
}	111 1101	7D	SYN	001 0110	16
~	111 1110	7E	US	001 1111	1F
ACK	000 0110	06	VT	000 1011	0B

a binary symbol with each voltage level. If we interpret the high level as a binary 1 and the low level as a binary 0, we are using a positive logic system. A negative logic system identifies the high level with binary 0 and the low level with binary 1. Most modern digital systems use positive logic or mixed logic systems.

Now suppose we want to represent a 4-bit binary code. Four binary circuit outputs are required. If we measure the output voltage levels of circuits A, B, C, and D to be low, high, high, and high, respectively, the code contained is 0111 if we are using positive logic. This assumes that A carries the most significant bit (MSB) of the code while D carries the least significant bit (LSB). If we want to represent a 32-bit binary code, 32 circuits are required. Thus, many circuits are necessary to contain reasonably large numbers, an obvious disadvantage of the binary system. This disadvantage is far outweighed by the natural correspondence of binary symbols to the two levels of digital circuits and the inherent reliability of the system.

For conventional digital systems, integrated circuits are used almost exclusively, with discrete digital circuits applied only to high-power or other special designs. There are four broad classifications of integration today, whose designations are based on circuit or system complexity.

Small-scale integration (SSI) is applied to a chip that performs at least one basic logic function. This chip could contain gates, flip-flops, level detectors, or other useful logic circuits. Using SSI, as many as four gates can be constructed on the same chip, totaling perhaps 20 to 40 components. The next step up in complexity is *medium-scale integration* (MSI). This level of integration represents an order of magnitude capability increase over SSI. Logic functions requiring 20 gates and 10 flip-flops can be integrated on a single MSI chip. *Large-scale integration* (LSI) utilizes very complex fabrication techniques to create, perhaps, 20,000 components on a single chip. Between 100 and 1000 logic gates that can perform the logic of an entire digital system may be included on a chip constructed by LSI techniques. *Very large-scale integration* (VLSI) has now entered the picture as electron-beam lithography and related techniques improve. More than 10 million devices may now be contained on a VLSI chip. Some refer to such circuits as ultra-large-scale integrated circuits.

All four levels of integration are widely used in the digital field. SSI is prominent in preliminary design work on systems that will later be fabricated using MSI or LSI circuits. It is also used in systems produced for a market too small to justify fabrication of MSI or LSI circuits. MSI is used for finished systems that do not require the smaller size of LSI and will be produced in relatively small quantities. LSI is directed toward finished products that require compactness or that have a large market potential. VLSI is used to implement complex microprocessor and large memory chips. An entire computer with main memory on a single chip has now become a reality. The cost per device on a VLSI chip is often less than one-thousandth of a cent, making these chips very cost effective in many applications.

1.4

LOGIC GATES

SECTION OVERVIEW The concept of a truth table to represent logic circuit behavior is introduced. The function table is also discussed. Logic gates are introduced, and their behavior is described in terms of function and truth tables.

The preceding section indicates that binary bits can be represented by the voltage level at a circuit output. We next consider what types of circuit are actually used in digital or logic systems.

A digital system uses a building block approach. Many small operational units are interconnected to make up the overall system. The most basic logical unit of the system is the gate circuit. There are several different types of gate with each type behaving in a different way. OR gates, AND gates, NOR gates, and NAND gates are some of the more important gates that will be considered in this section.

We now examine two important methods of characterizing logic gates. These circuits have one output and one or more inputs. The most basic description of operation is given by the function table. This table lists all possible combinations of inputs along with resulting outputs in terms of the two levels of voltage, high and low. Figure 1.1 shows a function table for a 2-input circuit. This table indicates that if both inputs are low or both are high, the output will be low. If one input is high while the other is low, a high level results on the output line.

As we deal with logic design, it is appropriate to use 1s and 0s rather than voltage levels. We must determine the proper relationship between the hardware voltage levels and the binary numbers, 0 and 1.

Positive Logic: Historically, the high voltage or more positive voltage level was associated with binary 1 while the low voltage or less positive level was associated with binary 0. If this choice is adhered to throughout an entire system of logic gates, it is called a positive logic system.

Negative Logic: If the high voltage level is associated with binary 0 while the low voltage level is associated with binary 1, we are using a negative logic scheme. Again, if an entire system of logic gates adheres to this choice, it is a negative logic system.

Mixed Logic: While both the positive and negative logic schemes have been popular at different times in the past three decades, the mixed logic scheme is now generally used. This scheme uses positive logic in some portions of the system while applying negative logic in other portions. Although the mixed logic scheme might minimize the number of gates required to implement a system, it introduces

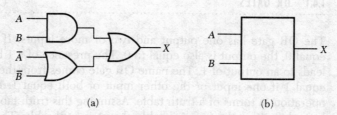

(a)   (b)

FIGURE 1.1 A function table for a logic circuit.

A	B	X
0	0	0
0	1	1
1	0	1
1	1	0

A	B	X
1	1	1
1	0	0
0	1	0
0	0	1

A	B	X
0	0	1
0	1	0
1	0	0
1	1	1

(a)   (b)   (c)

FIGURE 1.2 Truth tables for function table of Fig. 1.1: (a) positive logic, (b) negative logic, (c) positive logic inputs-negative logic output.

additional detail that must be addressed by the designer. We will consider the advantages and disadvantages of the mixed logic system after discussing individual gates.

Once the definition of logic levels is made, we use the function table to generate a truth table. The truth table describes inputs and outputs in terms of 1s and 0s rather than voltage levels. For a given circuit, the truth table for positive logic will be different than that for a negative logic choice. Figure 1.2 demonstrates this point for the function table of Fig. 1.1.

We note that for a positive logic definition, the circuit produces an output of 1 when one input is 0 while the other input is 1. For the negative logic definition, an output of 1 results when both inputs are either 0 or both are 1. Function and truth tables are used by manufacturers of logic gates to specify gate operation. The manufacturer conventionally defines gates in terms of positive logic rather than negative logic.

We could extend the discussion to a mixed logic definition for this individual gate. For example, the input levels could be chosen as a positive logic scheme while the output is defined as a negative logic scheme. The resulting truth table is shown in Fig. 1.2(c). It is certainly not obvious why this choice would be made, but this point will be clarified when assertion levels are discussed in the following chapter. A fourth option would be that of negative logic inputs and a positive logic output.

1.4.1   OR GATES

The OR gate has one output and two or more inputs. If all inputs are caused to equal 0, the output is also equal to 0. The presence of a 1 bit at one or more inputs leads to an output of 1. The name OR gate comes from the fact that the output will equal 1 if one input or the other input or both equal 1. Figure 1.3 describes this operation in terms of a truth table. Assuming this truth table results from a positive logic definition, the function table shown is applicable. The standard symbols for a 2-input OR gate and an 8-input OR gate are also shown.

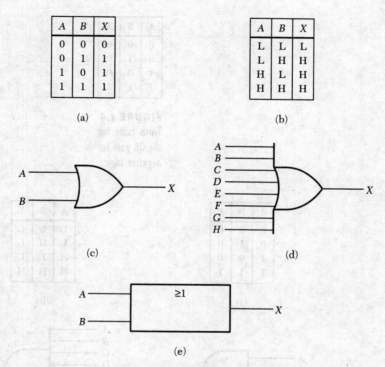

A	B	X
0	0	0
0	1	1
1	0	1
1	1	1

(a)

A	B	X
L	L	L
L	H	H
H	L	H
H	H	H

(b)

(c)

(d)

(e)

FIGURE 1.3 Truth tables and symbols for the OR gate: (a) truth table for positive logic, (b) function table, (c) symbol for 2-input OR gate, (d) symbol for 8-input OR gate, (e) IEEE/IEC logic symbol.

The symbol in Fig. 1.3(e) shows the OR gate logic symbol developed by the International Electrotechnical Commission (IEC). The logic symbols developed by the IEC were accepted as a standard in 1984 by the Institute of Electrical and Electronics Engineers as described in *IEEE Std 91-1984*. Both the older symbols of Fig. 1.3(c) and (d) along with the newer symbol of (e) are included in the logic handbooks. The IEEE/IEC notation ≥1 indicates the OR operation and means that the output will be asserted if one or more inputs are active.

If we choose to use negative logic with the OR gate of Fig. 1.3(c), it no longer performs the OR function. Figure 1.4 shows the truth table resulting from a negative logic definition. This truth table corresponds to that of an AND gate as we shall see in the following paragraphs. If a single gate can correspond to an OR gate or an AND gate depending on logic definition, we must take care to avoid confusion when specifying this gate. This problem is handled simply by naming the gate according to its positive logic function. All manufacturers refer to a gate having the function table of Fig. 1.3 as an OR gate. Although this circuit can be used as an AND gate with negative logic, the manufacturers' handbooks will always list it as an OR gate.

A	B	X
0	0	0
0	1	0
1	0	0
1	1	1

FIGURE I.4
Truth table for
the OR gate for
negative logic.

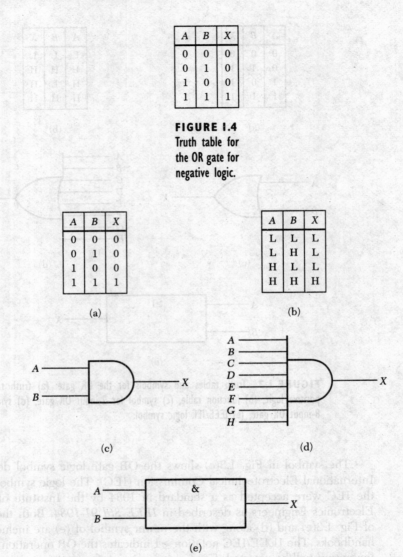

A	B	X
0	0	0
0	1	0
1	0	0
1	1	1

(a)

A	B	X
L	L	L
L	H	L
H	L	L
H	H	H

(b)

(c)

(d)

(e)

FIGURE I.5 Tables and symbols for the AND gate: (a) truth table for positive logic, (b) function table, (c) symbol for 2-input AND gate, (d) symbol for 8-input AND gate, (e) IEEE/IEC symbol.

I.4.2 AND GATES

The AND gate has one output and two or more inputs. The output will equal 0 for all combinations of input values except when all inputs equal 1. When each input is

1, the output will also equal 1. Figure 1.5 shows the AND gate, the function table, and the positive logic truth table. The IEEE/IEC symbol for the AND gate is the & as shown in Fig. 1.5(e). We could easily show that the AND gate will function as an OR gate for negative logic, but again the gate is named for its positive logic function.

1.4.3   THE INVERTER

The inverter performs the NOT or INVERT function. This logic element has one output and one input. The output level is always opposite to the input level. Figure 1.6 shows the function table, truth table, and symbols for the inverter.

The large triangle of the inverter symbol represents amplification (generally current amplification for logic circuits), while the small circle denotes an inversion of signals. We note that the inverter truth table for negative logic is the same as that for positive logic.

The 1 in the IEEE/IEC symbol indicates that the input must be active to achieve the inverted output. The circles in the older symbols and the triangle of the IEEE/IEC symbol are called qualifiers or qualifying symbols. The definitions of these qualifying symbols are shown in Fig. 1.7 [4]. The circles indicate logic negation. The circle at the input as in Fig. 1.7(a) indicates that an external applied 0 produces an internal 1. Figure 1.7(b) shows the qualifying symbol for logic negation

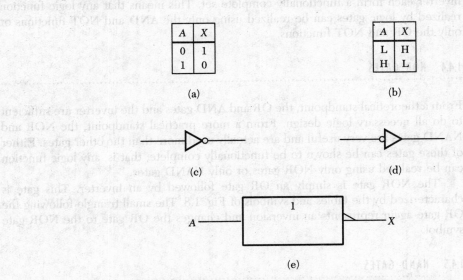

FIGURE 1.6 Tables and symbols for the inverter: (a) truth table, (b) function table, (c) inverter symbol, (d) alternate inverter symbol, (e) IEEE/IEC symbol.

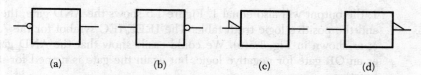

FIGURE 1.7 Qualifying symbols: (a) logic negation at input, (b) logic negation at output, (c) active-low level input, (d) active-low level output.

at the output; that is, an internal 1 produces an external 0 at the output. This symbol deals with the logical variables 1 and 0 much as a truth table does.

The triangle refers to voltage levels rather than to logic levels. The qualifying symbol of Fig. 1.7(c) indicates that this is an active-low input. A low voltage at this point is required to assert this input. The qualifying symbol of Fig. 1.7(d) designates an active-low output. When the logic circuit is asserted, this output will be driven low.

Because of the confusion resulting from the transition from old to new standard, this text will use the triangular qualifying symbol as defined in Figs. 1.7(c) and (d). This symbol will be used throughout the text. Rather than use this qualifier with the square blocks of the new standard, it will be used with the older gate and inverter symbols. The small circles will not be used, but the small triangle will be considered equivalent to an inversion.

At this points we are not yet aware of the kinds of logic functions that must be realized with gates. Nevertheless, we will now state a fact that will later become more meaningful, namely, that the AND gate and inverter or the OR gate and inverter each form a functionally complete set. This means that any logic function realized by logic gates can be realized using only the AND and NOT functions or only the OR and NOT functions.

1.4.4 NOR GATES

From a theoretical standpoint, the OR and AND gates and the inverter are sufficient to do all necessary logic design. From a more practical standpoint, the NOR and NAND gates are very useful and are actually used more than the other gates. Either of these gates can be shown to be functionally complete; that is, any logic function can be realized using only NOR gates or only NAND gates.

The NOR gate is simply an OR gate followed by an inverter. This gate is characterized by the tables and symbols of Fig. 1.8. The small triangle following the OR gate again represents an inversion and changes the OR gate to the NOR gate symbol.

1.4.5 NAND GATES

The NAND gate is an AND gate followed by an inverter. This gate is characterized by the tables and symbols of Fig. 1.9.

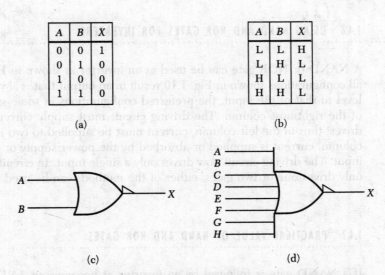

A	B	X
0	0	1
0	1	0
1	0	0
1	1	0

(a)

A	B	X
L	L	H
L	H	L
H	L	L
H	H	L

(b)

(c)

(d)

FIGURE 1.8 Tables and symbols for the NOR gate: (a) truth table for positive logic, (b) function table, (c) symbol for 2-input NOR gate, (d) symbol for 8-input NOR gate.

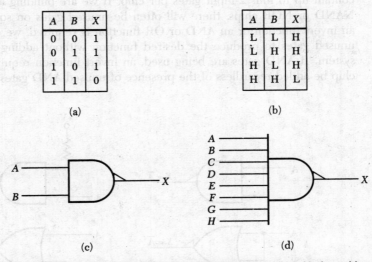

A	B	X
0	0	1
0	1	1
1	0	1
1	1	0

(a)

A	B	X
L	L	H
L	H	H
H	L	H
H	H	L

(b)

(c)

(d)

FIGURE 1.9 Tables and symbols for the NAND gate: (a) truth table for positive logic, (b) function table, (c) symbol for 2-input NAND gate, (d) symbol for 8-input NAND gate.

1.4.6   USING NAND AND NOR GATES FOR INVERTERS

A NAND or NOR gate can be used as an inverter as shown in Fig. 1.10. Although all configurations shown in Fig. 1.10 result in an output that is always at the opposite level to that of the input, the preferred configurations in some situations are those of the rightmost column. The driving circuit must supply current to each input it drives; thus in the left column, current must be supplied to two inputs. In the right column, current is supplied or absorbed by the power supply or by ground for one input. The driving circuit now drives only a single input. In circuits where an output only drives one or two gates, either of the methods can be used.

1.4.7   PRACTICAL VALUE OF NAND AND NOR GATES

If a NAND gate is followed by an inverter, it becomes an AND gate. Since both inverters and AND gates can be constructed from NAND gates, the NAND gate is seen to be a functionally complete set in itself. So also is the NOR gate, from which inverters and OR gates can be constructed.

From a practical standpoint, we note that the SSI chips containing gates may contain up to four 2-input gates per chip. If we are building a logic system using NAND or NOR chips, there will often be unused gates on some of the chips. If an invert function or an AND or OR function is required, we can often use these unused gates to produce the desired function without adding more chips to the system. If AND gates are being used, an invert function requires that an inverter chip be added regardless of the presence of unused AND gates in the system. The

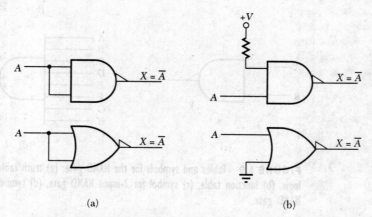

FIGURE 1.10   NAND and NOR gates used as inverters: (a) theoretical method, (b) practical method.

versatility of the NOR and NAND gates can often lead to a finished logic system requiring fewer chips than other implementations.

● ● ● **DRILL PROBLEMS** Sec. 1.4

1. Explain the difference between a truth table and a function table.

2. Construct a truth table for two series inverters; that is, the output of inverter *A* connects to the input of inverter *B*. The circuit input is the input of inverter *A*, while the output of the circuit is the output of inverter *B*.

3. Construct a truth table for a positive logic NOR gate followed by an inverter. What logic function does this circuit perform?

1.5
. . . .

LOGIC FAMILIES

SECTION OVERVIEW The important parameters of logic circuits that influence system performance are introduced in this section. Symbols representing required voltage levels, current levels, and switching times are defined. The effects of noise margin and fan-out are also considered.

It is unnecessary for a logic designer to understand the details of the electronic circuits that make up the gates. The manufacturers have designed these and other logic circuits as integrated circuits on silicon chips. These circuits can be fabricated in several different configurations and from several types of device. Those chips having in common a particular device and configuration are said to belong to a logic family.

Some examples of currently useful families are *transistor–transistor logic* (TTL), *emitter-coupled logic* (ECL), and *complementary MOS* (CMOS). These three families are discussed in more detail in the Appendix. Within each family of logic are several categories or types of circuit. For example, the TTL family includes the conventional TTL, low power, Schottky, low power Schottky, advanced Schottky (AS), and advanced low power Schottky (ALS) categories. All circuits of a given category must have compatible operating characteristics. The high-level voltage developed at the output of a gate must be sufficient to drive the input of any other gate to the same high level. The low-level output must pull the input of the next stage down to an acceptably low level. Certain current requirements must be met at each voltage level. Each category is compatible to some degree with other categories of the same family. For perfect compatibility, a single category is used for a given design. In general, an entire digital system, such as a computer, will use only one or two logic-circuit families. Hundreds to thousands of SSI or MSI logic elements are connected properly to form the required subsystems of the digital system.

Sometimes it is necessary to connect logic elements that are not of the same family. When this is done, an interface between the different elements may be required. An interface consists of circuits that translate the output signals from one family to the input signals required by the other family. Certain families can be combined without interface circuits. A family is said to be compatible with another family when both families can be interconnected without requiring interface circuits. Other combinations of families are popular enough that standard interface circuits are provided within the IC families.

As mentioned previously, this textbook emphasizes the TTL family. In addition to the popularity of this logic family, many other types of LSI or VLSI circuits are designed to be compatible with the TTL logic family. Several microprocessors and memories are implemented in MOS technology because more devices can be fabricated within a given silicon chip volume than can be fabricated with TTL. In general, these devices are designed to have compatible voltage levels with the TTL family and can easily be interfaced with TTL devices.

1.5.1 CURRENT AND VOLTAGE DEFINITIONS

Although it is not the purpose of this textbook to duplicate a data handbook for digital ICs, some definitions are pertinent to further discussion of logic circuits.

V_{IHmin} = *minimum input voltage that the logic element is guaranteed to interpret as the high logic level*

V_{ILmax} = *maximum input voltage that the logic element is guaranteed to interpret as the low logic level*

V_{OHmin} = *minimum high logic-level voltage appearing at the output terminal of the logic element*

V_{OLmax} = *maximum low logic-level voltage appearing at the output terminal of the logic element*

I_{IHmax} = *maximum current that will flow into an input when a specified high logic-level voltage is applied*

I_{ILmax} = *maximum current that will flow into an input when a specified low logic-level voltage is applied*

I_{OH} = *current flowing into the output when a specified high-level output voltage is present*

I_{OL} = *current flowing into an output when a specified low-level output voltage is present*

I_{OS} = *current flowing into an output when the output is shorted and input conditions are such to establish a high logic-level output.*

(Current flowing out of a terminal has a negative value.)

1.5.2 FAN-OUT

In a digital system, a given gate may drive the inputs to several other gates. The designer must be certain that the driving gate can meet the current requirements of the driven stages at both high and low voltage levels. The number of inputs that can be driven by the gate is referred to as the *fan-out* of the circuit. This figure is expressed in terms of the number of standard inputs that can be driven. Most circuits of a family will require the same input current, but a few may require more. If so, the specs for such a circuit will indicate that the input is equivalent to some multiple of standard loads. For example, a circuit may present an equivalent input of two standard loads. If fan-out of a gate is specified as 10, only five of these circuits could be safely driven. Figure 1.11 shows an AND gate loaded with 4 inputs, assuming each circuit presents one standard load to the AND gate output.

In several handbooks, the current requirements are given and fan-out can be calculated. For example, one TTL gate has the following current specs:

$$I_{IH} = 40\mu \text{ A} \qquad I_{IL} = -1.6 \text{ mA}$$
$$I_{OH} = -400\mu \text{ A} \qquad I_{OL} = 16 \text{ mA}$$

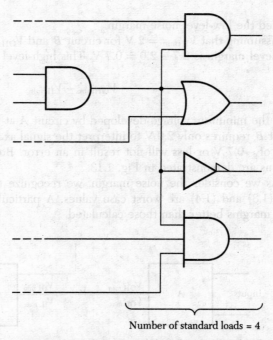

Number of standard loads = 4

FIGURE 1.11 Partial schematic demonstrating loading.

If this gate is to drive several other similar gates, we see that the output current capability of the stage is 10 times that required by the input. We note that the output stage can drive 400 μA into the following stages at the high level and sink 16 mA at the low level. The fan-out of this gate is 10.

1.5.3 NOISE MARGIN

Although current requirement is the major factor in determining fan-out, input capacitance or *noise margin* may further influence this figure. Noise margin specifies the maximum amplitude noise pulse that will not change the state of the driven stage. This assumes that the driving stage presents a worst case logic level to the driven stage. Noise margin can be evaluated from a consideration of the voltage levels V_{IHmin}, V_{ILmax}, V_{OHmin}, and V_{OLmax}. Figure 1.12 shows two logic circuits that are cascaded.

If we assume that $V_{\text{ILmax}} = 0.8$ V for circuit B, this means that the input must be less than 0.8 V to guarantee that circuit B interprets this value as a low level. If circuit A has a value of $V_{\text{OLmax}} = 0.4$ V, a noise spike of less than the difference $0.8 - 0.4$ V cannot lead to a level misinterpretation by circuit B. The difference

$$V_{\text{ILmax}} - V_{\text{OLmax}} \tag{1.3}$$

is called the low-level noise margin.

Assuming that $V_{\text{IHmin}} = 2$ V for circuit B and $V_{\text{OHmin}} = 2.7$ V for circuit A, the high-level margin is $2.7 - 2.0 = 0.7$ V. This high-level noise margin is found from

$$V_{\text{OHmin}} - V_{\text{IHmin}} \tag{1.4}$$

Since the minimum voltage developed by circuit A at the high level is 2.7 V, while circuit B requires only 2.0 V to interpret the signal as a high level, a negative noise spike of -0.7 V or less will not result in an error. Both low- and high-level noise margins are demonstrated in Fig. 1.13.

As we consider the noise margin, we recognize that the values calculated in Eqs. (1.3) and (1.4) are worst case values. A particular circuit could have actual noise margins better than those calculated.

FIGURE 1.12 Cascaded stages used to calculate noise margin.

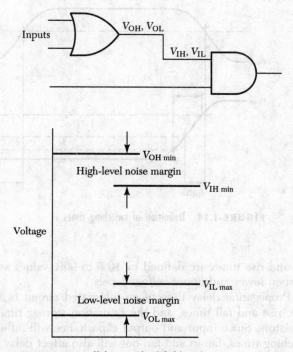

FIGURE 1.13 Noise margin definitions.

As more gate inputs are connected to a given output, the voltages generated at both high and low levels are affected as a result of increased current flow. Thus, fan-out is influenced by noise margin.

1.5.4 SWITCHING TIMES

Another quantity that is used to characterize switching circuits is the speed with which the device responds to input changes. For switching circuits, the graph of Fig. 1.14 is useful in defining delay times. This figure assumes an inverting gate. There is a finite delay between the application of the input pulse and the output response. A quantitative measure of this delay is the difference in time between the point where e_{in} rises to 50% of its final value and the time when e_{out} falls to its 50% point. This quantity is called leading-edge delay t_{pHL}. The trailing-edge delay t_{pLH} is the time difference between 50% points of the trailing edges of the input and output signals. The *propagation delay* is defined as the average of t_{pHL} and t_{pLH}, or

$$t_{pd} = \frac{t_{pHL} + t_{pLH}}{2}.$$

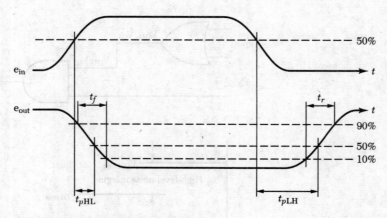

FIGURE I.14   Definition of switching times.

Fall and rise times are defined by 10% to 90% values as the output voltage swings between lower and upper voltage levels.

Propagation delay time of an integrated circuit is a function of passive delay time, rise and fall times, and the saturation storage time of the circuit's individual transistors. Since input and output capacitance will influence the integrated circuit switching times, fan-in and fan-out will also affect delay times. Switching times are sometimes specified by graphs showing the various times as functions of the number of standard input loads with specified driving conditions. Figure 1.15 shows a typical graph.

There will always be a certain amount of stray capacitance at the input and output terminals; thus, the careful designer will allow for the effect on switching time of these stray values. Generally a data sheet will specify switching times such as propagation delay times under specified loading conditions. A typical test load is a 1k Ω resistor in parallel with a 40 pF capacitor.

An understanding of the definitions given in the preceding paragraphs allows the designer to use logic gates as building blocks in digital systems. The TTL family has been the workhorse for many years in SSI and MSI applications. Fast and versatile, no other line offers as great a variety of circuits. The fabrication of resistors requires more chip volume than do transistors, and TTL chips use several resistors per gate. Consequently, applications in LSI circuits are somewhat limited.

CMOS has been used for all levels of integration from SSI through VLSI. The high-density chips so produced are ideally applied where large memories are required, although speed of operation sometimes limits performance. This family also exhibits higher input and output impedances than exhibited by TTL. While less current is required at the input of the CMOS circuit, the output cannot supply as much load current as a TTL circuit. The power dissipation of the CMOS gate is very small compared to the TTL gate and this technology is popular in systems that must minimize source power drain such as hand-held calculators.

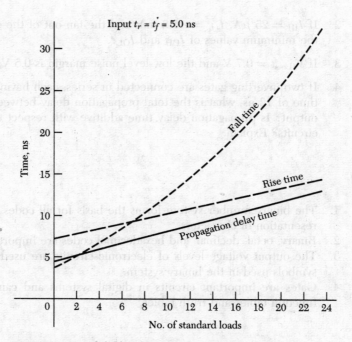

FIGURE 1.15 Switching times as a function of loading.

MOS circuitry is heavily used for such LSI circuits as microprocessors and large-memory chips. Earlier microprocessors were fabricated with p-MOS. Now most use n-MOS technology, but CMOS microprocessors are available also.

ECL logic is the highest speed family available. It does not offer as wide a variety of circuit types as TTL, but it is very popular in large computers (mainframe computers) and high-speed minicomputers. Since speed of operation of computers is important, most large computers use ECL technology to maximize performance.

The fundamental principles of logic systems are independent of the type of family used to implement the system. However, each family may influence the practical design of the system. When it is necessary to discuss a specific logic circuit, this text will consider the TTL family. This does not imply that other families are unimportant, but there is simply not enough space in the text to cover each family. Appendix 1 summarizes the basic characteristics of TTL, ECL, and CMOS logic and discusses some practical aspects of TTL gates.

●●● **DRILL PROBLEMS** Sec. 1.5

1. If the output of gate A drives an input to gate B, how should $V_{OHmin}(A)$ relate to $V_{IHmin}(B)$? How should $V_{OLmax}(A)$ relate to $V_{ILmax}(B)$?

2. If $I_{\text{IH}} = 25\ \mu\text{A}$, $I_{IL} = -1.2$ mA, and the fan-out of the circuit is 8, what are the minimum values of I_{OH} and I_{OL}?

3. If $V_{\text{ILmax}} = 0.7$ V and the low-level noise margin is 0.5 V, what is V_{OLmax}?

4. If two inverting gates are connected in series, each having a propagation delay time of 12 ns, what is the total propagation delay between circuit input and output? Is propagation delay time additive with respect to number of series circuits? Explain.

SUMMARY

1. The binary number system forms the basis for all codes used in machine representation of characters.
2. Binary, octal, decimal, and hexadecimal codes are important in digital systems.
3. The output voltage levels of electronic circuits are used to represent the two symbols used in the binary system.
4. Gates are important circuits in digital systems and can be characterized by function tables and truth tables.

CHAPTER I PROBLEMS

Note: Answers to starred problems may be found in Answers to Problems in the back of the book.

● **Sec. 1.1**

1.1 Explain why the binary number system is used for most digital systems.

°**1.2** Assume a circuit output can vary from 0 to 5 V. A detection circuit is capable of distinguishing levels that vary by 20 mV. How many different numbers can be represented by this system?

1.3 The drift in output voltage with temperature of a circuit is limited to ±10 mV. If the output voltage of this circuit is to accurately represent 1000 distinct numbers, what is the minimum size of the output voltage range?

● **Sec. 1.2.1**

°**1.4** Convert the following binary numbers to decimal.
  a. 11101.1101
  b. 10001.0001
  c. 10.1101101
  d. 110010110101.01

1.5 Convert the following decimal numbers to binary. Results should be accurate to within 0.01_{10}.
a. 584
b. 119.84
c. 33.45
d. 1.583

1.6 Convert 37.432_{10} to binary with an accuracy of 0.002_{10}.

°1.7 Convert 128.286_{10} to binary with an accuracy of 0.001_{10}.

Sec. 1.2.2

1.8 Convert the decimal numbers of Prob. 1.5 to octal and binary-coded octal with an accuracy of 0.01_{10}.

1.9 Convert the decimal number of Prob. 1.6 to octal and binary-coded octal with an accuracy of 0.002_{10}.

°1.10 Convert the decimal number of Prob. 1.7 to octal and binary-coded octal with an accuracy of 0.001_{10}.

1.11 Convert the binary numbers of Prob. 1.4 to octal and binary-coded octal.

1.12 Convert the binary numbers of Prob. 1.19 to octal and binary-coded octal.

Sec. 1.2.3

1.13 Convert the decimal numbers of Prob. 1.5 to hex and binary-coded hex with an accuracy of 0.01_{10}.

°1.14 Convert the decimal number of Prob. 1.6 to hex with an accuracy of 0.002_{10}.

1.15 Convert the decimal number of Prob. 1.7 to hex with an accuracy of 0.001_{10}.

1.16 Convert the binary numbers of Prob. 1.4 to hex and binary-coded hex.

1.17 Convert the binary numbers of Prob. 1.19 to hex and binary-coded hex.

Sec. 1.2.4

°1.18 Convert the following decimal numbers to BCD.
a. 748
b. 5668

1.19 Convert the following binary numbers to decimal and to BCD.
a. 11011101
b. 10001110
c. 1.001
d. 101.1011

1.20 Convert the binary numbers of Prob. 1.4 to binary-coded decimal.

● Sec. 1.2.5

1.21 Write the ASCII code for the message

   This Code Is Neat.

1.22 Write the ASCII code for the message

   Most microcomputers use ASCII code.

● Sec. 1.3

°**1.23** A positive logic system with logic high defined as any voltage that exceeds 2.4 V and logic low as any voltage under 0.8 V is used. Six successive voltages from most-significant bit to least-significant bit are 2.8 V, 3.4 V, 0.4 V, 3.1 V, 0.6 V, 2.5 V. What binary number is represented by these voltages? Explain the problem if one of these voltages measured 1.7 V.

● Sec. 1.4

1.24 Two inputs A and B are inverted and applied to a NOR gate. Construct the corresponding truth table for positive logic.

1.25 Two inputs A and B are inverted and applied to a NOR gate. The gate output is connected to an inverter. Construct a truth table for the inverter output for positive logic.

°**1.26** An input A is inverted and applied to a NOR gate. The other input B is applied directly to the gate. Construct the corresponding truth table for positive logic.

1.27 Two inputs A and B are inverted and applied to a NAND gate. Construct the corresponding truth table for positive logic.

1.28 An input A is inverted and applied to an AND gate. The other input B is applied directly to the gate. Construct the corresponding truth table for positive logic.

°**1.29** Construct a truth table for the circuit of Fig. P1.29.

1.30 Show how to connect a quad NOR gate chip to realize the circuit shown in Fig. P1.30 with a single chip. This quad NOR gate chip contains four, 2-input NOR gates.

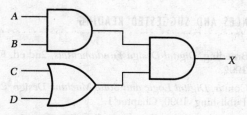

FIGURE PI.29

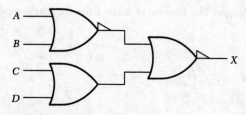

FIGURE PI.30

1.31 Show how to connect a quad 2-input NAND gate chip to realize the circuit shown in Fig. P1.31 with a single chip.

1.32 Show how to connect inverters to a positive logic OR gate to convert to a positive logic AND gate.

1.33 Show how to connect positive logic, 2-input AND gates to create a 3-input AND gate. Show how to connect any unused inputs.

● **Sec. 1.5**

°1.34 The current specs on the 7408-quad, 2-input AND gate are $I_{\text{IHmax}} = 40\ \mu\text{A}$, $I_{\text{ILmax}} = -1.6\ \text{mA}$, $I_{\text{OH}} = -800\ \mu\text{A}$, and $I_{\text{OL}} = 16\ \text{mA}$. How many similar inputs can one 7408 gate drive?

1.35 Repeat Prob. 1.34 for the 7428-quad, 2-input NOR buffer. The current specs are $I_{\text{IHmax}} = 40\ \mu\text{A}$, $I_{\text{ILmax}} = -1.6\ \text{mA}$, $I_{\text{OH}} = -2400\ \mu\text{A}$, and $I_{\text{OL}} = 48\ \text{mA}$.

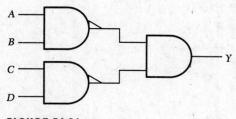

FIGURE PI.31

REFERENCES AND SUGGESTED READING

1. K. J. Breeding, *Digital Design Fundamentals,* 2nd ed. Englewood Cliffs, NJ: Prentice-Hall, 1992.
2. D. J. Comer, *Digital Logic and State Machine Design,* 2nd ed. New York: Saunders College Publishing, 1990, Chapter 1.
3. Engineering Staff, *TTL Logic.* Dallas, TX: Texas Instruments, 1988.
4. F. A. Mann, *Overview of IEEE Standard 91-1984.* Dallas, TX: Texas Instruments, 1984.
5. S. H. Unger, *The Essence of Logic Circuits.* Englewood Cliffs, NJ: Prentice-Hall, 1989.

Chapter 2

Boolean Algebra and Mapping Methods

L ogic circuits are subdivided into two types: combinational and sequential. Combinational circuits (sometimes called combinatorial circuits) have outputs that equal 1 only for certain combinations of input variables. Other input combinations cause the output to equal 0. The output is a direct function of the input values. There is no time dependence nor is there a dependence on some previously applied input combination that is no longer present. Sequential circuits have outputs that depend on present input, time, and past input history. These circuits are more complex than combinational circuits and use combinational circuits as building blocks. Before we can proceed to sequential design, we must understand combinational design. Thus, this and the following two chapters will be devoted to a study of combinational logic theory and application.

In designing any digital system there are three objectives that an effective design procedure should achieve. These objectives are to (1) build a system that operates within the given specifications, (2) build a reliable system, and (3) minimize resources. The principles to be discussed in this chapter allow objectives 1 and 2 to be met in a straightforward manner. Minimization of resources is a more complex issue and one that depends on several variables. For example, one might want to minimize the number of gates used in a circuit or the number of inputs. Often the number of IC chips used is a more important quantity to minimize than is the number of gates. There are other occasions when chip cost is the most significant quantity, and a system implemented with less chips may be more expensive due to higher individual chip cost. Other situations may call for a minimization of the designer's time rather than a minimum component cost. Still other cases may call for the entire circuit to be fabricated on a single chip since many firms own IC fabrication facilities or use custom fabrication companies. An important quantity to minimize in some instances is the time between start of design and availability of finished product. Minimization of this time often earns an edge over slower competitors in the same field.

Obviously, no design procedure can minimize all resources simultaneously. We will direct the following discussions toward the minimization of gates and inputs, which is a strength of Boolean algebra and Karnaugh map techniques. In later chapters, we will indicate alternate methods of logic design that can minimize other quantities listed previously.

2.1

BOOLEAN ALGEBRA

SECTION OVERVIEW Boolean algebra is introduced in this section followed by a discussion of several important algebraic relations. The basis for the reduction of logic functions to simpler forms is considered. The final topic of the section is that of assertion level and dependency notation.

This type of algebra for logic circuits was named after its inventor George Boole (1815–1864). Although available for several years, Boolean algebra was used little until the invention of the digital computer. It now serves as the basis for almost all design and analysis methods in digital systems.

We introduce the subject of Boolean algebra by considering a practical problem: that of identifying the presence of certain digital codes on a set of lines. A system may be transmitting 4 bits of binary information over 4 lines to another system. With 4 lines, 16 unique codes could be represented. Each code transmitted will exist for a certain length of time after which a new code can occur. The receiving system may need to identify the presence of certain transmitted codes. As an example, suppose the system is to identify the occurrence of the codes representing the decimal numbers 1, 4, 5, 9, 11, and 12. Each time one of these codes appears on the 4 lines, a circuit is to generate an output of 1. When any other code is present, the output should be 0. One method of implementing this system is shown in Fig. 2.1.

The four code lines are labeled with the symbols A, B, C, and D proceeding from the most significant bit (MSB) to the least significant bit (LSB). The output is designated F. Each of these lines represents a Boolean variable. A variable is an element that can take on a value of either 0 or 1. Variables are generally designated by alphabetic symbols. Often it is convenient to use a single letter for a variable symbol, although complex systems may use longer symbols for mnemonic purposes.

The input code for decimal 1 is $ABCD = 0001$. The only time we will have this combination of variables present simultaneously is when the code for 1 is on the lines. In order to check this simultaneous condition, we use the 4-input AND

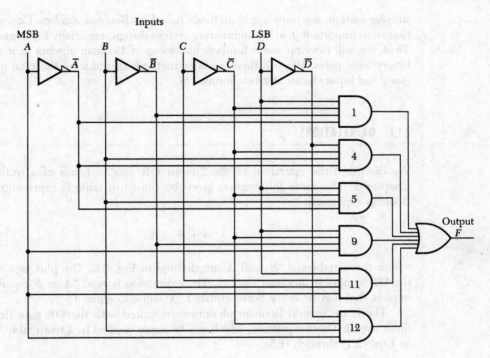

FIGURE 2.1   A logic system.

gate number 1, inverting all lines applied to the gate except the least significant bit. Each of the gate inputs will equal 1 only when 0001 is present on the lines. The remaining five AND gates perform the same function for the codes representing 4, 5, 9, 11, and 12. The OR gate causes the circuit output to equal 1 when any of these codes appears on the lines.

At any given time, each of these variables A, B, C, and D will equal either 0 or 1. When we assign a variable name to a line, we imply a value for the inverted variable. We create the inverted value by connecting an inverter to the line. The inverted variable is designated by the same symbol as the variable with the addition of an overbar. If A takes on a value of 0, $\bar{A}$ must equal 1. If A takes on a value of 1, $\bar{A}$ must equal 0. This allows us to write binary codes in terms of symbols. The code for decimal 9 is 1001. In terms of symbols we can write $A\bar{B}\bar{C}D$. Writing the code this way implies that when the code 1001 is present, an AND gate with inputs A, $\bar{B}$, $\bar{C}$, and D would produce an output of 1. We will expand on the use of this notation later.

While the problem of creating $F = 1$ whenever any of the codes for 1, 4, 5, 9, 11, or 12 appears on the line is solved by the circuit of Fig. 2.1, there are simpler methods of solving this problem. A later section will show that this system can be implemented with fewer gates and fewer connections. In order to implement the simpler system, we must apply methods based on Boolean algebra. Cost minimization is an important goal in engineering system design, especially for larger systems. Thus, we will develop some fundamental ideas of Boolean algebra as it applies to binary logic gates. We will then consider methods of reducing the total number of gates and inputs to an absolute minimum.

2.1.1   OR RELATIONS

We discussed the operation of the 2-input OR gate in terms of a truth table in Chapter 1. The same information given by the truth table is represented by the Boolean expression.

$$A + B = X$$

where the variables A, B, and X are defined in Fig. 2.2. The plus sign stands for the OR symbol in Boolean algebra. The expression is read "A or B equals X" and implies that if A or B (or both) equals 1, X will also equal 1.

There are several Boolean identities associated with the OR gate that can be quite useful. These equations, which can be easily verified by a truth table, are listed as Eqs. (2.1) through (2.5).

$$A + 0 = A \tag{2.1}$$

$$A + 1 = 1 \tag{2.2}$$

$$A + A = A \tag{2.3}$$

$$A + B = B + A \tag{2.4}$$

$$A + B + C = (A + B) + C = A + (B + C) \tag{2.5}$$

Equation (2.4) is the commutative law for the OR relation. Equation (2.5) is the associative law for the OR relation. The first equation can be proven from the

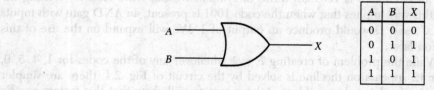

A	B	X
0	0	0
0	1	1
1	0	1
1	1	1

FIGURE 2.2   An OR gate with truth table.

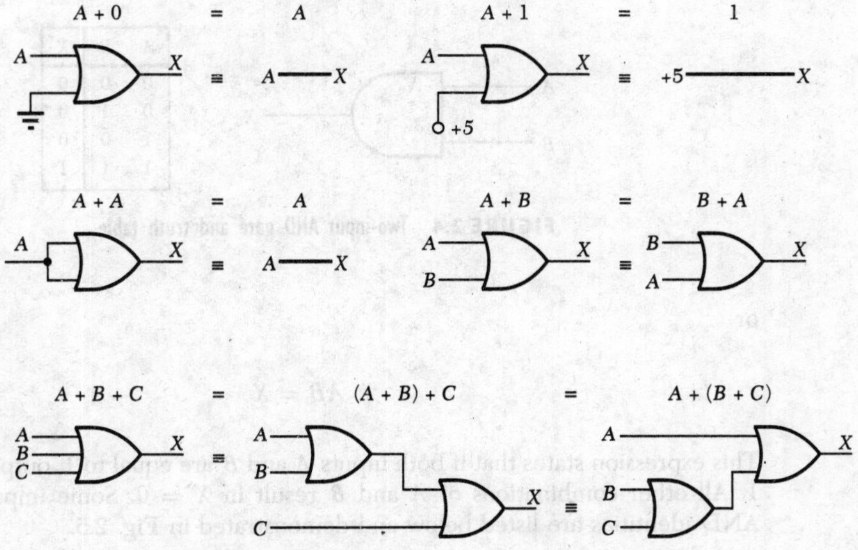

FIGURE 2.3 Implementation of OR relations.

truth table by setting $B = 0$. Rows 1 and 3 show that $X = A$ for this case. If $B = 1$, rows 2 and 4 prove that $A + 1 = 1$ as indicated by Eq. (2.2). Equation (2.3) is proven by considering rows 1 and 4 in which A and B are equal. If A and B values are exchanged in the truth table, we see that the output is unchanged, and thus Eq. (2.4) is demonstrated. The associative law can also be proven with a truth table. We should recognize the physical meaning of these OR relations. Figure 2.3 demonstrates the implementation of each equation using positive logic.

We note in Fig. 2.3 that the OR relations can be used in certain cases to minimize the physical circuit. For example, an OR gate with one input connected to ground and the other input connected to A can be replaced with a wire connected to A as shown. Three of the five equivalencies of Fig. 2.3 show instances of replacing an OR gate by a single conducting wire. The extension of these relations to a higher number of input variables is easily made.

2.1.2 AND RELATIONS

A 2-input AND gate is shown in Fig. 2.4 along with its truth table. The Boolean expression meaning the AND operation of A and B is

$$A \cdot B = X$$

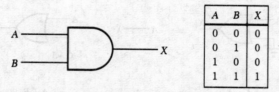

A	B	X
0	0	0
0	1	0
1	0	0
1	1	1

FIGURE 2.4   Two-input AND gate and truth table.

or

$$AB = X$$

This expression states that if both inputs A and B are equal to 1, output X will equal 1. All other combinations of A and B result in $X = 0$. Some important Boolean AND identities are listed below and demonstrated in Fig. 2.5.

$$A \cdot 0 = 0 \tag{2.6}$$

$$A \cdot 1 = A \tag{2.7}$$

$$A \cdot A = A \tag{2.8}$$

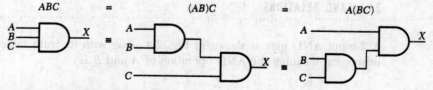

FIGURE 2.5   Implementation of AND relations.

$$A \cdot B = B \cdot A \tag{2.9}$$

$$A \cdot B \cdot C = (A \cdot B) \cdot C = A \cdot (B \cdot C) \tag{2.10}$$

Again, these identities can be proven by the AND gate truth table. Equation (2.9) is the commutative law for the AND relation. Equation (2.10) is the associative law for the AND relation. The physical implementations of these relations, shown in Fig. 2.5, demonstrate the reduction in circuitry that can result from using the equations.

2.1.3 OTHER RELATIONS

There are several other relations that are useful in logic design. Many of these pertain to the inversion of a variable. Inverting a variable is also called complementing a variable. If we invert or complement a variable twice, the result equals the input variable. That is,

$$\bar{\bar{A}} = A \tag{2.11}$$

A variable ORed with its complement equals 1, since one of these inputs will equal 1. We can write

$$A + \bar{A} = 1 \tag{2.12}$$

If a variable is ANDed with its complement, the result is always 0 since one of the inputs will equal 0. This gives the equation

$$A\bar{A} = 0 \tag{2.13}$$

Another form that is encountered in logic design is given by $A + \bar{A}B$ or $\bar{A} + AB$. The truth table of Fig. 2.6 can be used to show that

$$A + \bar{A}B = A + B \tag{2.14a}$$

A	B	$\bar{A}$	$A + \bar{A}B$	$A + B$	$\bar{A} + AB$	$\bar{A} + B$
0	0	1	0	0	1	1
0	1	1	1	1	1	1
1	0	0	1	1	0	0
1	1	0	1	1	1	1

FIGURE 2.6 Truth table to prove Eq. (2.14).

and

$$\bar{A} + AB = \bar{A} + B \tag{2.14b}$$

The last relation we will consider is called the distributive law. It can be expressed as

$$A(B + C) = AB + AC \tag{2.15}$$

Figure 2.7 shows the implementation of this relation. We note that one realization of the expression uses only two gates while the other form uses three. Figure 2.8 proves this relation with a truth table.

A very significant relation can be considered at this point. If we encounter the equation

$$F = A\bar{B} + \bar{A}\bar{B}$$

we can apply the distributive law to get

$$F = \bar{B}(A + \bar{A})$$

Equation (2.12) is then applied to result in

$$F = \bar{B}$$

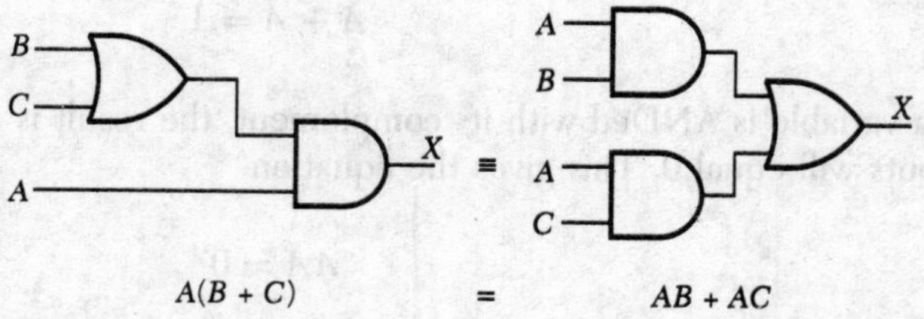

$$A(B + C) \qquad = \qquad AB + AC$$

FIGURE 2.7   Implementation of distributive law.

A	B	C	B + C	A(B + C)	AB	AC	AB + AC
0	0	0	0	0	0	0	0
0	0	1	1	0	0	0	0
0	1	0	1	0	0	0	0
0	1	1	1	0	0	0	0
1	0	0	0	0	0	0	0
1	0	1	1	1	0	1	1
1	1	0	1	1	1	0	1
1	1	1	1	1	1	1	1

FIGURE 2.8   Truth table to prove distributive law.

The starting equation contains two terms that are logically adjacent. The difference between $A\bar{B}$ and $\bar{A}\bar{B}$ lies in the variable A. In one term A appears, while $\bar{A}$ appears in the other term. The remaining variable in both terms is $\bar{B}$. If only a single variable changes between two terms, the terms are said to be logically adjacent. The expression

$$X = AB\bar{C}\bar{D} + AB\bar{C}D$$

demonstrates two 4-variable terms that are logically adjacent. The significance of logical adjacency will become apparent when we discuss the Karnaugh map later in the chapter.

We will see later that the Karnaugh map is more efficient than Boolean algebra in reducing a logic expression to its simplest form. We will show an example of a reduction problem at this point simply to illustrate the use of the preceding relations. Let us assume that the logic expression $F = A\bar{B} + \bar{A}\bar{B} + AB\bar{C} + \bar{A}B\bar{C}$ is to be reduced. We will tabulate the steps that can be used to reduce this expression.

$$F = A\bar{B} + \bar{A}\bar{B} + AB\bar{C} + \bar{A}B\bar{C}$$

$$= \bar{B}(A + \bar{A}) + B\bar{C}(A + \bar{A}) \qquad \text{by Eq. (2.15)}$$

$$= \bar{B} + B\bar{C} \qquad \text{by Eq. (2.12)}$$

$$= \bar{B} + \bar{C} \qquad \text{by Eq. (2.14b)}$$

This final result can be implemented by a single 2-input OR gate (ignoring inverters), whereas the original expression requires two 2-input AND gates, two 3-input AND gates, and one 4-input OR gate.

There are several manipulations that can be done in Boolean algebra that require either creativity or guess work. For example, suppose we are asked to prove the equality

$$AB + BC + \bar{A}C = AB + \bar{A}C$$

One approach to this is to AND the quantity $B + 1$ with the term $\bar{A}C$ appearing on the right side of the equation. Since $B + 1 = 1$ and $\bar{A}C1 = \bar{A}C$, the value of the equation is unchanged. We continue with

$$AB + \bar{A}C = AB + \bar{A}C(B + 1)$$

$$= AB + \bar{A}CB + \bar{A}C \qquad \text{by Eq. (2.15)}$$

$$= B(A + \bar{A}C) + \bar{A}C \qquad \text{by Eq. (2.15)}$$

$$= B(A + C) + \bar{A}C \qquad \text{by Eq. (2.14a)}$$

$$= AB + BC + \bar{A}C$$

The proper approach to proving a Boolean relationship is not always obvious as demonstrated by the previous proof. This type of ambiguity does not exist in Karnaugh mapping methods as we shall later see.

● ● ● **DRILL PROBLEMS Secs. 2.1.1 to 2.1.3**

1. Implement the expression $X = (A + B)(\bar{A} + \bar{B})$ with two OR gates, an AND gate, and two inverters.

2. Implement the expression X in Problem 1 using two AND gates, an OR gate, and two inverters.

3. Reduce the expression $X = \bar{A} + A\bar{B} + BC$ to simplest terms.

4. Reduce the expression $F = (A + \bar{B} + \bar{C})(\bar{A} + B + C)BC$ to simplest terms.

5. Prove that $C = AC + \bar{A}C$.

6. Prove that $AB + A\bar{B} + \bar{A}C = A + C$.

● **2.1.4 DEMORGAN'S LAWS**
....

These two laws allow us to convert between forms of logic equation and also to convert between types of gate used. DeMorgan's laws are given by

$$\overline{ABC \cdots N} = \bar{A} + \bar{B} + \bar{C} + \cdots + \bar{N} \qquad (2.16)$$

and

$$\overline{A + B + C + \cdots + N} = \bar{A}\bar{B}\bar{C} \cdots \bar{N} \qquad (2.17)$$

We can state Eq. (2.16) in words as follows:

When N variables are ANDed and then inverted, the only combination leading to a 0 output occurs when all N variables equal 1. If we invert the N variables and OR the result, the only combination leading to a 0 output again occurs when all N variables equal 1.

The second law can be stated as:

When N variables are ORed and then inverted, the only combination leading to a 1 output occurs when all N variables equal 0. If we invert the N variables and AND the result, the only combination leading to a 1 output again occurs when all N variables equal 0.

These word statements essentially prove the results without the necessity of using truth tables. A physical implementation of the two laws again leads to significant practical results. Figure 2.9 shows these equivalencies for three input variables.

In terms of gates, we see that a NAND gate is equivalent to an OR gate with inverted inputs and a NOR gate is equivalent to an AND gate with inverted inputs. As a memory aid we note that any gate, AND or OR, followed by an inverter is equivalent to the opposite type gate, OR or AND, with inverted inputs.

Alternate forms of DeMorgan's laws are found by complementing both sides of Eqs. (2.16) and (2.17) yielding

$$ABC \cdots N = \overline{\bar{A} + \bar{B} + \bar{C} + \cdots + \bar{N}} \tag{2.18}$$

and

$$A + B + C + ... + N = \overline{\bar{A}\bar{B}\bar{C} \cdots \bar{N}} \tag{2.19}$$

Application of these two laws along with the Boolean identities allows a designer more flexibility to implement a given expression in terms of the gates available. In constructing a system, several unused gates may be available on the IC chip used. If these are NANDs, for example, we can use them for other types of gates. Obviously, if an AND gate is required, we can follow a NAND by an inverter and produce this function. In this way, we minimize unused gates on chips that are already in the system rather than add new chips.

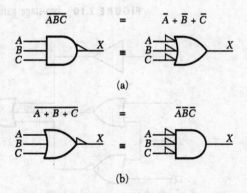

FIGURE 2.9 Implementations of DeMorgan's laws:
(a) first law, (b) second law.

Let us demonstrate the key points in this method. Suppose we have a function

$$X = A(B + CD)$$

that we must implement with NOR gates and inverters. The form of the function we should obtain is one that can be realized directly with NOR gates. An expression such as

$$Y = \overline{\overline{(R + S)} + \overline{(P + Q)}}$$

results from two stages of NOR gates as shown in Fig. 2.10. We can approach this form for the function X by applying DeMorgan's laws to obtain

$$X = \overline{\overline{A} + \overline{B + CD}} = \overline{\overline{A} + \overline{B}\,\overline{CD}} = \overline{\overline{A} + \overline{B}(\overline{C} + \overline{D})}$$

$$= \overline{\overline{A} + \overline{B}\overline{C} + \overline{B}\overline{D}} = \overline{\overline{A} + \overline{B + C} + \overline{B + D}}$$

The final expression can be realized as shown in Fig. 2.11.

Rather than striving to put an expression into NOR or NAND form, we can convert directly to the gates required. The expression $X = A(B + CD)$ can be realized as shown in Fig. 2.12. If we want to convert to NOR gates and inverters,

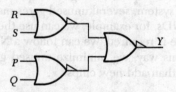

FIGURE 2.10   Two-stage gating.

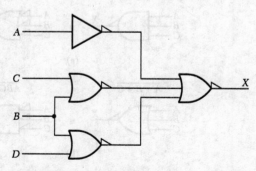

FIGURE 2.11   Implementation of $X = A(B + CD)$ using NOR gates.

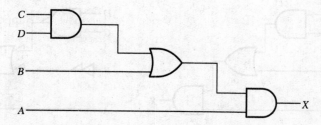

FIGURE 2.12   Realization of $X = A(B + CD)$.

we first note that the OR gate is similar to a NOR. By adding two inversions at the output of this OR, we can convert to a NOR gate without changing the function of the circuit as shown in Fig. 2.13(a). To change the AND gates to NOR gates, we must introduce inverters at the gate inputs. To preserve the logic function, we introduce two inverters at each desired point as shown in Fig. 2.13(b). Next, the input inverters are moved to the outputs of the two gates while the AND symbol is changed to the OR. The result is shown in Fig. 2.13(c). We note that this realization is not unique when compared with the earlier realization of the same function as shown in Fig. 2.11. Figure 2.14 shows the development of the NAND gate realization of the function $X = A(B + CD)$.

DeMorgan's laws can also be used to reduce logic expressions. For example, given the expression $X = ABC + \bar{A}CD + \bar{B}CD$, we could reduce as follows:

$$X = ABC + \bar{A}CD + \bar{B}CD$$

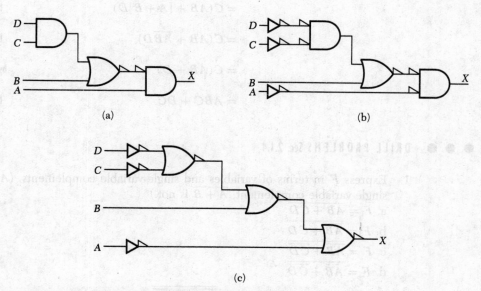

FIGURE 2.13   Steps in producing a NOR gate realization of X.

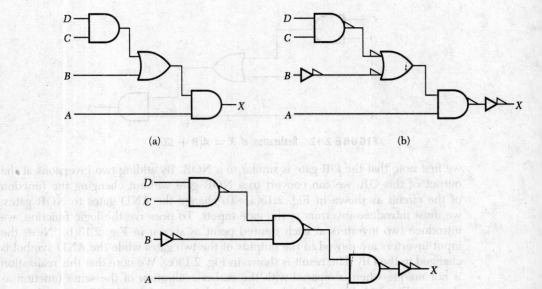

FIGURE 2.14 Steps in producing a NAND gate realization of *X*.

$$= C(AB + \bar{A}D + \bar{B}D) \qquad \text{by Eq. (2.15)}$$

$$= C(AB + [\bar{A} + \bar{B}]D) \qquad \text{by Eq. (2.15)}$$

$$= C(AB + \overline{AB}D) \qquad \text{by Eq. (2.16)}$$

$$= C(AB + D) \qquad \text{by Eq. (2.14a)}$$

$$= ABC + DC \qquad \text{by Eq. (2.15)}$$

● ● ●   **DRILL PROBLEMS Sec. 2.1.4**

1. Express F in terms of variables and single-variable complements. ($\bar{A}$ is a single-variable complement; $\overline{A+B}$ is not.)

    a. $F = \overline{AB + CD}$

    b. $F = \overline{\overline{AB} + CD}$

    c. $F = \overline{\overline{AB} + \overline{CD}}$

    d. $F = \overline{\bar{A}\bar{B} + \overline{CD}}$

2. Reduce to simplest form $X = \overline{\overline{\bar{A}\bar{B}\bar{C}} + \bar{A}B}$.

2.1.5 DEPENDENCY NOTATION

We have mentioned the IEEE standard for logic symbols earlier in Chapter 1. One of the fundamental ideas making the new standard more powerful is that of dependency notation. This notation provides the means of noting relationships between inputs, outputs, or inputs and outputs without showing all elements and interconnections involved. For a full understanding of the concept of dependency notation, the IEEE standard, *IEEE Std. 91-1984*, should be consulted.

We will present here only a few ideas and symbols related to this symbology. Logic polarity indicators are used to eliminate the need for specifying whether positive logic or negative logic definitions apply. This leads to an advantage in handling mixed logic systems. When used for an input, the triangular polarity indicator of Fig. 2.15 tells us that a low applied level will lead to an internal state of 1 for the device. When used for an output, the triangle indicates that the internal 1 state will produce a low external level.

The IEEE/IEC symbol for a quad 2-input NAND gate is shown in Fig. 2.16(a). We will not use this type of symbol for gates. Instead, the symbol of Fig. 2.16(b) will be used since we will often need to consider individual gates. The character & is used to indicate the AND operation, while ≥ 1 would indicate the OR operation for the standard symbols.

Figures 2.16(c) and (d) show two other symbols of importance to gate circuits. These are the open-collector output and the three-state output symbols which appear inside the gate outline near the output terminal when such gates are used.

Input	Output

FIGURE 2.15 Logic polarity indicator.

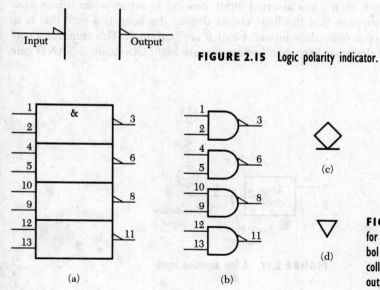

(a)   (b)   (c)

(d)

FIGURE 2.16 (a) IEEE/IEC symbol for quad 2-input NAND chip, (b) Symbol to be used in this text, (c) Open-collector output symbol, (d) Three-state output symbol.

2.1.6 ASSERTION LEVELS

The meaning of the infinitive "to assert" is to put in force or to cause a positive action. This term is used to describe the state of logic signals in digital systems. It is conventional to associate the binary value 1 with a signal that is asserted. It follows that a signal that is not asserted has a value of 0.

As we shall shortly see, a signal may be asserted when the voltage level of the signal is either high or low. Thus, in logic circuits we refer to high asserted or low asserted signals. If a terminal is a low assertion input, the circuit requires this input to be at the low voltage level to activate the circuit. High assertion inputs can activate or assert the circuit only for high-level input voltages. A low asserted output indicates that when the circuit is active, the output exists at the low voltage level.

There are two obvious situations that call for low assertion logic levels. The first occurs when the physical characteristics of some circuit require a low logic level for activation. The second occurs when it is convenient to change the logical function performed by a gate from that performed when positive logic is used. Several modern logic systems involve both high and low asserted signals and are therefore mixed logic systems rather than strictly positive or strictly negative.

The indicator light of Fig. 2.17 demonstrates a circuit with a low assertion input. The resistor-lamp combination will be active only when current flows through this circuit. Since the upper terminal is connected to +5 V, no current will flow if the input to this circuit is also at 5 V. Thus, when the logic circuit output is near the high level of 5 V, the indicator lamp is not asserted. In order to cause a voltage drop across the resistor and lamp, the logic circuit must drive point X to the 0 V level. The input to the lamp is then a low asserted input, causing an action when driven low.

Now let us suppose that the logic circuit driving the lamp is a gate that is to assert the lamp input only when inputs A and B are both high. This requires a gate having an output that goes low when both inputs are high. Obviously, a NAND gate

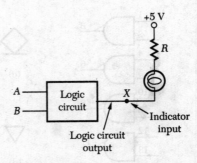

FIGURE 2.17 A low assertion input.

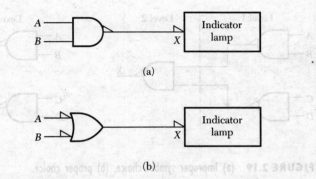

FIGURE 2.18 (a) A low assertion lamp circuit, (b) An equivalent of the circuit in (a).

will work in this circuit. The two NAND gate inputs are signals that are asserted high while the output is asserted low. The logic function that is being performed is an AND function, looking for the condition of $A = B = 1$. When this condition occurs, the gate causes a positive action, asserting the output. If we indicate low assertion level by the logic polarity indicator, we could represent the circuit of Fig. 2.17 as shown in Fig. 2.18(a).

The indicator at the lamp input does not represent an inverter. It simply indicates that this input must be driven low to assert the lamp. The triangle on the AND gate output can represent an inversion, converting the AND to a NAND. We would consider this circuit as a gate that performs the AND function on high assertion inputs to produce a low asserted output. Because the signal X is low asserted, $X = 1$ when a low voltage level appears on this line.

From DeMorgan's laws, we recognize that we could also represent the circuit as shown in Fig. 2.18(b). Although technically correct, the circuit of part (b) does not directly indicate what function the gate is actually performing. It also does not match the assertion levels of the variables A and B to the assertion levels of the gate inputs. Nor does it match assertion levels of the gate output to the indicator lamp input.

More information is conveyed by a schematic using a symbol choice that reflects the actual function being performed by each circuit. The key to choosing the correct gate symbols is to match assertion levels as far as possible. Figure 2.18(a) uses a gate with high assertion level inputs, matching the variables A and B. A low gate output assertion level matches the input level required to assert the lamp. When all assertion levels are matched, we can read from the logic symbol the function being performed. From Fig. 2.18(a) we see that A ANDed with B will assert the light. Figure 2.18(b) does not immediately express this information.

As another example of choosing symbols to match assertion levels, consider the logic circuits of Fig. 2.19. The two circuits are equivalent, but the first does not immediately indicate the function being performed while the second does. For the

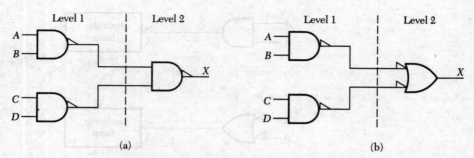

FIGURE 2.19   (a) Improper symbol choice, (b) proper choice.

first circuit, the two input gates perform the AND function with output asserted low. Since these outputs are asserted low, they should drive low assertion inputs to the last gate, which they do not. Figure 2.19(b) chooses the correct symbol for this gate and we can directly see that the OR function is performed with output asserted high. We can now write by inspection that

$$X = AB + CD$$

We could arrive at this same conclusion from the first circuit only after some algebraic manipulation. From an examination of the circuit, we would write

$$X = \overline{\overline{AB} \cdot \overline{CD}}$$

and then apply DeMorgan's first law to obtain the earlier result. Obviously, the method of matching assertion levels is preferable when dealing with logic diagrams, since the resulting functions are directly observable.

Figure 2.20 offers a more complex example of this method. We first note that gates 1 and 2 have low asserted outputs; thus, gate 4 is changed to a low asserted input gate. This results in an AND gate with a low asserted output. Since both gates 3 and 4 drive gate 5, we also want gate 3 to match the low asserted output of gate 4. The inputs to gate 3 are high assertion signals; thus, the NAND gate form is appropriate. After changing this gate to its NAND equivalent, we then match gate 5 to its driving signals to arrive at the circuit of Fig. 2.20(b). Since all assertion levels are now matched, we can write by inspection that

$$X = AB \cdot (C + D) + EF$$

It would be very nice indeed if we encountered only logic circuits that allowed matching of all assertion levels. Unfortunately, the situation sometimes occurs when the inputs to a gate are not all asserted at the same level. In these cases, it is not obvious which symbol should be used. The circuits of Fig. 2.21 demonstrate this situation.

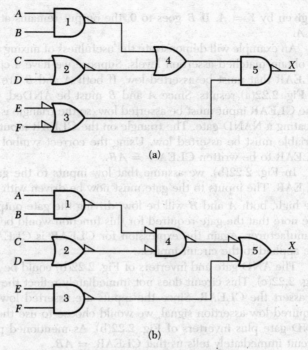

FIGURE 2.20 (a) Given logic circuit, (b) revised symbols.

Figure 2.21(a) demonstrates an inhibit circuit. If $B = 1$ and is a high asserted signal, the gate output will be unaffected by changes in A. For this case the effect of A on the output is inhibited. If B goes to 0, then $X = A$. The circuits represented in Fig. 2.21(b) are enable circuits. If B is asserted high or equal to 1, the output Y

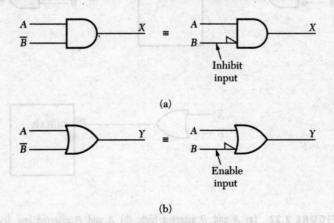

FIGURE 2.21 (a) An inhibit circuit, (b) an enable circuit.

is given by $Y = A$. If B goes to 0, the output remains at 1 regardless of the value of A.

An example will demonstrate the usefulness of mixing positive and negative logic to obtain matched assertion levels. Suppose we have a circuit with an input called CLEAR that must be asserted low. If both A and B are asserted high, the circuit of Fig. 2.22(a) results. Since A and B must be ANDed, the AND symbol is used. The CLEAR input must be asserted low; so the triangle is drawn at the gate output, creating a NAND gate. The triangle on the CLEAR input simply indicates that this variable must be asserted low. Using the correct symbol allows the expression for CLEAR to be written CLEAR $= AB$.

In Fig. 2.22(b), we assume that low inputs to the gate are required to assert CLEAR. The inputs to the gate must now be driven with $\bar{A}$ and $\bar{B}$. When A and B are high, both $\bar{A}$ and $\bar{B}$ will be low, driving the gate output low to assert CLEAR. We note that the gate required for this function would be called an OR gate by the manufacturer. Again the expression for CLEAR is CLEAR $= AB$, but $\bar{A}$ and $\bar{B}$ are applied to the circuit inputs.

The AND gate and inverters of Fig. 2.22(b) could be represented by the OR of Fig. 2.22(c). This circuit does not immediately reflect the single condition required to assert the CLEAR. Since the inputs are asserted low and CLEAR indicates a required low assertion signal, we would choose to use the circuit that includes the AND gate plus inverters of Fig. 2.22(b). As mentioned previously, this equivalent circuit immediately tells us that CLEAR $= AB$.

We will emphasize the fact that the manufacturers name their gates based on a positive logic definition. A NAND gate assumes that high assertion inputs

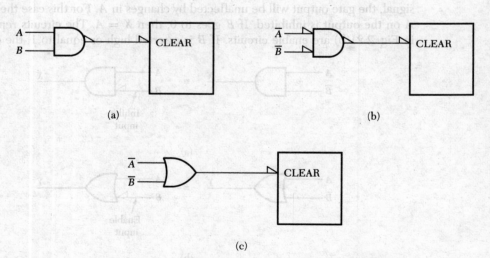

(a)   (b)

(c)

FIGURE 2.22   (a) A and B asserted high, (b) A and B asserted low, (c) An equivalent of the circuit in (b).

are required to produce the low-level output. We recognize from the preceding discussion that a redefinition of levels results in a new function performed by the gate. The general practice of basing the gate name on a positive logic definition removes any ambiguity that could result from other choices of assertion level. Of course, in dealing with circuits we will mix positive and negative logic to match assertion levels.

The concept of matching assertion levels is useful in determining the function performed by a given circuit. Often a particular output must be asserted with a specified level. In such a case, the output assertion level is first matched, then levels are matched working from the output gate toward the inputs. When matching is completed, the resulting schematic immediately indicates the logic function that creates the output. The lamp indicator of Fig. 2.17 demonstrates the requirement for a low assertion input. These required signals can be called defined assertion levels. All remaining undefined input and output lines are then manipulated using DeMorgan's laws to match the defined assertion-level lines. If all output levels match corresponding input levels, the function of the circuit can be immediately identified.

It is not necessary to match low assertion levels with complemented system input variables. Input variables that are not matched appear in the result in complemented form. Figure 2.23 is used to demonstrate the application of this method. Since the block requires a high assertion input, called X, gate 3 is converted to match its

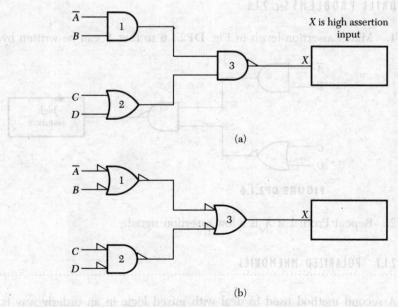

FIGURE 2.23 (a) A circuit with unmatched levels, (b) The matched equivalent.

output to input X. After this change, gate 3 has low assertion inputs; therefore gates 1 and 2 must be converted to match output levels to these inputs. This conversion results in the circuit of Fig. 2.23(b). All levels are now matched with the exception of the input variables. Once this matching has taken place, the function of each gate is indicated by its symbol. Gate 1 and gate 3 perform the OR function, and gate 2 performs the AND function. The expression for X is now written by inspection, complementing those variables that do not match input assertion level, namely, B, C, and D. The result is

$$X = (A + \bar{B}) + \bar{C}\bar{D}$$

We note that A appears in uncomplemented form in this expression even though the actual input is complemented. As mentioned previously, matched inputs appear in the expression as uncomplemented values, while unmatched inputs appear in complemented form.

While this method is useful in many situations, problems arise when a gate output must drive multiple inputs and at least one of these inputs has a different assertion level than the remaining inputs. This situation can be handled by creating a separate diagram for each gate with an input that does not match the other levels, but that can become complex for complex systems.

● ● ●   **DRILL PROBLEMS Sec. 2.1.6**

1. Match assertion levels in Fig. DP2.1.6 so that X can be written by inspection.

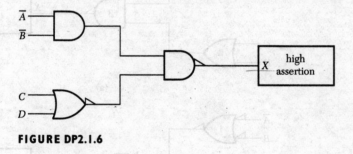

FIGURE DP2.1.6

2. Repeat Prob. 1 if X is a low assertion signal.

2.1.7   POLARIZED MNEMONICS

A second method used to deal with mixed logic in an orderly way is referred to as polarized mnemonics. This method is popular in industry and is formalized in Fletcher's text [2]. In this approach the letter H or L is appended to each variable, usually in parentheses, to indicate the assertion level of that variable. In addition, an appropriate mnemonic is often used for the variable name to prompt the designer's memory relative to the function of that variable. Examples of polarized mnemonics are STRCLK(H) and LDRG(L). The first polarized mnemonic might stand for a signal needed to start a gated clock and the (H) indicates that this signal must be asserted high to initiate the clock signal. The second example indicates that the signal level needed to load a register is low.

The method of polarized mnemonics leaves little room for doubt and does not require that each gate be matched in function to the assertion levels. Returning to the indicator lamp circuit of Fig. 2.18(a), the input variables A and B would be replaced by $A(H)$ and $B(H)$ while the output variable would be replaced by $X(L)$. This would indicate the levels that result in assertion or the levels defined as logic 1 for each variable.

Figure 2.24 shows an example of this method using positive logic symbols in a mixed logic system. Since in the polarized mnemonic scheme

$$F(L) = \bar{F}(H)$$

the output of the gate that drives $X(L)$ is given by

$$X(L) = \overline{A(H)B(H) + C(L)D(H)}$$

Using the method of matching assertion levels, the expression for X as a low asserted signal is found to be

$$X = (\bar{A} + \bar{B})(C + \bar{D})$$

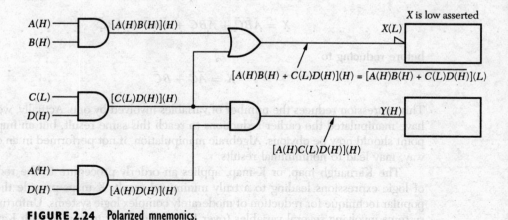

FIGURE 2.24 Polarized mnemonics.

The two expressions can be manipulated to show equivalence. Either approach can be used with positive, negative, or mixed logic circuits.

2.2

KARNAUGH MAPS

SECTION OVERVIEW The Karnaugh map can be used to minimize logic functions allowing the function to be implemented with the minimum number of gates. This method is considered in this section. The section is concluded with a discussion of static and dynamic hazards.

Boolean algebra allows us to express logic functions in a concise form and to reduce expressions to simpler equations. With more complex expressions, the result of algebraic manipulation may not lead to a unique reduced expression. In order to demonstrate this point, consider the equation

$$X = \bar{A}\bar{B}\bar{C} + \bar{A}B\bar{C} + AB\bar{C}$$

If the distributive law is applied to the first two terms, the expression reduces to

$$X = \bar{A}\bar{C}(\bar{B} + B) + AB\bar{C} = \bar{A}\bar{C} + AB\bar{C}$$

Another form of reduction is found by applying the distributive law to the second two terms, giving

$$X = \bar{A}\bar{B}\bar{C} + B\bar{C}(\bar{A} + A) = \bar{A}\bar{B}\bar{C} + B\bar{C}$$

The expression for X could be reduced slightly further by noting that $\bar{A}B\bar{C} = \bar{A}B\bar{C} + \bar{A}B\bar{C}$ from Eq. (2.3). We then expand X to be

$$X = \bar{A}\bar{B}\bar{C} + \bar{A}B\bar{C} + \bar{A}B\bar{C} + AB\bar{C}$$

before reducing to

$$X = \bar{A}\bar{C} + B\bar{C}$$

This expression reduces the number of variables involved by one. Actually, we could have manipulated the earlier reductions to reach this same result, but an important point should now be obvious. Algebraic manipulation, if not performed in an orderly way, may lead to nonminimal results.

The Karnaugh map, or K-map, applies an orderly procedure to the reduction of logic expressions leading to a truly minimized result. K-maps provide the most popular technique for reduction of moderately complex logic systems. Unfortunately, systems involving several variables (over five) are difficult to handle with Karnaugh methods. In these situations, the computer becomes necessary to minimize the logic function [2].

2.2.1 STANDARD SUM-OF-PRODUCTS FORM

In order to apply the K-map method, a function should be expressed in standard sum-of-product (SSOP) form. An expression such as

$$AB + BC + \bar{B}D$$

is in sum-of-product (SOP) form since the products (ANDs) of individual terms are first formed, followed by the summation (OR) of these products. Another form that can occur in logic expressions is the product-of-sums (POS). This form is exemplified by the expression

$$(A + B + C)(\bar{B} + \bar{C})$$

A POS expression can be converted to the SOP form by using the distributive law. The preceding expression would then become

$$A\bar{B} + B\bar{B} + C\bar{B} + A\bar{C} + B\bar{C} + C\bar{C} = A\bar{B} + C\bar{B} + A\bar{C} + B\bar{C}$$

Although the last equation is in SOP form, it is not in SSOP form. This form requires that each product term contain all variables involved. In the previous expression, the variables A, B, and C are involved; thus each product term should include these variables or complemented forms of these variables. To expand the SOP form into SSOP, we note that we can AND (or multiply) any term by 1 without changing the value of the term. Furthermore, since $X + \bar{X} = 1$, we could multiply any term with $C + \bar{C}$, $B + \bar{B}$, or $A + \bar{A}$ without affecting the value of the function. To put the earlier expression in SSOP form, we proceed as shown in the following equation:

$$A\bar{B}(C + \bar{C}) + C\bar{B}(A + \bar{A}) + A\bar{C}(B + \bar{B}) + B\bar{C}(A + \bar{A})$$

$$= A\bar{B}C + A\bar{B}\bar{C} + A\bar{B}C + \bar{A}\bar{B}C + AB\bar{C} + A\bar{B}\bar{C} + AB\bar{C} + \bar{A}B\bar{C}$$

$$= A\bar{B}C + A\bar{B}\bar{C} + \bar{A}\bar{B}C + \bar{A}B\bar{C} + AB\bar{C}$$

Each product term with a missing variable is ANDed with the sum of the variable and its complement. All duplicate product terms are eliminated since $X + X = X$ or

$$A\bar{B}C + A\bar{B}C = A\bar{B}C$$

If both B and C were missing, leaving only A as a term in an expression, $(C + \bar{C})$ and $(B + \bar{B})$ must multiply A. For example,

$$A + \bar{A}\bar{B}\bar{C} = A(B + \bar{B})(C + \bar{C}) + \bar{A}\bar{B}\bar{C}$$

$$= ABC + AB\bar{C} + A\bar{B}C + A\bar{B}\bar{C} + \bar{A}\bar{B}\bar{C}$$

The SSOP terms in an expression are called minterms. Minterms are numbered according to the decimal code they represent. The preceding expression contains minterms m_0, m_4, m_5, m_6, and m_7. The equation

$$F = \bar{A}\bar{B}C + \bar{A}BC + A\bar{B}\bar{C}$$

tells us that $F = 1$ when minterm m_1, m_3, or m_4 is present and implies that $F = 0$ when minterm m_0, m_2, m_5, m_6, or m_7 is present. We will use minterm notation as we discuss mapping of logic functions.

● ● ●   **DRILL PROBLEMS Sec. 2.2.1**

1. Put into SOP form $X = AB + (C + \bar{B})(AB + \bar{C})$.

2. Put X of Prob. 1 into SSOP form.

3. Write the expression for F if F is to equal 1 when any of the minterms m_0, m_3, m_4, or m_6 is present. Assume A is the most significant bit in the three-bit code.

● **2.2.2   CONSTRUCTING A K-MAP**

Now we turn our attention to the construction of K-maps. All possible terms of an expression can be represented by a location on a map. The two variables A and B have four possible combinations that can be represented by the maps of Fig. 2.25. The second map is more conventional than the first with each square corresponding

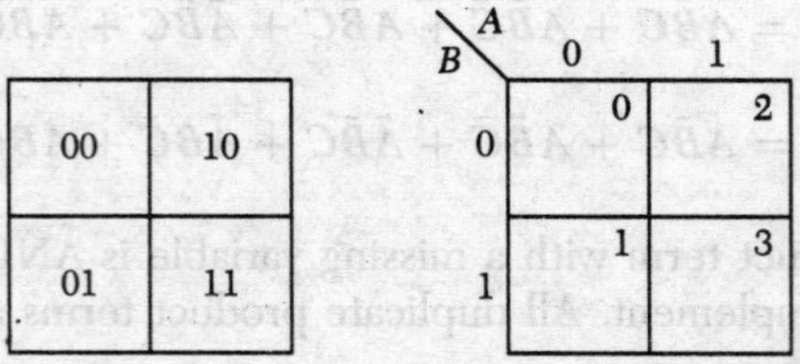

FIGURE 2.25 Maps showing all possible combinations of a two-variable expression.

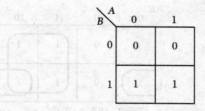

FIGURE 2.26 A Karnaugh map.

to a unique set of values for A and B. The top left square represents the values $A = 0$ and $B = 0$, the bottom left represents $A = 0$ and $B = 1$, the top right represents $A = 1$ and $B = 0$, and the bottom right represents $A = 1$ and $B = 1$. Each location represents a minterm and can be numbered as shown.

When plotting a function on a K-map we recognize that a logical function is specified in terms of those combinations of variables that lead to a value of 1 for the function. The meaning of the equation

$$F = AB + \bar{A}B$$

is that F will be 1 when $B = 1$ and $A = 1 (BA = 1)$ or when $B = 1$ and $A = 0$ $(B\bar{A} = 1)$. The equation also implies that $F = 0$ for the values $B = 0$ and $A = 0$ $(\bar{B}\bar{A} = 1)$ or $B = 0$ and $A = 1$ $(\bar{B}A = 1)$. Figure 2.26 plots this information.

● ● ●   **DRILL PROBLEMS Sec. 2.2.2**

1. Plot the K-map for $X = AB + A\bar{B}$.

2. Plot the K-map for $Y = AB + A\bar{B} + \bar{A}\bar{B}$.

3. Plot the K-map for $Z = AB + \bar{A}$.

●   **2.2.3   REDUCING AN EXPRESSION**

In reducing the expression F to the simplest terms, we look for adjacent locations of the map containing values of 1. These locations form a couple. A couple consists of two terms that are logically adjacent. As indicated earlier, this means the two terms would be identical except that a single variable appears in its complemented form in one term and in its uncomplemented form in the other term. The expression

$$F = AB + \bar{A}B$$

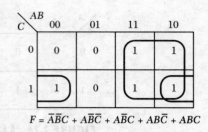

$$F = \bar{A}\bar{B}C + A\bar{B}\bar{C} + A\bar{B}C + AB\bar{C} + ABC$$

FIGURE 2.27 A three-variable K-map.

comprises a couple. From our experience with Boolean algebra, we recognize that a couple can be reduced; thus,

$$F = BA + B\bar{A} = B(A + \bar{A}) = B$$

A three-variable term has eight possible combinations leading to a K-map as shown in Fig. 2.27. Note the arrangement of the variables AB. Rather than counting from 00 to 11 in numerical sequence, the order is such that adjacent squares are always logically adjacent. Only one variable is complemented with the other two unchanged as we move from one square to an adjacent one. Furthermore, logical adjacency also exists between the extreme squares of a row or column. The location $A\bar{B}\bar{C}$ (top right) is adjacent to the location $\bar{A}\bar{B}\bar{C}$ (top left).

In the map of Fig. 2.27, there is a quad that contains four 1s, each logically adjacent to two others, and there is also a couple. This quad of four terms can be reduced to A. We can note this by factoring the four terms to give

$$A\bar{B}\bar{C} + A\bar{B}C + AB\bar{C} + ABC = A(\bar{B}\bar{C} + \bar{B}C + B\bar{C} + BC)$$

$$= A(\bar{B}[\bar{C} + C] + B[\bar{C} + C])$$

$$= A(\bar{B} + B) = A$$

The four terms in parentheses encompass all possible combinations of the variables B and C. This tells us that, regardless of the values of B and C, the function equals 1 when A is 1. Thus, the term reduces to A.

A simple way to arrive at the reduced expression without using algebra is to note which variable or variables of a group remain unchanged. From the K-map of Fig. 2.27, the quad shows both B and C changing from 0 to 1 as we move to different locations of the quad while A remains constant at 1. The couple shows A changing from 0 to 1 while B remains constant at 0 and C remains constant at 1. The reduced expression is then $A + \bar{B}C$.

Because of the arrangement of the map to achieve logical adjacency for physically adjacent squares, the minterm numbers do not proceed sequentially. Fig-

	AB			
C	00	01	11	10
0	0	2	6	4
1	1	3	7	5

FIGURE 2.28 Minterm numbering for a three-variable map.

ure 2.28 shows the three-variable map with numbered locations corresponding to minterms.

For expressions involving more terms, the K-map provides an orderly means of locating all terms that can be reduced in the manner described. Consider the four-variable function:

$$F = ABCD + A\bar{B}CD + A\bar{B}\bar{C}D + \bar{A}BCD + \bar{A}B\bar{C}D + \bar{A}\bar{B}CD + \bar{A}BC\bar{D}$$

With four variables there are 16 possible combinations; thus, the K-map must contain 16 locations. Figure 2.29 shows the standard arrangement of a four-variable K-map. After plotting F, we note several adjacent locations containing a 1. The row containing four ones tells us that a 1 occurs whenever $C = D = 1$, regardless of the values of A or B. The second column indicates that a 1 should also occur in the expression whenever $\bar{A}B\bar{D} = 1$. The fourth column indicates the occurrence of a 1 in the expression when $A\bar{B}D = 1$. Thus, the entire expression can be reduced to

$$F = CD + \bar{A}B\bar{D} + A\bar{B}D$$

	AB			
CD	00	01	11	10
00	0	1	0	0
01	0	0	0	1
11	1	1	1	1
10	0	1	0	0

$$F = ABCD + A\bar{B}CD + A\bar{B}\bar{C}D + \bar{A}BCD + \bar{A}B\bar{C}D + \bar{A}\bar{B}CD + \bar{A}BC\bar{D}$$

FIGURE 2.29 A four-variable K-map.

If F is implemented by hardware, seven 4-input AND gates plus a 7-input OR would be required for the original expression. Alternatively, the reduced expression could be implemented using two 3-input AND gates plus a 2-input AND gate, along with a 3-input OR gate. Obviously the reduced expression is much more economical.

The basic idea in reducing an expression consists of first plotting the K-map, then forming groups containing 2, 4, 8, or 16 adjacent 1s. A couple allows two original terms to be reduced to one smaller term, a quad allows four original terms to be reduced to one smaller term, and an octuplet reduces eight terms to a single smaller term. Given a choice we always form the largest group possible to achieve maximum reduction. A square containing a 1, surrounded by adjacent zero squares, indicates a term in the original expression that must be included in the final expression with no further reduction. We note that the two squares at opposite ends of a row or column are adjacent locations. Some examples of these ideas are demonstrated in Fig. 2.30. In the first map, the couple is formed from locations at opposite ends of a

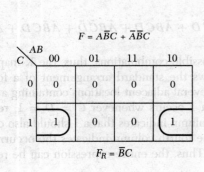

FIGURE 2.30 Some examples of reduction.

row. The second map shows a quad that can be represented by a single two-variable term. The third map also contains a quad, although it is not so obvious as in the preceding map. In addition, this map contains a four-variable term that cannot be reduced. The last map demonstrates that one square containing a 1 can be used to form more than one couple. With these basic ideas, a great deal of reduction can be performed.

● ● ● **DRILL PROBLEMS Sec. 2.2.3**

1. Reduce to simplest form $X = AB + A\bar{B}\bar{C} + \bar{A}\bar{B}\bar{C}$.

2. Reduce to simplest form $Y = ABC + A\bar{B}\bar{C}D + \bar{A}\bar{B}\bar{C}D$.

3. Reduce to simplest form $Z = ABC\bar{D} + AB\bar{C}D + A\bar{B}CD + \bar{A}BC\bar{D} + \bar{A}\bar{B}CD$.

● **2.2.4 A CLOSER LOOK AT MINIMIZATION**

It is possible that the application of the steps previously outlined may lead to redundant or unnecessary terms in the final expression. In order to obtain a true minimum expression, all redundant terms must be eliminated. The source of these redundant terms can be understood after a consideration of prime implicants.

When a K-map is plotted, all possible largest groups are circled to include all 1s on the map. These groups will contain 1, 2, 4, 8, or 16 mutually adjacent 1s for a four-variable map as shown in Fig. 2.31.

The term that represents each group is called a prime implicant. The prime implicants for the map of Fig. 2.31(a) are $\bar{C}\bar{D}$, $B\bar{D}$, $A\bar{B}\bar{C}$, $A\bar{B}D$, ACD, and ABC. For Fig. 2.31(b) the prime implicants are $\bar{A}BCD$, $A\bar{B}\bar{C}$, $A\bar{D}$, and $\bar{B}\bar{C}D$. Not all prime implicants are necessary in the final expression since certain prime implicants include 1s that have all been included as parts of other prime implicants. The prime implicant covering locations 8 and 9 in either map is an example of a redundant implicant. For absolute minimization of a function, all redundant implicants must be eliminated. The remaining prime implicants are called essential implicants. The final expression of a logic function will contain only essential implicants.

The map of Fig. 2.31(a) will result in a minimized expression of

$$F = \bar{C}\bar{D} + B\bar{D} + A\bar{B}D + ABC$$

Two prime implicants were redundant in this map. This result, although minimum, may not be unique. Two other possible expressions are

$$F = \bar{C}\bar{D} + B\bar{D} + A\bar{B}\bar{C} + ACD$$

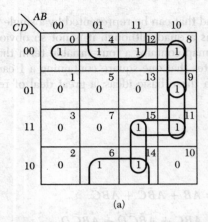
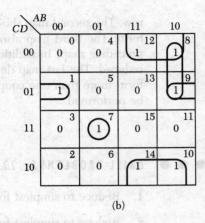

FIGURE 2.31   K-maps with groups that represent prime implicants.

and

$$F = \bar{C}\bar{D} + B\bar{D} + A\bar{B}D + ACD$$

We will leave it to the reader to verify that the groups chosen for the last two expressions include only essential implicants.

The map of Fig. 2.31(b) leads to a minimized expression of

$$F = A\bar{D} + \bar{B}\bar{C}D + \bar{A}BCD$$

There are no other options for this minimized expression.

We will now return to the example that opened the discussion of Boolean algebra in Sec. 2.1. This problem will be restated in the following example.

EXAMPLE 2.1
. . . .

Four variables are used to represent a 4-bit binary number. When certain numbers occur, a logical 1 should be generated. The numbers for which the output should indicate 1 are 1, 4, 5, 9, 11, and 12. Assume that each variable is also present in complemented form.

SOLUTION

Logical 1 is to be generated for any of the following combinations.

Decimal	Binary	Four-Variable Term
1	0001	$\bar{A}\bar{B}\bar{C}D$
4	0100	$\bar{A}B\bar{C}\bar{D}$
5	0101	$\bar{A}B\bar{C}D$
9	1001	$A\bar{B}\bar{C}D$
11	1011	$A\bar{B}CD$
12	1100	$AB\bar{C}\bar{D}$

Therefore, the expression is

$$F = \bar{A}\bar{B}\bar{C}D + \bar{A}B\bar{C}\bar{D} + \bar{A}B\bar{C}D + A\bar{B}\bar{C}D + A\bar{B}CD + AB\bar{C}\bar{D}$$

This expression was realized directly with six 4-input AND gates plus one six-input OR in Fig. 2.1.

In order to minimize this expression, the K-map of Fig. 2.32(a) is plotted. The three couples circled in the figure are the essential implicants necessary to represent the expression giving

$$F_R = B\bar{C}\bar{D} + \bar{A}\bar{C}D + A\bar{B}D$$

Note that the numbered minterm locations can simplify the plotting process.

This reduced expression is implemented in Fig. 2.32(b) and requires only 4 total gates and 12 total inputs, compared with 7 total gates and 30 total inputs to implement the original expression. It is left as an exercise for the student to realize this expression with NAND-NOR gate logic.   ●

"Don't Care" Conditions

In several applications, we can take advantage of "don't care" conditions to minimize an expression. Suppose we are given a problem of implementing a circuit to generate a logical 1 when a 2, 7, or 15 appears on a four-variable input. A logical 0 should be generated when 0, 1, 4, 5, 6, 9, 10, 13, or 14 appears. The input conditions for the numbers 3, 8, 11, and 12 never occur in this system. When we know that certain combinations never occur, we don't care whether these inputs generate a one or a zero in the K-map. We can interpret these "don't care" conditions as either ones or zeros, whichever allows us to further minimize the reduced function. A Θ can be used to represent the "don't care" combinations as shown in the map of Fig. 2.33. The two "don't care" conditions in the third row are taken as 1s, while the other two are taken as 0s. This allows the reduced expression to be written

$$F_R = CD + \bar{A}\bar{B}C$$

Had all "don't care" conditions been interpreted as 0s, the expression would have

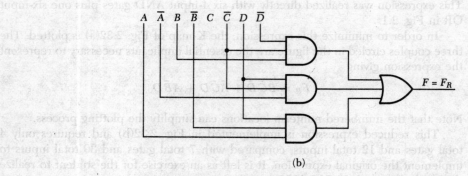

FIGURE 2.32 (a) K-map for Example 2.1. (b) Implementation of the reduced expression.

been

$$F = BCD + \bar{A}\bar{B}C\bar{D}$$

Obviously, we must consider these conditions in order to truly minimize the system.

Factoring

After the reduced expression is obtained, it may be possible to further minimize the number of gates used. Consider an expression such as

$$X = A\bar{B} + C\bar{B}$$

This can be factored to give

$$X = \bar{B}(A + C)$$

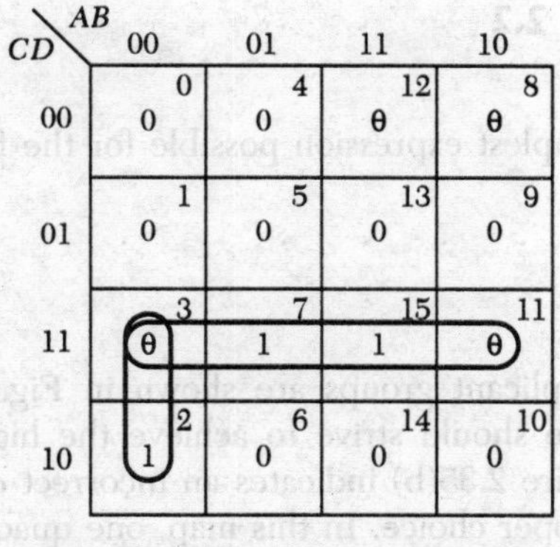

FIGURE 2.33 K-map with "don't care" conditions.

Implementing the first form of X requires two 2-input AND gates and a 2-input OR. The second expression can be realized with one 2-input OR gate followed by a 2-input AND. Thus, we should consider factoring the reduced expression prior to implementation of the circuit (see Prob. 2.32).

We will emphasize that, in reducing expressions with K-maps, we should always include any logical 1 within the largest group of 1s possible. For example, consider the maps of Fig. 2.34.

The expression for the first map is

$$F = \bar{C}D + BCD + A\bar{B}\bar{C}\bar{D}$$

By including each 1 in the largest group possible, as in the second map, the expression becomes even simpler. It is

$$F_R = \bar{C}D + BD + A\bar{B}\bar{C}$$

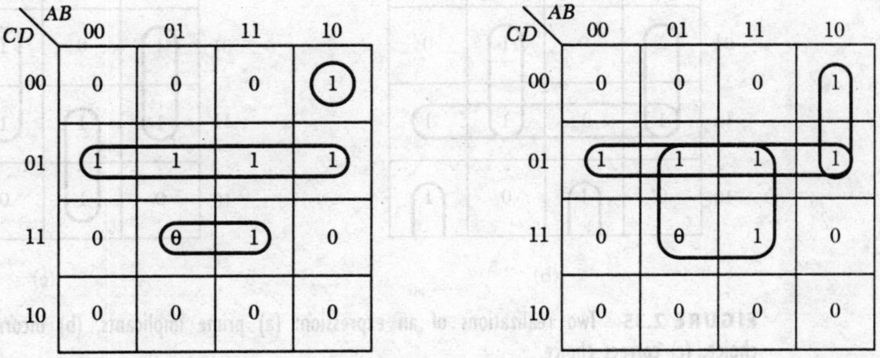

FIGURE 2.34 Map reductions using different groupings.

EXAMPLE 2.2

Write the simplest expression possible for the function represented by the map of Fig. 2.35(a).

SOLUTION

All prime implicant groups are shown in Fig. 2.35(a). In selecting the essential implicants, we should strive to achieve the highest number of redundant groups possible. Figure 2.35(b) indicates an incorrect choice of groups, while Fig. 2.35(c) shows the proper choice. In this map, one quad becomes redundant since all 1s of

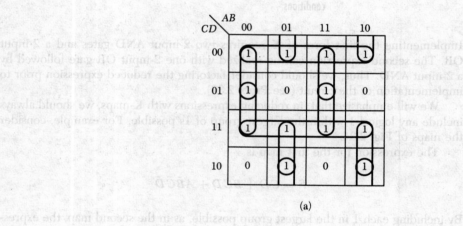

(a)

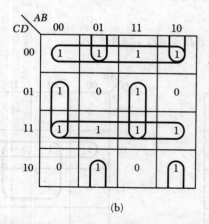

(b)

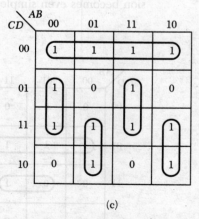

(c)

FIGURE 2.35 Two realizations of an expression: (a) prime implicants, (b) incorrect implicant choice, (c) correct choice.

this group are included in other necessary groups. The resulting expression is

$$F = \bar{C}\bar{D} + \bar{A}\bar{B}D + \bar{A}BC + ABD + A\bar{B}C$$

●

Realizing $\bar{F}$

Another check that should be made before selecting a final expression is the possibility of realizing the complemented function rather than the function. Each location of the map that contains a 0 indicates that $F = 0$ for this term. This is equivalent to saying that F is not 1 or $F = \bar{1}$. Complementing both sides of this equation results in $\bar{F} = 1$. Thus, $\bar{F} = 1$ results from the terms corresponding to $F = 0$. We can write an expression for $\bar{F}$ in terms of the 0 locations of the map using the same ideas to group the 0s as we did previously to group the 1s.

Returning to Fig. 2.35, we can write

$$\bar{F} = \bar{A}B\bar{C}D + A\bar{B}\bar{C}D + \bar{A}BC\bar{D} + ABC\bar{D}$$

Since we ultimately want to realize F rather than $\bar{F}$, both sides of the equation are complemented, giving

$$\bar{\bar{F}} = F = \overline{\bar{A}B\bar{C}D + A\bar{B}\bar{C}D + \bar{A}BC\bar{D} + ABC\bar{D}}$$

In some instances, the expression resulting from this method may be simpler than that resulting from realizing F directly. Groups of zeros can be used to minimize the $\bar{F}$ expression just as groups of ones are used to reduce F.

EXAMPLE 2.3

Minimize the expression

$$F = \bar{A}B\bar{C}\bar{D} + \bar{A}B\bar{C}D + \bar{A}BCD + \bar{A}BC\bar{D} + A\bar{B}\bar{C}\bar{D} + A\bar{B}\bar{C}D$$
$$+ A\bar{B}C\bar{D} + A\bar{B}CD + AB\bar{C}\bar{D} + ABC\bar{D}$$

SOLUTION

The Karnaugh map is shown in Fig. 2.36. Grouping the ones as shown results in

$$F_1 = \bar{A}B + A\bar{B} + A\bar{D}$$

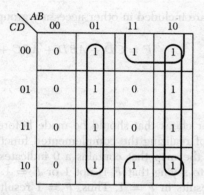

FIGURE 2.36   K-map for Example 2.3.

If we use the two groups of zeros, we obtain

$$\bar{F}_2 = \bar{A}\bar{B} + ABD$$

and

$$F_2 = \overline{\bar{A}\bar{B} + ABD}$$

which is the simplest possible realization for F.

The function F_2 can be realized with one 3-input gate and two 2-input gates. F_1 requires three 2-input gates and one 3-input gate.   ●

A five-variable map is shown in Fig. 2.37. In this situation, couples or quads can be formed in corresponding locations of the two parts of the map. Generally, expressions involving seven or more variables are reduced by means of computer programs.

Tabulation Methods of Reduction

We have considered algebraic methods and K-map methods of logic function reduction to this point. Algebraic manipulation is reliable only for functions involving two or three variables. The use of the K-map is much more accurate for expressions that involve up to about six variables. This approach is useful for manual reduction of expressions because it is based on a graphical approach.

When the computer is used to minimize logic functions, graphical methods are not appropriate. Furthermore, K-maps become overly complex for expressions containing many variables. Most computer reduction programs are based on tabulation methods such as the Quine-McCluskey procedure. While it is not the intent of this book to cover all reduction methods, a simple example of the use of the Quine-McCluskey procedure will now be given.

$$F = \bar{A}BC\bar{D}\bar{E} + AB\bar{C}\bar{D}\bar{E} + \bar{A}B\bar{C}D\bar{E} + \bar{A}BC\bar{D} + \bar{A}BC\bar{D}E + ABC\bar{D}\bar{E} + AB\bar{C}DE + \bar{A}B\bar{C}DE$$

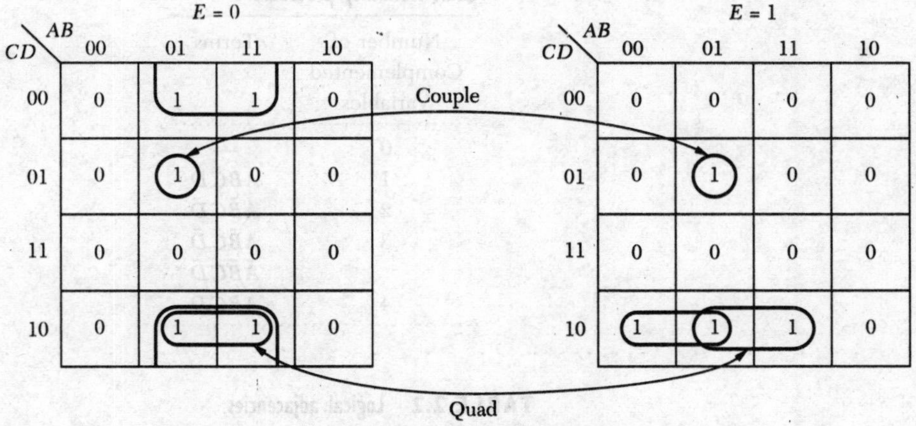

FIGURE 2.37 A five-variable map.

This procedure consists of two main steps [4]:

1. Eliminate as many terms as possible using logical adjacencies. The resulting terms will be prime implicants.
2. Select the appropriate prime implicants to minimize the final expression.

The expression

$$F = AB\bar{C}D + A\bar{B}\bar{C}\bar{D} + A\bar{B}C\bar{D} + \bar{A}BC\bar{D} + \bar{A}B\bar{C}\bar{D}$$

reduced earlier by K-map in Fig. 2.30, will be reduced by the Quine-McCluskey procedure to demonstrate the key points.

Logical adjacency between two 4-variable terms can only exist if three variables appear the same in both terms while the fourth variable is complemented in one term and uncomplemented in the other. Thus, logically adjacent terms must differ by only one complemented variable. This suggests an ordering of terms as shown in Table 2.1. We now check the logical adjacency of each term in a group with all terms in the next group. In this expression, the term in the first group is not logically adjacent to any other term, but there are several other adjacencies as shown in Table 2.2. We note that every term in the original expression has been included in these reduced terms. The procedure is continued to find additional logical adjacencies as indicated in Table 2.3. Again, all of the reduced terms of Table 2.2 are included in the original terms. Since there are no other possible adjacencies, the procedure is finished and the reduced expression is

$$F_R = AB\bar{C}D + \bar{B}\bar{D}$$

TABLE 2.1 Starting point for Quine-McCluskey procedure.

Number of Complemented Variables	Terms
0	—
1	$AB\bar{C}D$
2	$A\bar{B}C\bar{D}$
3	$A\bar{B}\bar{C}\bar{D}$
	$\bar{A}\bar{B}C\bar{D}$
4	$\bar{A}\bar{B}\bar{C}\bar{D}$

TABLE 2.2 Logical adjacencies.

Original Terms	Reduced Terms
$AB\bar{C}D$	$AB\bar{C}D$
$A\bar{B}C\bar{D} + A\bar{B}\bar{C}\bar{D}$	$A\bar{B}\bar{D}$
$A\bar{B}C\bar{D} + \bar{A}\bar{B}C\bar{D}$	$\bar{B}C\bar{D}$
$A\bar{B}\bar{C}\bar{D} + \bar{A}\bar{B}\bar{C}\bar{D}$	$\bar{B}\bar{C}\bar{D}$
$\bar{A}\bar{B}C\bar{D} + \bar{A}\bar{B}\bar{C}\bar{D}$	$\bar{A}\bar{B}\bar{D}$

TABLE 2.3 Additional logical adjacencies.

Original Terms	Reduced Terms
$AB\bar{C}D$	$AB\bar{C}D$
$A\bar{B}\bar{D} + \bar{A}\bar{B}\bar{D}$	$\bar{B}\bar{D}$
$\bar{B}C\bar{D} + \bar{B}\bar{C}\bar{D}$	$\bar{B}\bar{D}$

In Table 2.3, the column of reduced terms shows a repeat of the $\bar{B}\bar{D}$ term. Obviously, this term need only be included once in the reduced expression. In a complex expression containing many variables, several prime implicants will result. It is not always obvious when one prime implicant will cover another prime implicant. For this reason, a prime implicant chart is necessary to determine the essential implicants and the minimal expression. The interested reader is referred to the literature [4] for more information on this topic.

For manual reduction of logic functions, the K-map approach is very useful. Because of the ability of computers to tabulate and sort, the tabulation methods are more effectively used in automatic reduction methods.

● ● ● DRILL PROBLEMS Sec. 2.2.4

1. Reduce to simplest form

$$X = \bar{A}B\bar{C}\bar{D} + \bar{A}B\bar{C}D + AB\bar{C}\bar{D} + AB\bar{C}D + A\bar{B}\bar{C}D$$

$$+ ABC\bar{D} + \bar{A}BCD + \bar{A}\bar{B}\bar{C}\bar{D}$$

2. A system sends BCD code over four lines A (MSB), B, C, and D. Design a minimal circuit to generate a 1 when the codes for 1, 5, 6, 7, or 9 are present. *Hint:* Remember to use "don't cares."

3. Reduce X from Prob. 1 by reducing $\bar{X}$ and then complementing.

4. Repeat Prob. 2 using tabulation methods.

2.2.5 HAZARD COVERS

In our previous discussions, we have assumed ideal switching characteristics for all gates and inverters. In practice, there is a propagation delay associated with each logic element as explained in Sec. 1.5. Each waveform also has finite rise and fall times as indicated by the output waveform of Fig. 2.38. In certain situations this delay can lead to undesirable results as outlined in the following paragraphs.

In order to demonstrate the problem of propagation delay time in combinational circuits, we will consider the K-map of Fig. 2.39. For the input conditions of $A = B = 1$ and C changing from 1 to 0, we can plot the timing chart of Fig. 2.40. Although the equation for F indicates that F should remain equal to 1 as C switches, we see that in reality F may contain a short excursion toward zero. This unwanted "glitch" is called a static hazard because the signal was supposed to remain static at the 1 level. The change in $\bar{C}$ occurs slightly later than the change in C due to the propagation delay time of the inverter. Since C drops from 1 to 0 prior to $\bar{C}$ rising from 0 to 1, there is a short period during which neither AND condition is satisfied.

When C switches from 0 to 1, if the propagation delay of the upper gate is less than the total delay through the inverter plus the lower gate, the anomaly does not

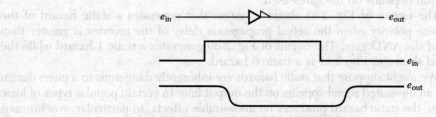

FIGURE 2.38 Practical output waveform.

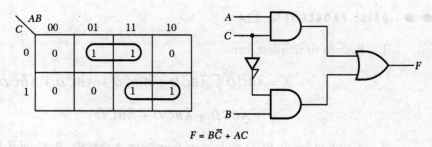

FIGURE 2.39 K-map and logic circuit.

$$F = B\overline{C} + AC$$

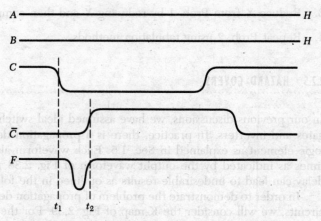

FIGURE 2.40 Timing chart for the circuit of Fig. 2.39.

occur. In order to ensure that a hazard does not occur in this situation, the worst case propagation delay times must be considered. If the maximum propagation delay time of the upper gate is less than the minimum propagation delay of the inverter plus the minimum propagation delay of the lower gate, the hazard will not take place. With the considerable variation in minimum and maximum propagation delay times of gates and inverters of the same family, the occurrence of a hazard in this situation depends on the gates used.

The circuit of Fig. 2.41 shows a system that generates a static hazard of the opposite polarity when the actual propagation delay of the inverter is greater than that of the AND gate. The circuit of Fig. 2.39 generates a static 1 hazard while the hazard shown in Fig. 2.41 is a static 0 hazard.

We might suppose that static hazards are inherently dangerous to a given design since an unwanted signal appears on the output line. In certain popular types of logic design, the static hazard produces no undesirable effects. In particular, synchronous sequential circuits (Chap. 6) are designed to ignore the static hazards that may be

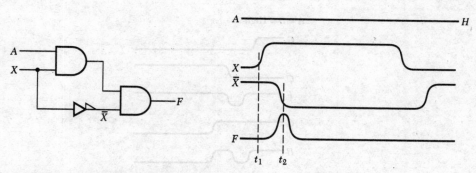

FIGURE 2.41 A logic circuit with a static hazard.

present. Other types of systems can produce erroneous outputs when these hazards occur, however, and in these instances the hazard must be eliminated.

In order to eliminate the hazard in the circuit of Fig. 2.39, we note that the glitch occurs as C changes from 1 to 0 when $A = B = 1$. We can make the output independent of the value of C, for these values of A and B, by including the redundant implicant or couple as shown in Fig. 2.42. This couple is referred to as a hazard cover and results in one more required gate in the actual circuit. Any adjacent couples must be overlapped by a hazard cover to eliminate the possibility of a static hazard. Although the system is no longer a minimal system, it can now operate reliably when static hazards are significant.

A dynamic hazard differs from a static hazard in that a transition is made from 1 to 0 or 0 to 1 accompanied by two or more unwanted transitions. Figure 2.43 demonstrates two dynamic hazards. If the circuit of Fig. 2.39 is followed by a gate, a dynamic hazard might result as indicated in Fig. 2.44.

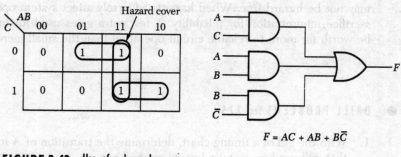

$$F = AC + AB + B\bar{C}$$

FIGURE 2.42 Use of a hazard cover.

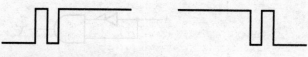

FIGURE 2.43 Dynamic hazards.

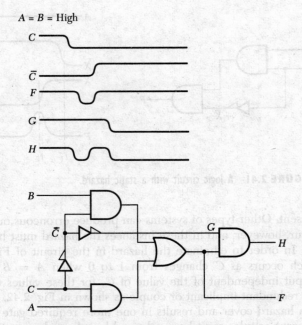

FIGURE 2.44 Circuit with dynamic hazard.

It can be shown that a dynamic hazard can only exist as a result of static hazards. Therefore, if hazard covers are used to eliminate all static hazards, we need not be concerned with dynamic problems.

Although we will not be concerned further with hazards until we consider asynchronous state machines, it is mentioned here to emphasize that minimal circuits may not be hazard-free. When hazards adversely affect system operation, we must sacrifice minimization for reliability. A few extra gates used for hazard covers will be worth far more to reliable circuit operation than the small increase in cost.

● ● ●   DRILL PROBLEMS Sec. 2.2.5

1. With the aid of a timing chart, determine the transition of A in Fig. DP2.2.5-1 that will produce a static hazard on X.

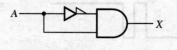

FIGURE DP2.2.5-I

2.3

VARIABLE-ENTERED MAPS

SECTION OVERVIEW The variable-entered map is closely related to the K-map but is more useful in some cases. The construction of the map and its use in minimization is covered in this section.

The K-map is a useful tool in logic function minimization if fewer than six variables are involved. When the number of variables exceeds six, computer programs can be used to solve this problem. The variable-entered map can be used to plot an n-variable problem on an $n - 1$ variable map. It is actually possible to reduce the map dimension by two or three in some cases, but a reduction of one is a more typical situation. A second advantage of the variable-entered map occurs in design problems involving multiplexers, which can be solved more readily by this approach than by K-maps.

2.3.1 PLOTTING THE VARIABLE-ENTERED MAP

We will introduce the map-entered variable (MEV) approach by considering the three-variable function

$$X = \bar{A}\bar{B}\bar{C} + AB\bar{C} + A\bar{B}\bar{C} + ABC$$

Of course, this function could be reduced by means of a three-variable K-map. On the other hand, we could plot the function on a two-variable map if we consider the value of X to be a function of map location and the variable C. This possibility is demonstrated in Fig. 2.45.

The first map lists the conditions required of C at each location of A and B to result in $X = 1$. The only exception to this is location $\bar{A}B$, which yields $X = 0$, regardless of the value of C. The second map conveys the same information in a more efficient manner. This map is called a variable-entered map (VEM) and

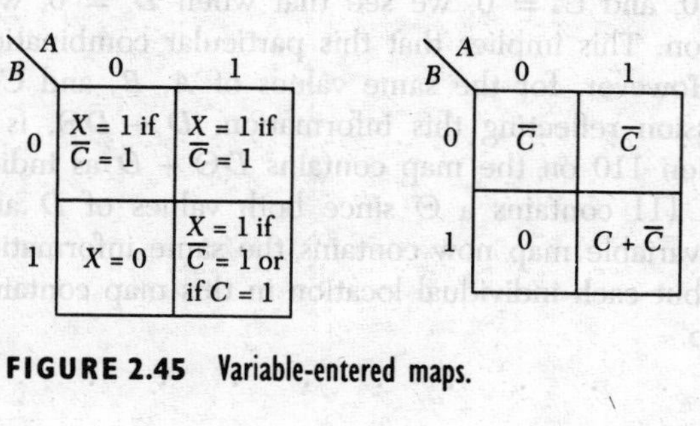

FIGURE 2.45 Variable-entered maps.

Inputs A B C D (MEV)	Output X
0 0 0 0	0
0 0 0 1	1
0 0 1 0	0
0 0 1 1	0
0 1 0 0	1
0 1 0 1	1
0 1 1 0	1
0 1 1 1	0
1 0 0 0	θ
1 0 0 1	1
1 0 1 0	0
1 0 1 1	0
1 1 0 0	1
1 1 0 1	θ
1 1 1 0	θ
1 1 1 1	θ

(a)

C \ AB	00	01	11	10
0	D	1 or $D + \bar{D}$	$D\theta + \bar{D}$	$D + \bar{D}\theta$
1	0	$\bar{D}$	θ	0

X

(b)

FIGURE 2.46 (a) Truth table for X. (b) Variable-entered map for X.

contains the same information as a three-variable K-map. In this case, C is called the map-entered variable.

Figure 2.46 demonstrates the procedure used to plot the variable-entered map. For $A = 0$, $B = 0$, and $C = 0$, we see from the table that X follows the value of D. In the corresponding map location, a D is entered. This location then generates a 1 in the map when $D = 1$ and a 0 when $D = 0$. For $A = 0$, $B = 0$, and $C = 1$, the output X is 0 regardless of the value of D. When $A = 0$, $B = 1$, and $C = 0$, X is 1 for either value of D; thus, a 1 is entered in this location of the map. We note that a 1 is equivalent to $D + \bar{D}$. For $A = 0$, $B = 1$, and $C = 1$, the table shows that when $D = 0$, $X = 1$, and when $D = 1$, $X = 0$. The corresponding map location then contains a $\bar{D}$. When $A = 1$, $B = 0$, and $C = 0$, we see that when $D = 0$, we do not care what value X takes on. This implies that this particular combination of variables will never occur. However, for the same values of A, B, and C, $X = 1$ when $D = 1$. The expression reflecting this information, $D + \bar{D}\Theta$, is then entered into the map. Location 110 on the map contains $D\Theta + \bar{D}$ as indicated by the truth table. Location 111 contains a Θ since both values of D are "don't care" conditions. A three-variable map now contains the same information as does a four-variable K-map, but each individual location in this map contains more information than the K-map.

2.3.2 REDUCING EXPRESSIONS WITH THE VARIABLE-ENTERED MAP

In order to reduce an expression using a VEM, we again choose groups of similar terms to cover all nonzero terms appearing in the map, except "don't care" terms that are taken as 0s. Figure 2.47 shows some examples of VEMs with the appropriate groups. In these maps, 1s are written as the MEV plus its complement. The first

$$F = A\overline{B}CD + \overline{A}B\overline{C}D + A\overline{B}C\overline{D} + AB\overline{C}D + \overline{A}\overline{B}CD$$

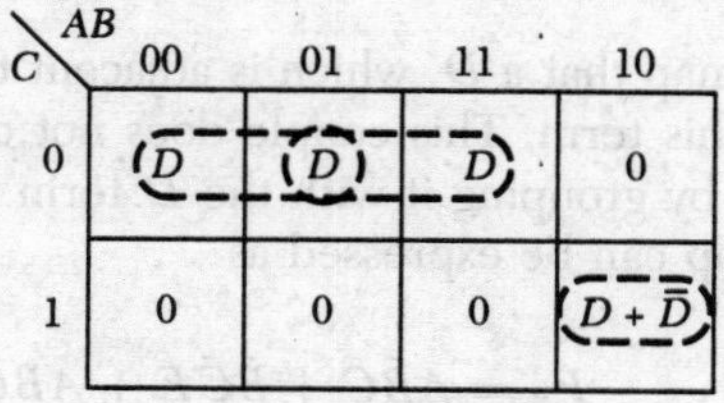

$$F = \overline{A}\overline{B}\overline{C}D + \overline{A}B\overline{C}D + \overline{A}B\overline{C}\overline{D} + \overline{A}\overline{B}C\overline{D} + A\overline{B}\overline{C}D + A\overline{B}CD + AB\overline{C}\overline{D}$$

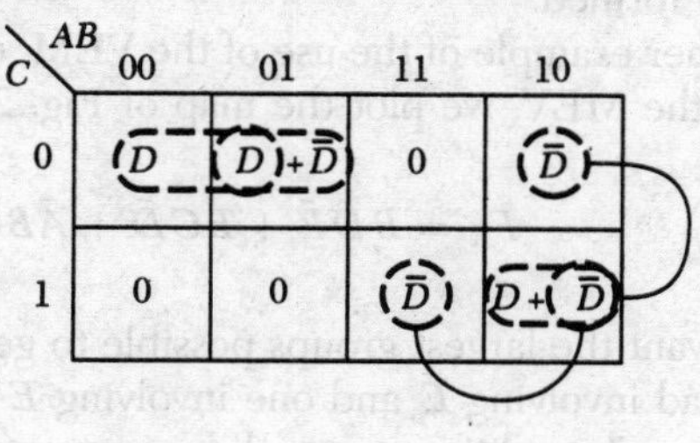

$$F = \overline{A}\overline{B}\overline{C}\overline{D}E + \overline{A}\overline{B}C\overline{D}E + \overline{A}BC\overline{D}\overline{E} + \overline{A}BC\overline{D}E + A\overline{B}\overline{C}\overline{D}E + A\overline{B}C\overline{D}E + A\overline{B}\overline{C}D + \overline{A}BCD\overline{E}$$

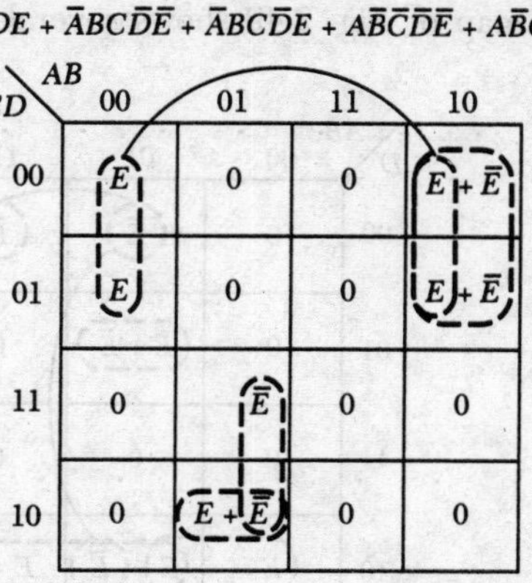

FIGURE 2.47   Variable-entered maps.

map of Fig. 2.47 reduces to

$$F_R = \bar{A}\bar{C}D + B\bar{C}D + A\bar{B}C$$

All three groups in this map are couples, including the one in location 101. This term involves $D + \bar{D}$ and consequently will not contain D when reduced. The second map reduces to

$$F_R = \bar{A}\bar{C}D + \bar{A}B\bar{C} + AC\bar{D} + A\bar{B}\bar{D} + A\bar{B}C$$

We see in this map that a D, which is adjacent to a 1 or $D + \bar{D}$, can form a couple with the D of this term. This couple does not cover the $\bar{D}$ term, but this variable can be covered by grouping it with the D term in the same map location.

The last map can be expressed as

$$F_R = A\bar{B}\bar{C} + \bar{B}\bar{C}E + \bar{A}BC\bar{E} + \bar{A}BC\bar{D}$$

Here we see that, when a term such as $E + \bar{E}$ is adjacent to an identical term, a quad can be formed.

As another example of the use of the VEM, consider the expression of Fig. 2.37. Using E as the MEV, we plot the map of Fig. 2.48. The reduced expression is

$$F_R = B\bar{D}\bar{E} + BC\bar{D} + \bar{A}B\bar{C}D + \bar{A}C\bar{D}E$$

We always want the largest groups possible to get maximum reduction. In Fig. 2.48, we see a quad involving $\bar{E}$ and one involving $E + \bar{E}$.

"Don't care" conditions serve the same purpose in reducing VEMs as in K-map reduction. The map of Fig. 2.49 shows several "don't care" conditions. There are

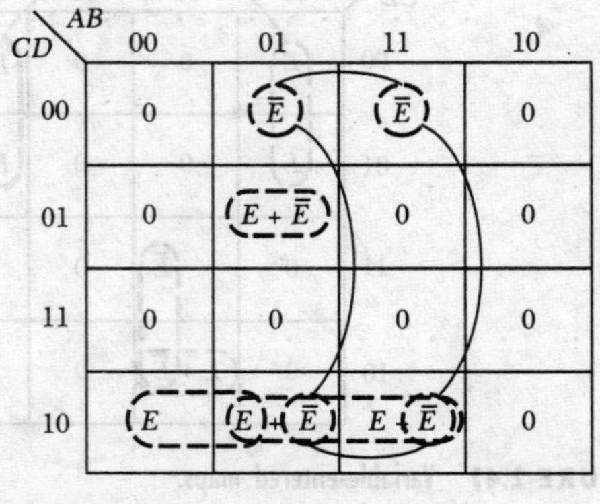

FIGURE 2.48   A five-variable VEM.

CD \ AB	00	01	11	10
00	0	Θ	E	0
01	0	Θ	Θ	$E\Theta + \bar{E}$
11	0	0	Θ	$E + \bar{E}\Theta$
10	0	0	0	0

FIGURE 2.49 "Don't care" conditions.

two types of "don't care" situations possible. The first is a location containing only a Θ. This location can then be taken as a 0, 1, E, or $\bar{E}$, whichever is appropriate. The second type of "don't care" condition is exemplified by $E\Theta + \bar{E}$. In this case, the "don't care" can be taken as a 0 or 1 to give either $\bar{E}$ or $E + \bar{E}$. In the map of Fig. 2.49, the Θs in the second column are taken as an E, the Θs in column 3 are taken as $E + \bar{E}$, and each Θ in column 4 is taken as a 1. This allows two quads to cover all terms, giving

$$F_R = B\bar{C}E + AD$$

EXAMPLE 2.4

Construct a VEM from the truth table of Fig. 2.50 using E as the MEV. Write the reduced expression for X.

SOLUTION

A five-variable MEV requires a four-variable map to implement. Figure 2.51 shows the map for this example. As we group terms we attempt to create the largest nonredundant groups possible. The map shown in Fig. 2.51 takes all Θs within the groups as 1s and all exterior Θs as 0s. The expression reduces to

$$X = \bar{A}\bar{B}\bar{D} + \bar{B}DE + AB\bar{D}$$

One disadvantage of the VEM is the difficulty in realizing a logic function by first finding the minimum form of $\bar{F}$ and then complementing this result. The K-map is much easier to use for this purpose.

A	B	C	D	E	X
0	0	0	0	0	θ
0	0	0	0	1	1
0	0	0	1	0	0
0	0	0	1	1	1
0	0	1	0	0	1
0	0	1	0	1	1
0	0	1	1	0	θ
0	0	1	1	1	θ
0	1	0	0	0	θ
0	1	0	0	1	θ
0	1	0	1	0	0
0	1	0	1	1	0
0	1	1	0	0	0
0	1	1	0	1	0
0	1	1	1	0	0
0	1	1	1	1	0

A	B	C	D	E	X
1	0	0	0	0	0
1	0	0	0	1	0
1	0	0	1	0	0
1	0	0	1	1	θ
1	0	1	0	0	θ
1	0	1	0	1	0
1	0	1	1	0	0
1	0	1	1	1	1
1	1	0	0	0	1
1	1	0	0	1	θ
1	1	0	1	0	0
1	1	0	1	1	0
1	1	1	0	0	1
1	1	1	0	1	1
1	1	1	1	0	θ
1	1	1	1	1	θ

FIGURE 2.50 Truth table for Example 2.4.

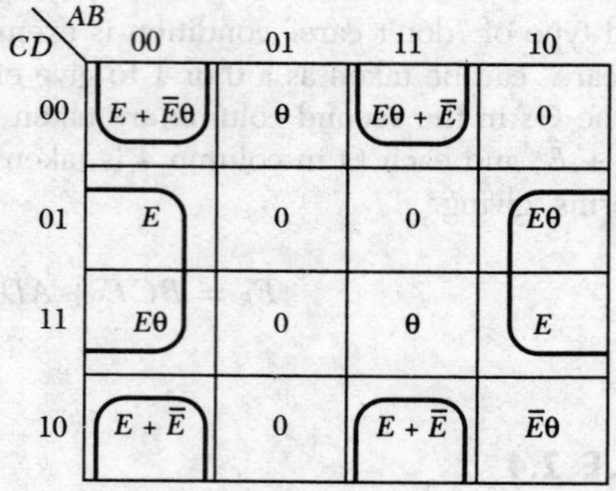

FIGURE 2.51 VEM for Example 2.4.

● ● ●   **DRILL PROBLEMS** Sec. 2.3.1 and 2.3.2

1. Using C as the MEV, reduce $X = \bar{A}B\bar{C} + \bar{A}BC + AB\bar{C} + A\bar{B}\bar{C}$.

2. Using D as the MEV, reduce $Y = \bar{A}\bar{B}\bar{C}\bar{D} + \bar{A}\bar{B}C\bar{D} + A\bar{B}\bar{C}\bar{D} + A\bar{B}\bar{C}D + A\bar{B}CD + A\bar{B}C\bar{D}$.

3. Repeat Drill Prob. 2 from Sec. 2.2.4 with D as the MEV.

2.3.3   THE VARIABLE-ENTERED MAP WITH TWO MAP-ENTERED VARIABLES

There are occasions when it is useful to reduce the dimension of the K-map by 2. This can be done by plotting two MEVs rather than one. Let us plot the VEM for

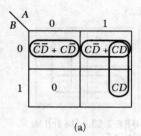
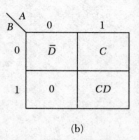

(a)   (b)

FIGURE 2.52 A VEM with two MEVs: (a) original map, (b) simplified map.

$X = \bar{A}\bar{B}\bar{C}\bar{D} + \bar{A}\bar{B}C\bar{D} + ABCD + A\bar{B}CD + A\bar{B}C\bar{D}$ using C and D as MEVs. We first factor the expression into terms involving the map locations $\bar{A}\bar{B}$, $\bar{A}B$, $A\bar{B}$, and AB. This results in

$$X = \bar{A}\bar{B}(\bar{C}\bar{D} + C\bar{D}) + A\bar{B}(C\bar{D} + CD) + ABCD$$

The map is shown in Fig. 2.52. After plotting the map of Fig. 2.52(a), we note that the terms in the upper two locations can be simplified, leading to the map of Fig. 2.52(b). This type of map is useful when combining gates with multiplexers to realize logic functions. We will consider this combination of elements in the next chapter.

The reduction of the map is more complex than the single MEV case. Figure 2.52(a) can be used for this reduction. With the groups shown, the resulting expression is

$$X_R = ACD + A\bar{B}C + \bar{A}\bar{B}\bar{D}$$

EXAMPLE 2.5

. . . .

Plot the VEM for the expression

$$Y = \bar{A}\bar{B}\bar{C}\bar{D} + \bar{A}\bar{B}\bar{C}D + \bar{A}\bar{B}C\bar{D} + \bar{A}BCD + ABCD + A\bar{B}CD$$

using C and D as MEVs. Write the reduced expression for Y.

SOLUTION

The expression is first written in factored form as

$$Y = \bar{A}\bar{B}(\bar{C}\bar{D} + \bar{C}D + C\bar{D}) + \bar{A}BCD + A\bar{B}CD + ABCD$$

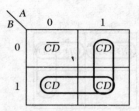

FIGURE 2.53 The VEM for Example 2.5.

The VEM is then plotted in Fig. 2.53. We note that $\bar{C}\bar{D} + \bar{C}D + C\bar{D}$ is also equal to not CD. When either $\bar{C}\bar{D}$, $\bar{C}D$, or $C\bar{D}$ occurs, the term CD cannot be true; thus not CD is equivalent to the sum of these three terms.

The reduced expression is

$$Y_R = \bar{A}\bar{B}\bar{C} + \bar{A}\bar{B}\bar{D} + BCD + ACD$$ ●

● ● ● **DRILL PROBLEMS** Sec. 2.3.3

1. Plot the function

$$F = A\bar{B}\bar{C}DE + A\bar{B}CD E + \bar{A}\bar{B}\bar{C}DE + \bar{A}\bar{B}C D\bar{E}$$

$$+ AB\bar{C}DE + ABCD\bar{E} + \bar{A}BCDE + \bar{A}\bar{B}\bar{C}\bar{D}E + A\bar{B}\bar{C}D\bar{E} + \bar{A}\bar{B}\bar{C}\bar{D}\bar{E}$$

on a VEM with D and E as MEVs.

2. Plot and reduce

$$X = A\bar{B}CD + AB\bar{C}D + ABC\bar{D} + \bar{A}BCD + A\bar{B}C\bar{D}$$

$$+ ABCD + \bar{A}\bar{B}\bar{C}D + AB\bar{C}\bar{D}$$

using a VEM with C and D as MEVs.

2.4
. . . .

REALIZING LOGIC FUNCTIONS WITH GATES

SECTION OVERVIEW This section considers some practical aspects of implementing logic functions with gates. The applications of the exclusive OR and coincidence gates are also discussed.

2.4.1 GATES

During the 1950s and 1960s, gates were used exclusively to implement logic functions. In some cases, these realizations took the form of diode matrices, which were essentially equivalent to gate circuits. In the late 1960s and early 1970s, as MSI circuits became available, different forms of function realization were developed. The multiplexer, the decoder, the read-only memory, and the programmable logic array now allow alternative forms of function implementation. These devices will be considered in Chapter 3 and Chapter 8.

Although newer methods are available, gate realization is still used. Minimization techniques along with gate implementations generally lead to minimal component cost of a finished circuit. This section notes a few practical aspects involved in using gate circuits.

When SSI circuits are used to construct a system, IC chips with 1, 2, 3, or 4 gates will be utilized. There may be several logic functions to implement in a system; consequently, when one function is realized with chips, unused gates often result. These gates can then be used in constructing some other logic function of the system. Furthermore, it is often appropriate to convert to NOR or NAND gates to fully utilize the several gates of the same type contained on each IC chip. For example, the expression $X = AB + CD$ could be realized with two 2-input AND gates and a 2-input OR requiring two chips. A single quad 2-input NAND could implement this expression as shown in Fig. 2.54. The application of DeMorgan's laws is quite useful in converting functions to implementations involving a single type of gate, resulting in a lower chip count.

Another practical problem often encountered is that of unused inputs. Suppose we have an expression $Y = \overline{A\bar{B}C}$ to implement and have a 4-input NAND available. If we connect these three variables to the gate inputs, we must determine how to connect the fourth input. Remembering that $A\bar{B}C \cdot 1 = A\bar{B}C$ dictates that the fourth input be tied high in this case. In TTL circuits, a floating input is interpreted as a logic one; hence, we might be tempted to allow the fourth input to float. An open input exhibits a reasonably high impedance and becomes less immune to noise. We should always tie unused inputs to either ground or the +5 V supply, whichever is appropriate, rather than allowing the input to float and possibly introduce noise problems. If an input is tied to +5 V, a series resistance of 1 to 10 kΩ should be used between the source and gate input to limit current.

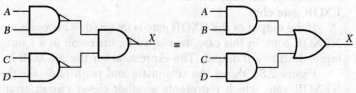

FIGURE 2.54 NAND gate realization.

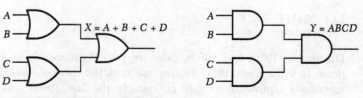

FIGURE 2.55 Expansion of inputs.

A 4-input OR gate can be used to implement a three-variable OR expression, but in this case we note that $A + B + C + 0 = A + B + C$. Thus, here we want the fourth input connected to ground rather than to +5 V.

It might also be convenient to construct functions requiring a larger number of inputs with 2- or 3-input gates. Figure 2.55 shows two such examples. Four-input OR gates or AND gates can be based on 2-input devices. If unused gates are available, this method may conserve chip count even though 4-input gates are included in the TTL line.

A final point that should be emphasized is that a NAND or NOR gate can be used as an inverter if needed. Utilization of a spare gate may avoid the addition of an inverter chip. All inputs can be tied together to convert a NAND or NOR to an inverter, but the driving circuit is loaded more heavily than necessary in this configuration. Instead of driving all inputs in parallel, the practical method shown in Fig. 1.10(b) is generally preferred if minimal current loading of the driving gate is appropriate.

2.4.2 THE EXOR GATE IN FUNCTION REALIZATION

The exclusive OR function obeys the truth table of Fig. 2.56. The symbol for the EXOR gate and the equation it satisfies are also shown in the figure. A 1 is generated at the gate output only if one input equals 1 while the other input is 0. If both inputs are equal, the output is 0. The K-map for the EXOR function is indicated in Fig. 2.57. In writing the expression for this function, we see that three 2-input gates are required since

$$X = A \oplus B = A\bar{B} + \bar{A}B \tag{2.20}$$

Fortunately, the function is implemented on a chip. The 7486 is a quad 2-input EXOR gate chip.

If the output of the EXOR gate is inverted, it becomes a coincidence circuit or a NEXOR gate. In this case, two equal inputs result in a 1 output while two different inputs lead to a 0 output. The expression for the NEXOR is $X = AB + \bar{A}\bar{B}$.

Figure 2.58 shows the schematic and truth table for the 74135 Quad EXOR/ NEXOR gate, which represents a rather clever circuit arrangement. A and B are the inputs to an EXOR gate. The output of the first EXOR gate connects to the

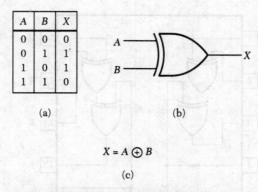

A	B	X
0	0	0
0	1	1
1	0	1
1	1	0

(a)

(b)

$$X = A \oplus B$$

(c)

FIGURE 2.56   The exclusive OR gate: (a) truth table, (b) symbol, (c) equation.

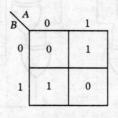

FIGURE 2.57   K-map for the EXOR.

input of a second EXOR gate. If the C input to this gate is low, the output equals the signal applied to the other input. Therefore, when C is low, the circuit output Y represents the EXOR of inputs A and B.

In order to produce a NEXOR gate, input C is taken high. For this condition, the EXOR function of gate 1 is inverted by gate 2 resulting in a NEXOR function for Y. Each input C drives two gate circuits in order to conserve pins on the chip.

Since EXOR and NEXOR gates are available on chips, their use in logic function realization should be considered. In certain cases, these gates can implement functions more efficiently than conventional gates. The map of Fig. 2.59 reflects one such function. There is no reduction possible, and the resulting expression is

$$X = \bar{A}B\bar{C} + \bar{A}\bar{B}C + ABC + A\bar{B}\bar{C}$$

A total of 5 gates and 17 inputs are required to implement this expression.

The expression for X can be factored to yield

$$X = \bar{A}(B\bar{C} + \bar{B}C) + A(BC + \bar{B}\bar{C})$$

The expression in the first set of parentheses is an EXOR relation while that in the second set of parentheses is a NEXOR relation. This function could be implemented

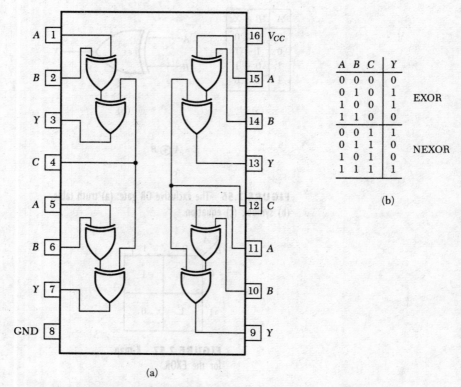

A	B	C	Y	
0	0	0	0	
0	1	0	1	EXOR
1	0	0	1	
1	1	0	0	
0	0	1	1	
0	1	1	0	NEXOR
1	0	1	0	
1	1	1	1	

(b)

(a)

FIGURE 2.58 The 74135 EXOR/NEXOR gate: (a) schematic, (b) truth table.

with two AND gates, an OR gate, an EXOR gate, and a NEXOR gate requiring 10 inputs. On the other hand, we can note that X can also be expressed as

$$X = A \oplus (B \oplus C)$$

We can implement this expression with two 2-input EXOR gates as indicated in Fig. 2.60.

C \ AB	00	01	11	10
0	0	1	0	1
1	1	0	1	0

X

FIGURE 2.59 A K-map.

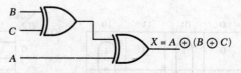

FIGURE 2.60 Implementation of the map of Fig. 2.59.

The map of Fig. 2.59 represents a special case that allows the EXOR to be applied much more efficiently than conventional gates. This map contains four 1s with no adjacencies. Any map with more nonadjacent 1s than adjacent 1s is a candidate for realization using EXOR circuitry. We should note in the map of Fig. 2.59 that the diagonal 1s represent a nonadjacent pair as also do the pairs of 1s separated by a single 0.

A theory can be developed for absolute minimization using EXOR circuits, but a rather simple method can be used to decrease gate count. This method consists simply of reducing the circuit by conventional methods, then looking for EXOR or NEXOR relations within the reduced expression. We can demonstrate this method with the maps of Fig. 2.61. The expression for X in Fig. 2.61(a) is reduced to

$$X = \bar{A}B + A\bar{B}$$

This expression obviously reduces to

$$X = A \oplus B$$

The expression for Y in Fig. 2.61(b) is reduced to

$$Y = \bar{B}\bar{C} + BC$$

which can be realized with a NEXOR gate.

The map of Fig. 2.61(c) gives an expression

$$Z = \bar{A}\bar{B}\bar{D} + \bar{A}BD + A\bar{C}\bar{D} + A\bar{B}CD$$

Although the first two terms could be combined in a NEXOR relation to reduce total inputs, the number of gates would not be reduced. Unless an unused NEXOR gate happens to be available in the system, conventional gate realization would probably lead to a lower chip count.

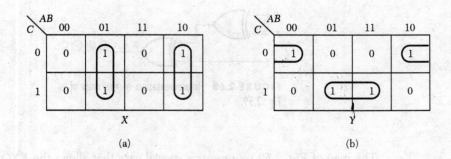

(a)   (b)

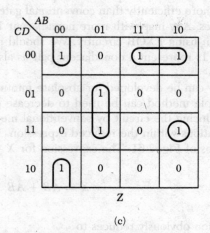

(c)

FIGURE 2.61 K-map examples implemented with EXOR gates.

● ● ●   DRILL PROBLEMS Sec. 2.4.2

1. Minimize the number of gates used to implement the expression
   $X = \bar{A}\bar{B}\bar{C} + \bar{A}\bar{B}C + AB\bar{C} + ABC$ using conventional gates only. Repeat if
   EXOR/NEXOR gates are allowed.

2. Repeat Prob. 1 if

$$X = \bar{A}\bar{B}\bar{C} + \bar{A}\bar{B}C + \bar{A}BC + AB\bar{C} + ABC + A\bar{B}C$$

2.5

. . . .

COMBINATIONAL DESIGN PROCEDURE AND EXAMPLES

SECTION OVERVIEW This section proposes a design procedure for combinational logic circuits and
demonstrates the design of practical combinational logic systems.

2.5.1   A DESIGN PROCEDURE

The introductory section of this textbook discussed the notion of applying top-down design methods to the development of digital systems. While many combinational problems are so straightforward that a design procedure is unnecessary, it is useful to consider a procedure that can later be extended to more complex problems.

The top-down approach begins with a broad, abstract solution that is then successively refined to add the required detail to construct the system. For most combinational problems, the following procedure is appropriate.

1. Determine, from a thorough study of the specifications, the goals of the system.
2. Draw a block diagram of the system showing all input and output lines. Each different function of the system should show a separate block or functional module. Assign names or labels to all variables.
3. Express the relationship of each output variable to the input variables. This can be done by tables or by equations.
4. Construct truth tables for all output variables.
5. Reduce output variable expressions to simplest terms.
6. Check for any necessary hazard covers.
7. Implement circuits eliminating any redundant gates.

The last step is appropriate in multifunction realizations wherein the same terms are implemented for more than one function.

At the risk of "overkill" we will apply this procedure to some simple combinational logic design problems.

2.5.2   A COMBINATIONAL FUNCTION GENERATOR

A circuit is required to produce an output X for various combinations of the inputs A, B, and C. The chart of Fig. 2.62 plots all combinations of the inputs along with the corresponding values of the output X.

The block diagram reflects the fact that this circuit is driven by three inputs to produce a single output. Since the table expresses the relationship of inputs to the output variable, the K-map can be immediately constructed. Figure 2.63 includes the truth table, the block diagram, and the K-map for this circuit.

The reduced expression for X is

$$X = \bar{A}\bar{C} + AC$$

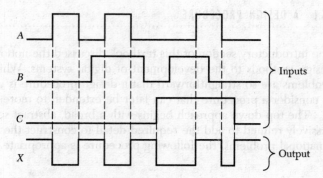

FIGURE 2.62 An input-output chart.

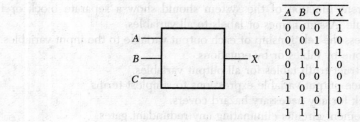

A	B	C	X
0	0	0	1
0	0	1	0
0	1	0	1
0	1	1	0
1	0	0	0
1	0	1	1
1	1	0	0
1	1	1	1

C \ AB	00	01	11	10
0	1	1	0	0
1	0	0	1	1

FIGURE 2.63 Truth table and K-map for the information of Fig. 2.62.

Although the specifications did not mention whether hazards are to be eliminated, a quick inspection of the K-map indicates that hazards will not be generated for this system. This expression can be implemented by three 2-input gates or by a single NEXOR gate.

This rather simple design problem demonstrates the basic procedure for combinational circuit design. More complex problems use the same tools, but require more decision making as the next example shows.

2.5.3   A NUMBER COMPARATOR

Design a binary number comparator that accepts the numbers A and B, each two bits long, and compares the numerical values of these numbers. The result of the comparison is to be presented on three output lines. These lines are labeled G, E, and L. The line G should be asserted if the number A is larger than B. If the numbers are equal, E should be asserted. If A is less than B, the line L should be asserted. The block diagram is indicated in Fig. 2.64. The bits of the two input numbers are designated A_1 A_0 and B_1 B_0. Hazards are unimportant in this design.

A truth table for all possible inputs is also shown in Fig. 2.64. This table expresses the relationships between all output and input variables.

A K-map is then constructed for each output from the truth table information. These maps appear in Fig. 2.65.

The expressions for G, E, and L are reduced to

$$G = A_1\bar{B}_1 + A_0\bar{B}_1\bar{B}_0 + A_1A_0\bar{B}_0$$

$$E = \bar{A}_1\bar{A}_0\bar{B}_1\bar{B}_0 + \bar{A}_1A_0\bar{B}_1B_0 + A_1A_0B_1B_0 + A_1\bar{A}_0B_1\bar{B}_0$$

$$L = \bar{A}_1B_1 + \bar{A}_1\bar{A}_0B_0 + \bar{A}_0B_1B_0$$

These expressions could be realized with gates to complete the design. Before doing so, however, it is worthwhile to examine the expressions and K-maps carefully to see if further reduction can be accomplished.

An examination of the K-map for E indicates the horizontal relationship that is characteristic of EXOR or NEXOR gates. A modest amount of manipulation allows E to be written as

$$E = (A_1B_1 + \bar{A}_1\bar{B}_1)(A_0B_0 + \bar{A}_0\bar{B}_0)$$

A_1	A_0	B_1	B_0	G	E	L
0	0	0	0	0	1	0
0	0	0	1	0	0	1
0	0	1	0	0	0	1
0	0	1	1	0	0	1
0	1	0	0	1	0	0
0	1	0	1	0	1	0
0	1	1	0	0	0	1
0	1	1	1	0	0	1
1	0	0	0	1	0	0
1	0	0	1	1	0	0
1	0	1	0	0	1	0
1	0	1	1	0	0	1
1	1	0	0	1	0	0
1	1	0	1	1	0	0
1	1	1	0	1	0	0
1	1	1	1	0	1	0

FIGURE 2.64 Truth table for comparator.

K-map (a): output G

B_1B_0 \ A_1A_0	00	01	11	10
00	0	1	1	1
01	0	0	1	1
11	0	0	0	0
10	0	0	1	0

(a)

K-map (b): output E

B_1B_0 \ A_1A_0	00	01	11	10
00	1	0	0	0
01	0	1	0	0
11	0	0	1	0
10	0	0	0	1

(b)

K-map (c): output L

B_1B_0 \ A_1A_0	00	01	11	10
00	0	0	0	0
01	1	0	0	0
11	1	1	0	1
10	1	1	0	0

(c)

FIGURE 2.65 K-maps for the two-bit comparator: (a) output G, (b) output E, (c) output L.

This allows E to be implemented with two 2-input NEXOR gates and a 2-input AND gate.

Further examination of the maps shows that if E and L are ORed, the zeros of the resulting K-map occupy the same locations as the ones of the map for G. This allows G to be expressed as

$$G = \overline{E + L}$$

The final system is shown in Fig. 2.66.

This example demonstrates that the basic tools such as the truth table and K-map are good starting points for a design. It is generally necessary to inject human reasoning into the process to achieve a good design. In this example, some rather obvious observations led to a simpler overall design than would have been obtained directly from the K-maps.

As a matter of fact, a simpler design for the comparator is possible and is discussed in Chapter 3.

The design procedure for combinational problems will be expanded for the design of sequential circuits in Chapter 7.

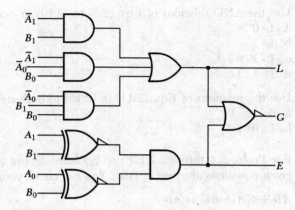

FIGURE 2.66 Number comparator.

SUMMARY

● ● ● ● ● ● ● ● ● ● ● ● ● ● ●

1. Boolean algebra can be used to express combinational logic problems in a concise manner.
2. Boolean algebra can be used to reduce the number of gates required to implement a logic function and can be used to convert the implementation to a particular type of gate.
3. The K-map is based on Boolean algebra and can be used to minimize logic functions.
4. Hazards may result in a minimized circuit. If it is important to eliminate these transient signals, hazard covers can be used.
5. The variable-entered map can also be used to minimize logic circuits. An advantage of this map over the K-map is its smaller size. We will later see that the VEM approach is useful in multiplexer design also.
6. The traditional method of function realization uses gates. This method minimizes component cost.
7. The EXOR or NEXOR gate can sometimes be used to reduce chip count over conventional gate implementation.
8. A simple procedure is useful for combinational circuit design.

CHAPTER 2 PROBLEMS

● ● ● ● ● ● ● ● ● ● ● ● ● ● ●

● Secs. 2.1.1 to 2.1.4

2.1 Use the OR relations of Eqs. (2.1) to (2.5) to evaluate
   a. $1 + 0$
   b. $1 + 1$
   c. $1 + 0 + 1$

2.2 Use the AND relations of Eqs. (2.6) to (2.10) to evaluate
a. $1 \cdot 0$
b. $1 \cdot 1$
c. $1 \cdot 1 \cdot 0$
d. $1 \cdot 1 \cdot 1$

2.3 Use the relations of Eqs. (2.11) to (2.15) to evaluate
a. $0 + 1 \cdot 1$
b. $1 \cdot (0 + 1)$

For Probs. 2.4 through 2.12, use Boolean algebra to determine whether the given equations are true or false. You can check your results with truth tables.

°2.4 $\bar{A}B + C\bar{A} + \bar{B}C = \bar{A}B$

2.5 $\bar{A}\bar{B}\bar{C} + \bar{A}BC = \bar{A}$

2.6 $\bar{A}\bar{B}\bar{C}\bar{D} + D = \overline{A + B + C + D}$

2.7 $AB\bar{C} + A\bar{B}C + \bar{A}BC = \overline{\bar{A}\bar{B} + \bar{A}\bar{C} + \bar{B}\bar{C} + ABC}$

°2.8 $AB + CD + A\bar{C} = \bar{A} + B + \bar{C} + D$

2.9 $\bar{A} + ABC = \overline{A\bar{B} + A\bar{C}}$

2.10 $CFG + C\bar{D}\bar{E} + EFG + DFG = DFG + EFG + C(\overline{D + E})$

2.11 $AB + \bar{A}\bar{C}\bar{D} + \bar{B}\bar{C}\bar{D} = AB + \bar{C}\bar{D}$

°2.12 $AB + BC + A\bar{C} + \bar{A}BC = AB\bar{C} + AB + A\bar{C}$

2.13 Realize, using NAND gates only, $X = \overline{A + \overline{B + C} + \overline{B} + D}$.

°2.14 Realize, using NOR gates only, $X = A(\bar{B} + \bar{C})(A + D)$.

2.15 Realize the expression of Prob. 2.13 using only NOR gates.

2.16 Realize the expression of Prob. 2.14 using only NAND gates.

2.17 Realize, using NAND gates only, $Y = \bar{A}C + A\bar{B}$.

°2.18 Use algebraic methods to express F as a function only of variables and single-variable complements:

$$F = \overline{\overline{(A + \bar{C})B + D} + E}$$

● Sec. 2.1.6

2.19 Write the equation for F in Fig. P2.19. Then convert the gates to reflect proper assertion level and write the expression for F directly from the circuit.

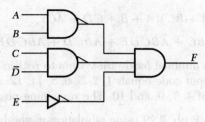

FIGURE P2.19

2.20 Repeat Prob. 2.19 for the circuit shown in Fig. P2.20.

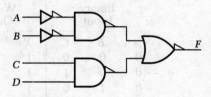

FIGURE P2.20

2.21 Repeat Prob. 2.19 for the circuit shown in Fig. P2.21.

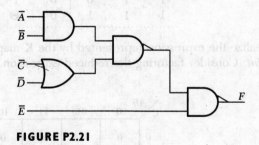

FIGURE P2.21

● Sec. 2.2.1

*2.22 Write the expression for F in Prob. 2.20 and put in SSOP form.

2.23 Write the expression for F in Prob. 2.19 and put in SSOP form.

● Secs. 2.2.2 to 2.2.4

Minimize the following expressions using K-maps.

2.24 $F = ABC + AB\bar{C} + A\bar{B}\bar{C}$

*2.25 $F = \bar{A}\bar{B}\bar{C}D + \bar{A}B\bar{C}D + \bar{A}\bar{B}CD + ABC\bar{D} + AB\bar{C}D$

2.26 $F = (AB + C\bar{D})(A\bar{C} + BD)$

2.27 $F = (A + \bar{B}C)(A + B + CD + \bar{A}C)$

*2.28 $F = ABC + \bar{A}\bar{B}\bar{C}DE + \bar{A}\bar{B}CD + A\bar{B}C\bar{D}\bar{E} + A\bar{B}\bar{C}\bar{D}E$

2.29 Build a minimal hardware system to realize a function F that equals 1 when a 4-bit input code equals 1, 2, 5, 6, 8, 11, 12, and 14. F should equal 0 for input codes of 4, 7, 9, and 10. The remaining possible codes will never occur.

2.30 Repeat Prob. 2.29 using tabulation methods to reduce the expression for F.

2.31 Build a code converter using gates and minimal design techniques to convert the binary code to the fictitious W-code shown.

Binary Input			W-Code Output		
A	B	C	X	Y	Z
0	0	0	1	1	1
0	0	1	1	Θ	1
0	1	0	1	1	0
0	1	1	0	1	1
1	0	0	1	0	0
1	0	1	0	0	1
1	1	0	0	1	0
1	1	1	0	Θ	0

*2.32 Realize the expression represented by the K-map with two gates (Fig. P2.32). *Hint:* Consider factoring the reduced expression.

C \ AB	00	01	11	10
0	0	1	0	0
1	0	θ	1	0

FIGURE P2.32

● Sec. 2.2.5

2.33 If F is realized as shown in Fig. P2.33, use a hazard cover to eliminate any static hazards. Explain hazards with a timing chart. Under what conditions of R, S, and X would you expect a glitch?

2.34 Instead of using a hazard cover, can you add a noninverting delay circuit at an appropriate point to eliminate the hazard in Prob. 2.33?

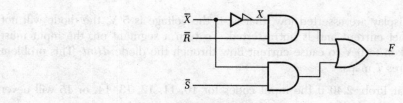

FIGURE P2.33

● Sec. 2.3.1 and 2.3.2

2.35 Work Prob. 2.24 using *C* as a map-entered variable.

*2.36 Work Prob. 2.24 using *B* as a map-entered variable.

2.37 Work Prob. 2.25 with *D* as a map-entered variable.

2.38 Work Prob. 2.27 with *A* as a map-entered variable.

*2.39 Work Prob. 2.28 with *E* as a map-entered variable.

2.40 A seven-segment LED display is shown in Fig. P2.40. Using VEMs, design the gate circuitry necessary to produce the display shown. Note that the inputs to

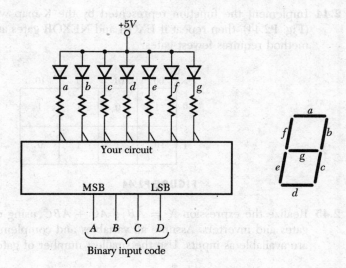

FIGURE P2.40

the display are asserted low, that is, if the voltage is 5 V, the diode will not conduct current and is not asserted. To turn a segment on, the input must be taken to 0 V to cause current flow through the diode. *Hint:* This problem requires 7 maps.

2.41 Repeat Prob. 2.40 if the input codes for 10, 11, 12, 13, 14, or 15 will never occur.

2.42 Design a code converter using gates to convert 4-bit binary code to 5-bit BCD, that is, 1 bit for the tens column and 4 bits for the ones column.

● Sec. 2.4

°2.43 Write the expression X realized by the circuit shown in Fig. P2.43. Use a K-map to implement this same function in a minimal way with no EXOR gates.

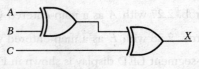

FIGURE P2.43

2.44 Implement the function represented by the K-map with conventional gates (Fig. P2.44); then repeat if EXOR and NEXOR gates are also allowed. Which method requires fewest gates?

C \ AB	00	01	11	10
0	1	0	1	1
1	0	1	1	0

FIGURE P2.44

2.45 Realize the expression $X = \bar{A}B + A\bar{C} + ABC$ using only two-input NAND gates and inverters. Assume all variables and complements of these variables are available as inputs. Use the smallest number of gates possible.

● Sec. 2.5

°2.46 Given four input lines that can carry any hex code from 0 to F. Design a circuit to generate a 1 on output line X when the input code is 2, 6, 7, 9, B, E, or F.

2.47 Repeat Prob. 2.46 if the circuit is to produce a 1 on line X for inputs of 1, 3, 5, 8, A, D, or E. The codes for B and F will never occur.

2.48 Design a 2-bit binary adder to add A_1A_0 to B_1B_0. The result should be presented on the lines $S_2S_1S_0$.

REFERENCES AND SUGGESTED READING

1. K. J. Breeding, *Digital Design Fundamentals*, 2nd ed. Englewood Cliffs, NJ: Prentice-Hall, 1992.
2. W. I. Fletcher, *An Engineering Approach to Digital Design*. Englewood Cliffs, NJ: Prentice-Hall, 1980.
3. M. M. Mano, *Digital Design*, 2nd ed. Englewood Cliffs, NJ: Prentice-Hall, 1991.
4. C. H. Roth, Jr., *Fundamentals of Logic Design*, 4th ed. St. Paul, MN: West, 1993.
5. S. H. Unger, *The Essence of Logic Circuits*. Englewood Cliffs, NJ: Prentice-Hall, 1989.

REFERENCES AND SUG...

1 Lee, Introduction Digital Circuit Fundamentals. Anta.j. Englewood Cliffs, N.J.: Prentice-Hall, 1967.

Chapter 3

Logic Function Realization with MSI Circuits

The traditional method of using gates to implement combinational logic functions has been applied for over three decades. Since this method often results in minimum component cost for many combinational systems, it continues to represent a popular approach. There are several newer circuits now available for function realization that lead to other advantages over gate implementations. This chapter will consider digital systems that use multiplexers, decoders, and other MSI circuits along with applicable design methods. Advantages of each approach will also be considered.

3.1

COMBINATIONAL LOGIC WITH MSI CIRCUITS

SECTION OVERVIEW This section considers two MSI circuits that can be used to implement combinational logic functions. The multiplexer and decoder are discussed and design methods are developed. Advantages of each device are covered.

3.1.1 MULTIPLEXERS

The multiplexer (MUX) is also called a data selector and was originally designed to time-division multiplex several lines of parallel information onto a single line. More recently the MUX has become popular in combinational logic circuit applications. The MUX has n select lines, 2^n input lines, and a single output line. It can be visualized as a switching circuit with the binary code on the select lines determining which input data line is connected to the output, hence the name data selector. Figure 3.1 shows the symbol for a 4:1 MUX and the switching circuit used to explain its operation.

The EN input enables the circuit when low. When high, the EN input either forces the output to remain at the low level or puts the output into the high-impedance state, depending on the MUX used. For example, the 74151 is a MUX that forces the output low while the 74251 presents a high output impedance when the EN input is high [2]. If the MUX is enabled, values of $S_1(\text{MSB}) = 1$ and $S_0 = 1$ cause input line 3 to be connected to the output. If $S_1 = 1$ and $S_0 = 0$, input line 2 is connected to the output, and so on. Of course, the actual MUX does not use a mechanical switch to selectively connect input to output. Instead, AND gates are used as shown in Figure 3.2. A particular line is gated to the output only when EN is asserted and S_1 and S_0 contain the code to open the corresponding AND gate.

The IEEE/IEC symbol requires some explanation. The upper block is a common control block. It is understood that an input to this block represents an input to all other common blocks of the symbol. The G indicates an AND dependency, and G_3^0 indicates there are four AND dependencies of lines 0 through 3 in the circuit. The two control inputs select a particular AND gate to allow data to pass through this gate to the output.

The MUX realizes logic functions by setting up a one-to-one correspondence between input line number and minterm number of a logic expression. This is demonstrated in Figure 3.3 where a two-variable K-map is shown with the minterm

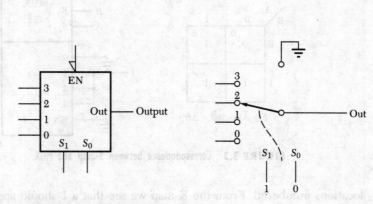

FIGURE 3.1 The multiplexer.

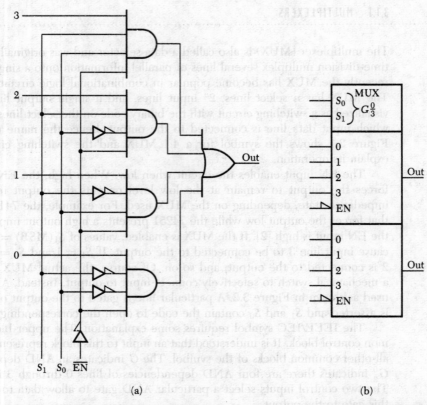

FIGURE 3.2 Implementation of 4:1 MUX: (a) actual circuit, (b) IEEE/IEC logic symbol for dual 4:1 MUX chip.

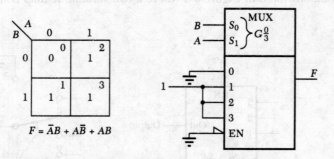

$$F = \bar{A}B + A\bar{B} + AB$$

FIGURE 3.3 Correspondence between K-map and MUX.

locations numbered. From the K-map we see that a 1 should appear at the output when minterm 1, 2, or 3 is present. A zero should result when minterm zero occurs. By connecting lines A and B to the select lines of the MUX, and then applying logic

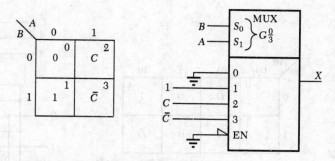

FIGURE 3.4 A three-variable realization.

1 to lines 1, 2, and 3, the correct function appears on the output line of the MUX. Logic zero is connected to the zero input line.

It is quite easy to extend the 4:1 MUX to a three-variable problem by using the map-entered variable approach. Let us assume we wish to synthesize the circuit to result in

$$X = AB\bar{C} + A\bar{B}C + \bar{A}B$$

The VEM and MUX implementation is shown in Figure 3.4.

When $AB = 00$, the VEM shows that a 0 output should result. The MUX implements this requirement by tying input line 0 to ground. When $AB = 01$, the VEM indicates that X should equal 1. The MUX produces a 1 at the output for this combination of A and B by tying input line 1 to logic 1 (+5 V). For $AB = 10$, X should equal 1 only if $C = 1$. Connecting C to input line 2 generates the correct value at the MUX output. Connecting $\bar{C}$ to input line 3 completes the implementation of X by producing a 1 at the output only if $A = 1$, $B = 1$, and $\bar{C} = 1$.

In general, an n-variable function requires a $2^{n-1}:1$ MUX to implement. For example, a four-variable design can be implemented by an 8:1 MUX having three select lines. Figure 3.5 demonstrates such a design.

It is possible to implement a system involving more variables without increasing the inputs to the MUX under certain conditions. One instance is that of seldom-used variables, a case that occurs often in sequential system design. A seldom-used variable is one that appears only occasionally in a logic expression. Each term of the expression

$$Y = A\bar{B}C + A\bar{B}\bar{C}D + \bar{A}BC\bar{D} + \bar{A}\bar{B}\bar{C}E + \bar{A}B\bar{C}$$

contains the variables A, B, and C, while D and E appear fewer times and do not appear simultaneously in the same term. Again the variables A, B, and C can be used to select the minterms or input lines, while both D and E can be used as inputs. This function is implemented in Figure 3.6.

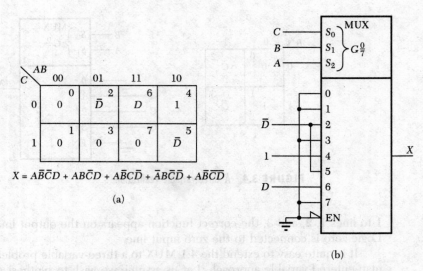

$$X = A\bar{B}\bar{C}D + AB\bar{C}D + A\bar{B}C\bar{D} + \bar{A}B\bar{C}D + A\bar{B}\bar{C}\bar{D}$$

(a)

(b)

FIGURE 3.5 A four-variable realization: (a) VEM, (b) 8:1 MUX realization.

Another method of increasing the number of variables without using a larger MUX is to use gating at the inputs. To demonstrate this method, let us consider the expression

$$X = A\bar{B}C\bar{D}E + AB\bar{C}\bar{D}\bar{E} + A\bar{B}C\bar{D}\bar{E} + \bar{A}B\bar{C}DE$$

$$+ \bar{A}B\bar{C}\bar{D}\bar{E} + \bar{A}B\bar{C}D\bar{E} + ABCDE$$

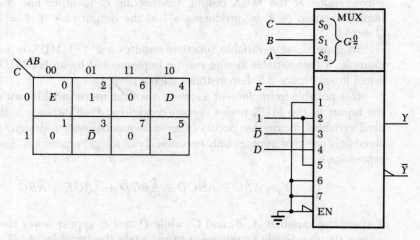

FIGURE 3.6 A seldom-used variable implementation (74LS151).

C \ AB	00	01	11	10
0	DE [0]	$D\bar{E} + \bar{D}E$ [2]	$\bar{D}\bar{E}$ [6]	0 [4]
1	0 [1]	0 [3]	DE [7]	$\bar{D}E + \bar{D}\bar{E}$ [5]

C \ AB	00	01	11	10
0	DE	$\bar{E}$	$\bar{D}\bar{E}$	0
1	0	0	DE	$\bar{D}$

(a) (b)

FIGURE 3.7   (a) A VEM. (b) Simplified VEM.

This five-variable expression can be realized with a three-select line MUX and additional gates. If variables A, B, and C are connected to the select lines, these same variables should be used to create a VEM. The remaining two variables, D and E, both become MEVs in this map. The original and simplified VEMs are shown in Figure 3.7. Since the variables A, B, and C act as minterm as well as MUX line selectors, the expression at each minterm location is implemented on the corresponding MUX line using gates. Figure 3.8 shows the finished circuit, which requires only two gates in addition to the 8:1 MUX.

There is a pin limitation on IC chips that generally restricts the size of the circuit that can be fabricated on the chip. A very complex circuit could be fabricated on a small chip, but the number of pins to the outside world ultimately determines what can be practically included on the chip. Most SSI or MSI circuits use 14 to 24 pins. Thus, a 16:1 is the largest available MUX on a chip. It is possible, however, to use several MUXs to expand to a larger multiplexing circuit. Figure 3.9 shows a 64:1

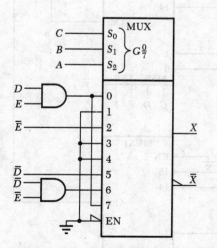

FIGURE 3.8   Implementation of a five-variable function with an 8:1 MUX.

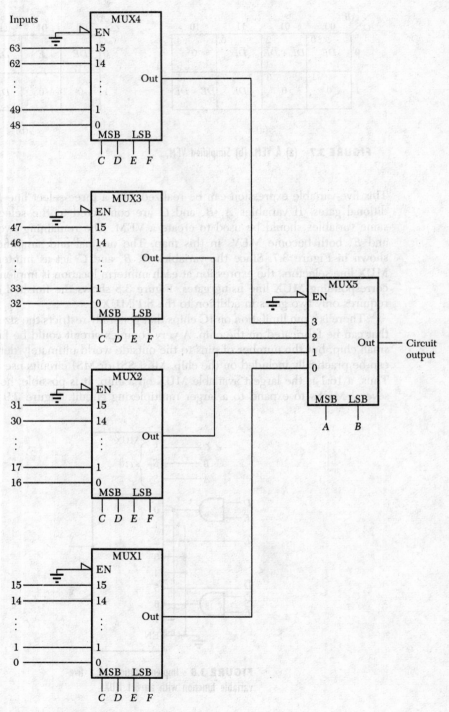

FIGURE 3.9   A 64:1 multiplexing circuit.

circuit. Six variables, designated A, B, C, D, E, and F, are required to address 64 input lines. The four least significant bits are connected in parallel to the select lines of four 16:1 MUXs. The two most significant bits connect to the select lines of a 4:1 MUX. A given input code will address one line of each of the 16:1 MUXs, but the 4:1 MUX will gate only one of the outputs of the larger MUXs to the circuit output. If the code happens to be $A\bar{B}C\bar{D}\bar{E}F$, which corresponds to 101001_2 or 41_{10}, each 16:1 MUX connects the ninth input line ($C\bar{D}\bar{E}F$) to its output. These four outputs connect to the input lines of the 4:1 MUX, which addresses input line 2 for this case ($A\bar{B}$). Thus, the input line that then connects to the circuit output is the ninth input line of MUX 3. This line corresponds to the forty-first circuit input line.

In chips containing more than one MUX, the select lines will normally appear in parallel for all MUXs. The 74LS153 is a dual 4:1 MUX chip, but select lines S_1 for both MUXs are connected to a single pin as also are lines S_0. Consequently, the MUXs cannot be addressed independently. The same is true of the 74LS157 quad 2:1 MUX chip. The select line S_0 is common to all 4 MUXs. Although this arrangement sacrifices some flexibility, the IC pin number requirement is minimized. Both the 74LS153 and the 74LS157 are 16 pin chips. The 74150 is a 24 pin 16:1 MUX.

EXAMPLE 3.1

A dual 4:1 MUX chip has common select lines for both MUXs in order to preserve pins on the chip. Show how to use this chip to implement the two functions

$$X = ABC + A\bar{B}\bar{C} + \bar{A}B\bar{C} + \bar{A}\bar{B}C$$

and

$$Y = \bar{A}\bar{B}C + \bar{A}B\bar{C} + A\bar{B}C$$

SOLUTION

One MUX will be used to implement X and the other will implement Y. If the variables A and B are used to drive the two select lines, the inputs for the two functions are given by

	S_1	S_0	X	Y
$\bar{A}\bar{B}$	0	0	0	1
$\bar{A}B$	0	1	$\bar{C}$	0
$A\bar{B}$	1	0	C	C
AB	1	1	C	0

The circuit of Figure 3.10 shows the actual realization of the functions.

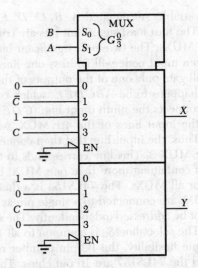

FIGURE 3.10 Implementation of X and Y.

In comparing MUX design of logic circuits to gate design, we note that map minimization techniques are not very significant in the MUX circuit. Less design time is then required for the MUX design. Another practical advantage of this circuit is that debugging and troubleshooting are easier than for gate circuits. If a MUX circuit malfunctions, the values of the variables connected to the select lines are measured. The corresponding input line and the output line are then checked to see if correct levels are present. In general, the problem can be readily isolated in this way. This is not the case for a complex gate circuit with several levels of gates. Pinpointing a malfunctioning gate may take considerably longer.

A third advantage is that chip count can often be minimized when using MUXs for function realization. Figure 3.11 compares MUX and gate design for the function

$$X = \bar{A}\bar{B}\bar{C}\bar{D} + \bar{A}B\bar{C}\bar{D} + \bar{A}B\bar{C}D + AB\bar{C}D + \bar{A}\bar{B}C\bar{D} + ABC\bar{D}$$

The gate realization, after minimization, requires three 3-input AND gates, one 4-input AND gate, and one 4-input OR gate. Converting to NAND gates allows the implementation to be accomplished with one triple 3-input NAND chip and one dual 4-input NAND chip. The MUX realization requires only one 8:1 MUX chip.

The chip costs of the two realizations are comparable, but the smaller number of inputs that must be wired may lead to a lower in-circuit cost for the MUX than for the gate circuit.

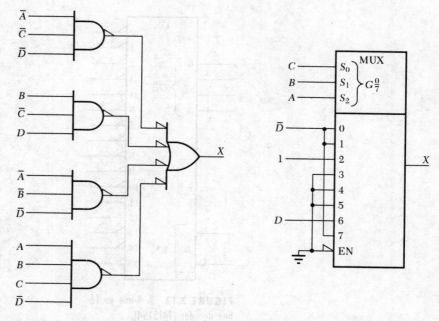

FIGURE 3.11 Gate and MUX realization of X.

● ● ● DRILL PROBLEMS Sec. 3.1.1

1. Realize the expression $X = \bar{A}\bar{B}C + A\bar{B}\bar{C} + \bar{A}B\bar{C} + ABC + \bar{A}\bar{B}\bar{C}$ with an 8:1 MUX.

2. Realize X from Prob. 1 with a 4:1 MUX.

3. Realize X from Prob. 1 with a 2:1 MUX and a minimal number of gates preceding the MUX inputs.

3.1.2 DECODERS

A binary decoder has n input lines and 2^n output lines. The output lines are numbered in accordance with the decimal equivalent of the input binary code. When the decoder is enabled, a code on the input asserts the corresponding output line while all other outputs remain unasserted. Figure 3.12 shows a 4-line to 16-line decoder. This device decodes all possible input codes with the outputs being asserted low. In order to implement a logic function, the outputs corresponding to all minterms that equal 1 in the function are ORed.

We will demonstrate the use of the decoder by implementing the function of Figure 3.5. This is shown in Figure 3.13. In decimal notation, the function can be

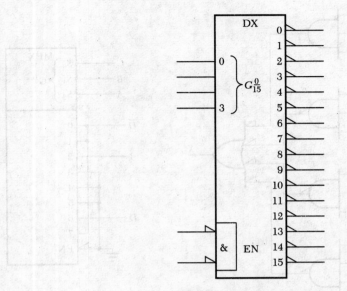

FIGURE 3.12   A 4-line to 16-line decoder (74LS154).

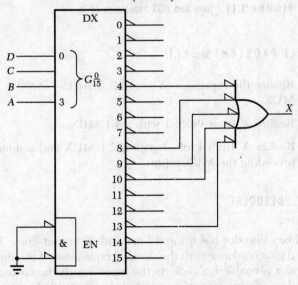

FIGURE 3.13   Decoder implementation of logic function.

written as

$$X = 4 + 8 + 9 + 10 + 13$$

These output lines are ORed to form the desired function. Since the decoder outputs are asserted low, a low-assertion OR gate (NAND gate) is required. When any of the codes 4, 8, 9, 10, or 13 is present at the input, X will be asserted.

The implementation of a binary decoder on the IC chip takes the form of 2^n NAND gates, each having $n + 1$ inputs. This configuration is demonstrated in Figure 3.14 for a 3-line to 8-line decoder, the 74LS138.

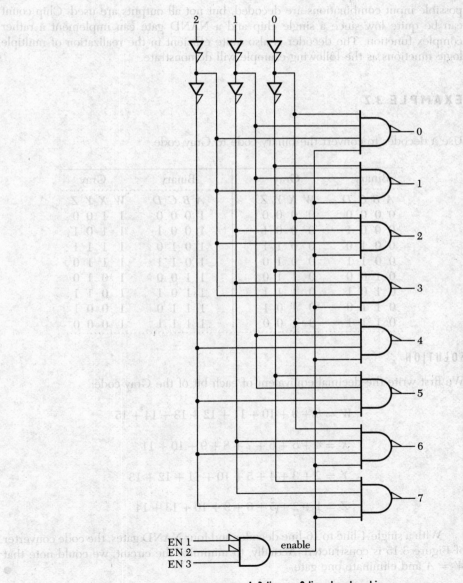

FIGURE 3.14 A 3-line to 8-line decoder chip.

Each possible input minterm is implemented on the chip. Each unique input code will assert a different output line, when the enable gate output is asserted. We note that when the enable gate output is not asserted, the 8 output lines of the decoder are forced to the high logic level. Unlike the three-state gates used for interfacing to the bus of a microprocessor, the enable input of the decoder cannot cause a high output impedance condition.

From the standpoint of efficiency, the decoder is somewhat wasteful in that all possible input combinations are decoded, but not all outputs are used. Chip count can be quite low since a single chip and a NAND gate can implement a rather complex function. The decoder is also quite efficient in the realization of multiple logic functions as the following example will demonstrate.

EXAMPLE 3.2

Use a decoder to convert the binary code to Gray code.

Binary	Gray	Binary	Gray
$A\ B\ C\ D$	$W\ X\ Y\ Z$	$A\ B\ C\ D$	$W\ X\ Y\ Z$
0 0 0 0	0 0 0 0	1 0 0 0	1 1 0 0
0 0 0 1	0 0 0 1	1 0 0 1	1 1 0 1
0 0 1 0	0 0 1 1	1 0 1 0	1 1 1 1
0 0 1 1	0 0 1 0	1 0 1 1	1 1 1 0
0 1 0 0	0 1 1 0	1 1 0 0	1 0 1 0
0 1 0 1	0 1 1 1	1 1 0 1	1 0 1 1
0 1 1 0	0 1 0 1	1 1 1 0	1 0 0 1
0 1 1 1	0 1 0 0	1 1 1 1	1 0 0 0

SOLUTION

We first write the decimal equivalent of each bit of the Gray code:

$$W = 8 + 9 + 10 + 11 + 12 + 13 + 14 + 15$$

$$X = 4 + 5 + 6 + 7 + 8 + 9 + 10 + 11$$

$$Y = 2 + 3 + 4 + 5 + 10 + 11 + 12 + 13$$

$$Z = 1 + 2 + 5 + 6 + 9 + 10 + 13 + 14$$

With a single 4-line to 16-line decoder and four NAND gates, the code converter of Figure 3.15 is constructed. Actually, to minimize the circuit, we could note that $W = A$ and eliminate one gate.

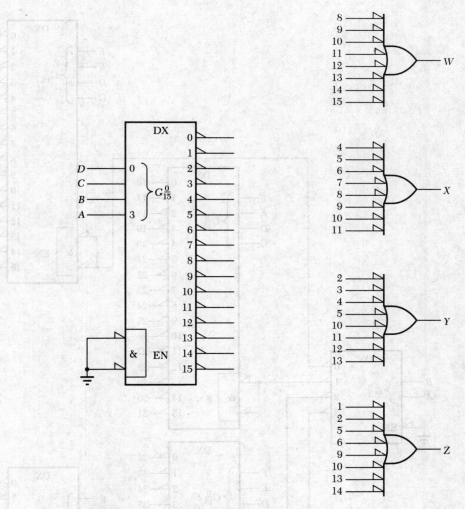

FIGURE 3.15 Binary to Gray code converter.

It is possible to use several chips to expand the number of possible input codes to a decoding circuit. Let us assume that a 6-bit binary code is to be decoded. This would result in 64 possible input combinations requiring 64 output lines. Figure 3.16 shows a method of implementing this circuit using one 2-line to 4-line decoder and four 4-line to 16-line decoders.

The two most significant bits of the input are used to enable the proper 4-line to 16-line decoder through the 2-line to 4-line device. If $AB = 00$, for example, the upper decoder is active. The four least significant bits, $CDEF$, determine which of the 16 lines of this decoder is asserted. The upper decoder has one output line

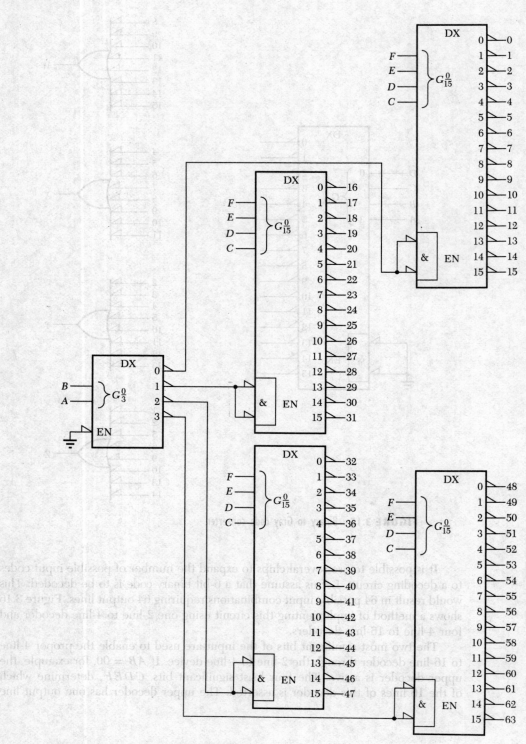

FIGURE 3.16 Expanded decoding circuit.

asserted for each input code between 000000 and 001111. The next lower decoder is active for input codes of 010000 to 011111. The third decoder is active for input codes of 100000 to 101111, and the last is active for input codes in the range of 110000 to 111111.

While the circuit of Figure 3.16 is useful in explaining the concept involved in decoder expansion, it is not an efficient realization of the 6-line to 64-line decoder. With two separate enable lines, inverters can be used to eliminate the 2-line to 4-line decoder as shown in Figure 3.17.

The 74154 chip is a 4-line to 16-line decoder packaged in a 24-pin circuit. The 74155 is a 16-pin chip that contains two 2-line to 4-line decoders with common input address lines. By using the separate enable lines, a 3-line to 8-line decoder can be constructed with the 74155 chip as shown in Figure 3.18. The MSB of the input code is used to enable decoder 1 when low and decoder 2 when high. If decoder 1 is enabled while decoder 2 is disabled, all outputs of decoder 2 are high and a single output of decoder 1 will be asserted low. An input code of 010 will assert circuit output line 2, while the code 110 will assert output line 6.

As mentioned earlier, the decoder may not be as efficient as the MUX in realizing a single function since a NAND chip is required. The real advantage of the decoder occurs in realizing several functions simultaneously. For comparison purposes, we will implement the same function realized by the gates and the MUX of Figure 3.11. In terms of decimal notation, this function is

$$X = 0 + 2 + 4 + 5 + 13 + 14$$

Figure 3.19 shows the decoder implementation of this function. The component cost for the decoder circuit is typically twice that of the MUX or gate realization.

EXAMPLE 3.3

The function

$$F = A\bar{B}CD + AB\bar{C}D + AB\bar{D} + \bar{A}BC\bar{D} + \bar{B}\bar{C}D$$

is to be realized. Implement this function with (a) gates, (b) a MUX, and (c) a decoder with gates.

SOLUTION

(a) The K-map of Figure 3.20 allows the minimization of the function F. The resulting reduced expression can be expressed as $F_R = A\bar{B}D + AB\bar{C} + \bar{B}\bar{C}D + BC\bar{D}$. The implementation of F_R is shown in Figure 3.21.

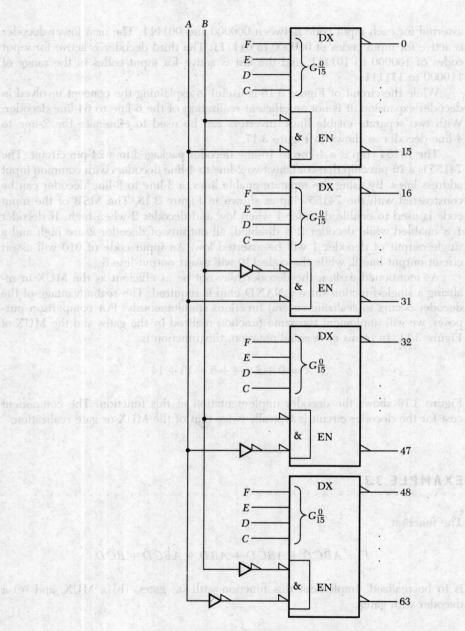

FIGURE 3.17 Practical realization of 6-line to 64-line decoder.

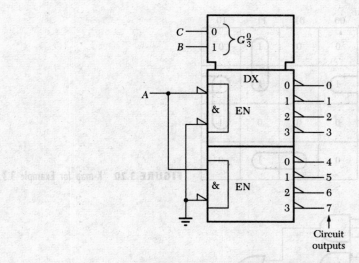

FIGURE 3.18   Dual 2-line to 4-line decoders used for a 3-line to 8-line decoder.

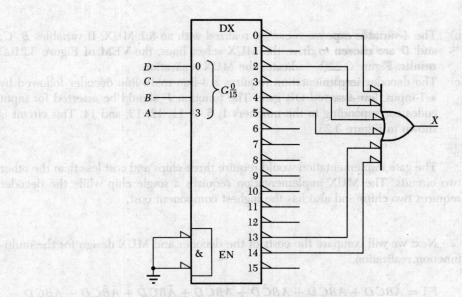

FIGURE 3.19   Decoder implementation of X.

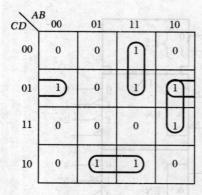

FIGURE 3.20   K-map for Example 3.3.

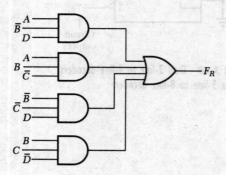

FIGURE 3.21   Realization of F with gates.

(b) The 4-variable expression can be realized with an 8:1 MUX. If variables B, C, and D are chosen to drive the MUX select lines, the VEM of Figure 3.21(a) results. Figure 3.22(b) indicates the MUX realization.

(c) The decoder implementation requires a 4-line to 16-line decoder followed by a 7-input, low-asserted OR gate. The function F should be asserted for input codes corresponding to the numbers 1, 6, 9, 11, 12, 13, and 14. This circuit is shown in Figure 3.23.

The gate implementation would require three chips and cost less than the other two circuits. The MUX implementation requires a single chip while the decoder requires two chips and also has the highest component cost.

Next we will compare the costs of the decoder and MUX design for the multi-function realization

$$F1 = \bar{A}BCD + A\bar{B}\bar{C}D + AB\bar{C}\bar{D} + \bar{A}B\bar{C}D + \bar{A}BC\bar{D} + \bar{A}\bar{B}CD + ABCD$$

$$F2 = \bar{A}\bar{B}\bar{C}D + A\bar{B}\bar{C}\bar{D} + A\bar{B}CD + ABC\bar{D} + ABCD + \bar{A}\bar{B}\bar{C}\bar{D}$$

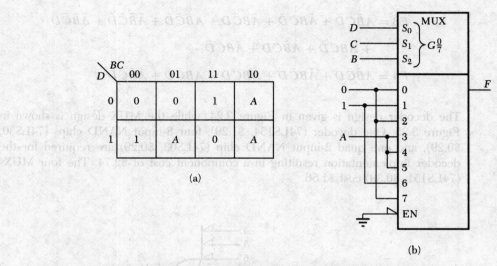

(a)

(b)

FIGURE 3.22 (a) VEM for MUX realization. (b) MUX implementation of *F*.

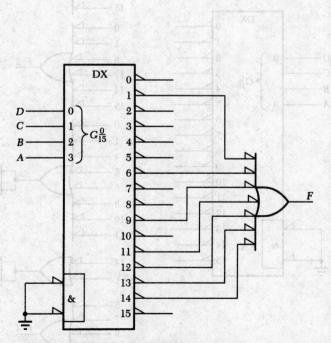

FIGURE 3.23 Decoder implementation of *F*.

$$F3 = A\bar{B}C\bar{D} + \bar{A}BC\bar{D} + \bar{A}B\bar{C}D + A\bar{B}\bar{C}\bar{D} + \bar{A}\bar{B}\bar{C}\bar{D} + \bar{A}\bar{B}\bar{C}D$$

$$+ ABC\bar{D} + A\bar{B}CD + \bar{A}B\bar{C}\bar{D}$$

$$F4 = A\bar{B}\bar{C}D + \bar{A}B\bar{C}D + \bar{A}BCD + AB\bar{C}\bar{D} + A\bar{B}C\bar{D}$$

The decoder design is given in Figure 3.24, while the MUX design is shown in Figure 3.25. One decoder (74LS154, $1.29), four 8-input NAND chips (74LS30, $0.29), and one quad 2-input NAND chip (74LS03, $0.29) are required for the decoder implementation resulting in a component cost of $2.74. The four MUXs (74LS151, $0.39) cost $1.56.

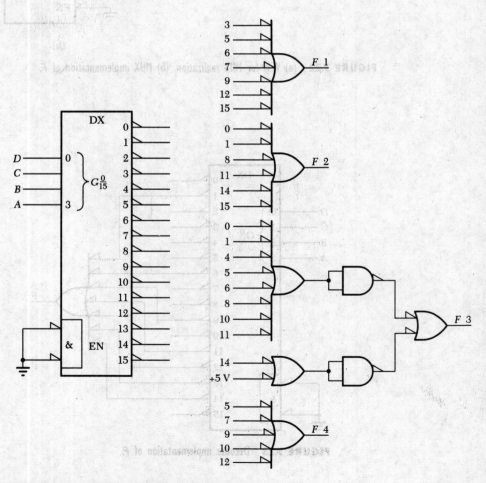

FIGURE 3.24 Multifunction realization with decoder.

EXAMPLE 3.4

Show how to construct a 5-line to 32-line decoder from standard MSI decoders.

SOLUTION

The 74154 4-line to 16-line decoder can be used to implement this larger decoder. Two such chips can be connected as shown in Figure 3.26 with the enable line connected to the MSB of the input code.   ●

● ● ● DRILL PROBLEMS Sec. 3.1.2

1. Realize the expression $X = \bar{A}\bar{B}C + A\bar{B}\bar{C} + \bar{A}B\bar{C} + ABC + \bar{A}\bar{B}\bar{C}$ with a 3-line to 8-line decoder and an 8-input NAND gate.

2. Realize the expressions $F1 = A\bar{B}C + AB\bar{C} + \bar{A}$, $F2 = \bar{A}\bar{B}C + \bar{A}B\bar{C} + A\bar{B}\bar{C}$, and $F3 = \bar{A}\bar{B}\bar{C} + ABC$ with a decoder followed by gates.

3. Show the pin connections necessary to implement X from Prob. 1 using a 74155 decoder chip along with an 8-input NAND gate.

3.2

STANDARD LOGIC FUNCTIONS WITH MSI CIRCUITS

SECTION OVERVIEW The encoder converts the assertion of a numbered line into the corresponding binary code. This circuit allows the depression of a key, which asserts an input line, to generate a binary code that can be used to communicate with a digital system.

The digital comparator examines two binary numbers and asserts an output that identifies which of the two numbers is the largest.

The light-emitting diode (LED) is very useful in logic circuits, serving as a visual indication of output voltage levels. This device is also used for the popular 7-segment display that forms numeric and other symbols. The 7-segment LED driver chip accepts a binary input code and produces the signals necessary to drive the display.

When binary codes are transmitted from one location to another, it is possible for noise to obscure a bit of the code. The received word would then differ from the transmitted word. It is possible to detect this type of error using a concept called parity.

This section will consider MSI chips that accomplish these four functions.

There are several logic functions that are often encountered in the digital field. Many of these functions have been implemented on single chips as MSI circuits. This section will consider the use of such MSI circuits as the encoder, the LED driver, the comparator, and the parity checker/generator circuit:

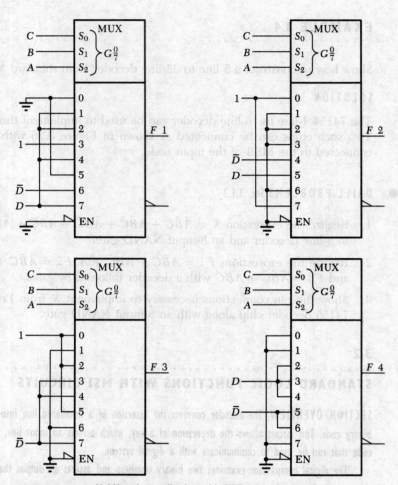

FIGURE 3.25 Multifunction realization with MUXs.

3.2.1 THE ENCODER

A binary encoder has n output lines and 2^n input lines. The input lines are numbered consecutively from 0 to $2^n - 1$. When a single input line is asserted, the binary code corresponding to the line number appears at the output.

The priority encoder is an extension of the binary encoder. This device allows several input lines to be asserted simultaneously while the output presents the binary code of the highest-numbered, asserted input line. For example, if input lines 3, 5, and 8 are asserted, the output code generated is 1000. The priority encoder was specifically designed to be used in conjunction with priority interrupt systems for

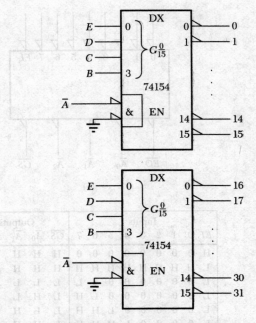

FIGURE 3.26   A 5-line to 32-line decoder.

computers. We note that if only one input line is asserted at a time, the priority encoder will function as a standard encoder.

The 74148 is an 8-input priority encoder that can be used in several applications. This particular device has low assertion input and output signals. The symbol for the 74148 priority encoder is shown in Figure 3.27 along with its function table. This chip has an enable input, EI, which enables the outputs when asserted low. If EI is taken high, all outputs are forced to the high level, regardless of the assertion of the input lines.

The other two output pins are the output enable, $\overline{EO}$, and the group signal, $\overline{GS}$. If EI is low, but all inputs are high, $\overline{EO}$ is low. This is the only condition for which this enable output is low. As long as the chip is enabled and $\overline{EO}$ is high, the output binary code will indicate the number of the highest numbered line asserted. The output $\overline{GS}$ is asserted low when any input line is active. Figure 3.28 shows the circuit diagram for the 74148 chip.

When an input line is asserted low, the corresponding AND gates that generate the proper binary code are asserted. This assumes that the enable input has been asserted low. There are several additional AND gates that are present to create a priority encoder. For example, if input lines 6 and 5 are asserted, output A_2 is asserted through AND gate 2, and output A_1 is asserted through AND gate 6.

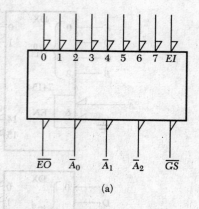

$$\overline{EO} \quad \overline{A_0} \quad \overline{A_1} \quad \overline{A_2} \quad \overline{GS}$$

(a)

Inputs									Outputs				
EI	0	1	2	3	4	5	6	7	$\overline{GS}$	$\overline{A_0}$	$\overline{A_1}$	$\overline{A_2}$	$\overline{EO}$
H	θ	θ	θ	θ	θ	θ	θ	θ	H	H	H	H	H
L	H	H	H	H	H	H	H	H	H	H	H	H	L
L	θ	θ	θ	θ	θ	θ	θ	L	L	L	L	L	H
L	θ	θ	θ	θ	θ	θ	L	H	L	L	L	L	H
L	θ	θ	θ	θ	θ	L	H	H	L	L	H	L	H
L	θ	θ	θ	θ	L	H	H	H	L	H	H	L	H
L	θ	θ	θ	L	H	H	H	H	L	L	L	H	H
L	θ	θ	L	H	H	H	H	H	L	H	L	H	H
L	θ	L	H	H	H	H	H	H	L	L	H	H	H
L	L	H	H	H	H	H	H	H	L	H	H	H	H

(b)

FIGURE 3.27   (a) Logic symbol for 74148 encoder.
(b) Function table.

Input line 5 would normally assert output lines A_2 and A_0; however, AND gate 10 is inhibited when input line 6 is asserted. This type of inhibition is performed to ensure that when more than one input line is asserted simultaneously, only the highest-numbered line will cause an output.

The additional outputs $\overline{GS}$ and $\overline{EO}$ allow the 74148 to be used to create larger encoders. A 16-line to 4-line encoder with high asserted outputs can be constructed as shown in Figure 3.29. When an input line numbered from 0 to 7 is asserted, chip 1 will present the corresponding binary code to the OR gates. This will drive Out_0, Out_1, and Out_2 while $\overline{EO}$ remains low since no input line is asserted on this chip. If an input line from 8 to 15 is asserted, the $\overline{EO}$ of chip 2 goes high to disable chip 1. This output also provides the MSB of the output code. Thus, if line 13 is asserted, chip 2 develops the code 101 and drives the OR gates to produce $Out_0 = 1$, $Out_1 = 0$, and $Out_2 = 1$. The value of Out_3 is provided by $\overline{EO}$, which is now high, generating a 1 for the MSB.

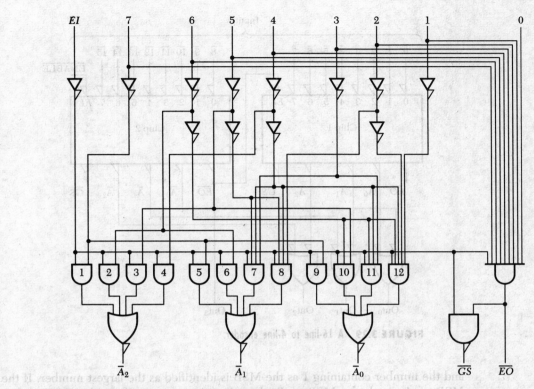

FIGURE 3.28 Circuit diagram for the 74148 encoder.

One application of a 16-line to 4-line encoder is in generating the proper code for a hex keypad. The 16 keys on the pad each represent a hex number. When depressed, each key closes a corresponding set of switch contacts. When the switch contacts close, the encoder input is driven low. The encoder then generates the binary code corresponding to the depressed key. Figure 3.30 shows this arrangement.

● ● ● **DRILL PROBLEM** Sec. 3.2.1

Show how to connect the 74147, 10-line-to-4-line priority encoder, to a decimal keypad.

3.2.2 THE COMPARATOR

It is often necessary to compare two binary numbers to determine which is larger. The basic method of comparison starts by comparing the MSB of input A to the MSB of input B. If one of these bits is 1 and the other 0, the process is completed

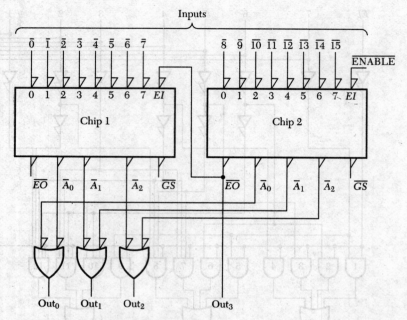

FIGURE 3.29 A 16-line to 4-line encoder.

and the number containing 1 as the MSB is identified as the largest number. If the MSB of A equals the MSB of B, then the next most significant bits of A and B are compared. This process continues until a bit of one number differs from the corresponding bit of the other. Rather than make these bit comparisons successively, the comparator is implemented so that all comparisons are made after only two gate delays.

There are three outputs for a comparator. One is asserted if $A = B$, another is asserted if $A > B$, and the other is asserted when $A < B$. The circuit of Figure 3.31 indicates the basic concept of a comparator. If $A_2 = 1$ while $B_2 = 0$, the $A > B$ output will be asserted. Since A_2 is unequal to B_2, the $A_2 = B_2$ line will disable the other four AND gates in the circuit. For either of these conditions on A_2 and B_2, no other input is significant in determining the output.

If $A_2 = B_2$, then A_1 and B_1 must be considered to generate the proper output. The output of the NEXOR gate driven by A_2 and B_2 will now enable the two AND gates associated with the A_1 and B_1 inputs. If A_1 and B_1 are unequal, one of these two AND gates will drive the appropriate output.

Only if $A_2 = B_2$ and $A_1 = B_1$ will the two AND gates associated with the inputs A_0 and B_0 be enabled. For this condition, the values of these two least significant bits determine the output. If all three bits of word A are equal to the corresponding bits of word B, all inputs to the output AND gate will be asserted to generate a 1 on the $A = B$ output line.

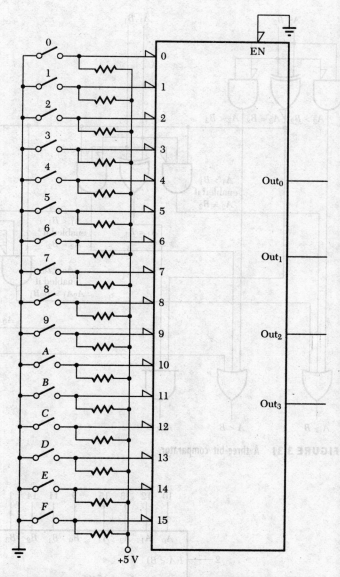

FIGURE 3.30 An encoder used to generate binary-coded hex for a hex keypad.

An actual 4-bit comparator, the 7485, also has three inputs to allow expansion to larger word comparison. The 74LS85 4-bit comparator is shown in Figure 3.32. The inputs $I(A < B)$, $I(A = B)$, and $I(A > B)$ are considered inputs from a comparison of four bits of lesser significance. In a 4-bit word comparison, $I(A < B)$

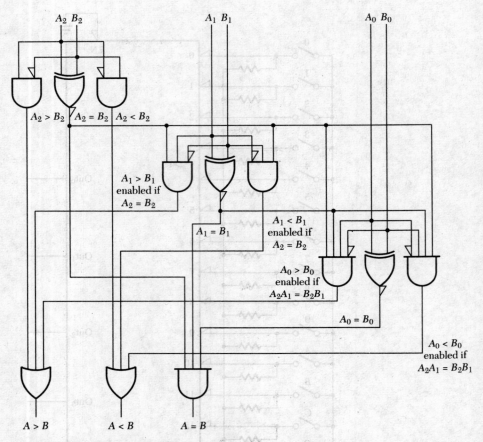

FIGURE 3.31 A three-bit comparator.

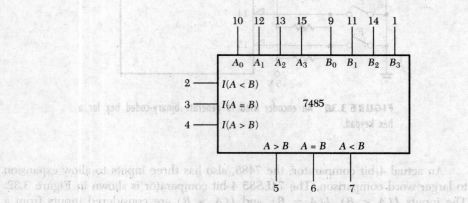

FIGURE 3.32 The 7485 four-bit comparator.

An actual 4-bit comparator, the 7485, also has three inputs to allow expansion to larger-word comparison. The 7LS85 4-bit comparator is shown in Figure 3.32. The inputs $I(A < B)$, $I(A = B)$, and $I(A > B)$ are considered inputs from a comparison of numbers of lesser significance than a 4-bit word comparison ($A < B$).

and $I(A > B)$ should be tied low while the $I(A = B)$ input should be tied high. Figure 3.33 shows an 8-bit comparator based on the 7485.

IC number 2 compares the most significant 4 bits, or nibble. If this nibble shows an unequal value between X and Y, the correct output is immediately determined. If X and Y are identical in the 4 most significant bits, the input from the 4 least significant bit comparison determines the output. When large words are compared, the method of Figure 3.33 suffers from a time-delay problem. Four comparators can only present a valid result after an overall propagation delay equal to four times the delay of an individual comparator. This figure is a "worst case" value based on the 3 higher-order nibbles being equal while the nibble of lowest significance determines the result. One IC delay time is required before comparator 2 receives an input from comparator 1. Two delay times are completed before the inputs to comparator 3 are stable. Three delay times are spent before comparator 4 receives stable inputs. The output of comparator 4 cannot be considered valid until 4 propagation delay times have been expended.

A parallel expansion scheme is recommended for comparison of larger words when high speed is required. Comparison of two 24-bit code words can be accomplished by the configuration of Figure 3.34. The output comparison is completed in this case after two IC propagation delay times.

The output comparator is driven by the 5 parallel comparators. The output of the upper comparator is the most important result and consequently drives the most significant bits of the output comparator. If $A_{23}A_{22}A_{21}A_{20}A_{19} < B_{23}B_{22}B_{21}B_{20}B_{19}$, then the $A < B$ output of comparator 5 will drive the B_3 input of comparator 6, resulting in an overall output of $A < B$. If the A_3 input of comparator 6 is driven, the overall output is $A > B$. When comparator 5 produces an output of $A = B$, neither A_3 nor B_3 is asserted at the input of comparator 6. In this case, comparator 4 will determine the circuit output assuming $A_{18}A_{17}A_{16}A_{15}A_{14}$ is not equal to $B_{18}B_{17}B_{16}B_{15}B_{14}$. We note that the cascading inputs $I(A < B)$, $I(A = B)$, and $I(A > B)$ are used as normal inputs for each comparator except comparator 1.

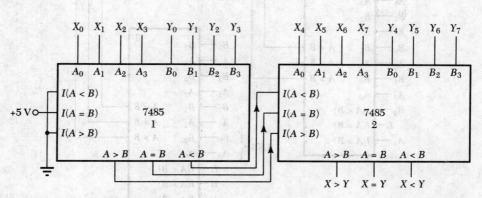

FIGURE 3.33 An eight-bit comparator.

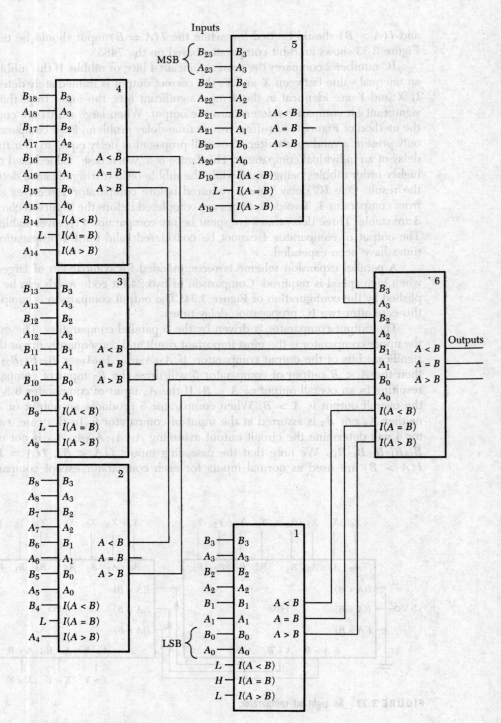

FIGURE 3.34   A 24-bit comparator.

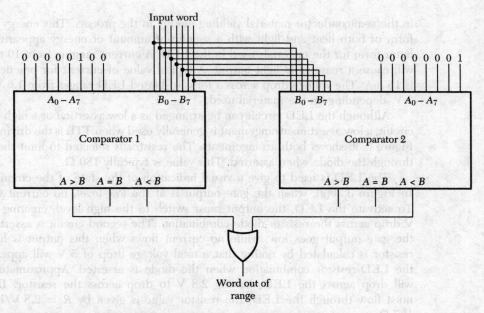

FIGURE 3.35 Circuit for determining if input word is out of range.

These inputs determine the output in exactly the same way a fifth bit comparison would; that is, they do not affect the output unless the higher four A and B inputs are equal.

There are several applications of comparators in digital systems. One common use is that of determining if a digital code or word falls within a certain range of values. The system of Figure 3.35 determines if an input word is within the range 00100000 to 10000000.

If the input word is below the boundary 00100000, the $A > B$ output of comparator 1 asserts the system output. When the input word is above the boundary 10000000, the $A < B$ output of comparator 2 drives the system output. For input words equal to or between the boundary values, the system output is not asserted.

● ● ●   **DRILL PROBLEM** Sec. 3.2.2

Show how to connect two 7485 chips to construct a 6-bit comparator.

3.2.3   THE SEVEN-SEGMENT LED DRIVER

The LED is a semiconductor diode constructed from gallium arsenide or gallium phosphide. When current flows through the diode, electrons recombine with holes in the semiconductor material yielding energy in the process. This energy is in the form of both heat and light with a significant amount of energy appearing in the latter form for the materials used in the LED. A current ranging from 10 to 40 mA will cause a reasonable light output. A typical value of current for practical LEDs is 15 mA. The voltage drop across a forward-biased LED ranges from 1.6 V to over 3 V, depending on the material used.

Although the LED circuit can be arranged as a low assertion or a high assertion circuit, a low assertion arrangement is generally used when TTL is the driving circuit. Figure 3.36 shows both arrangements. The resistor is selected to limit the current through the diode when asserted. This value is typically 180 Ω.

The LED is used to give a visual indication of the state of the circuit output. In Figure 3.36(a), when the gate output is at the low level, no current will flow. To activate the LED, the output must switch to the high level, creating a 4 to 5 V drop across the resistor-diode combination. The second circuit is asserted when the gate output goes low, while no current flows when this output is high. The resistor is calculated by noting that a total voltage drop of 5 V will appear across the LED-resistor combination when the diode is asserted. Approximately 2.2 V will drop across the LED, leaving 2.8 V to drop across the resistor. If 15 mA must flow through the LED, the resistor value is given by $R = 2.8$ V/15 mA = 187 Ω.

Another application of the LED is in the 7-segment display that is used to form numbers or characters. This display is popular in calculators, digital watches and clocks, computer readouts, digital meters, and other digital instrumentation. The 7-segment display uses either a common cathode or a common anode arrangement as shown in Figure 3.37. The common anode configuration of Figure 3.37(a) is a low assertion circuit while the common cathode is a high assertion circuit. A current-limiting resistor must be placed in series with each LED. TTL gates can sink considerably more current at the low output voltage level than they can

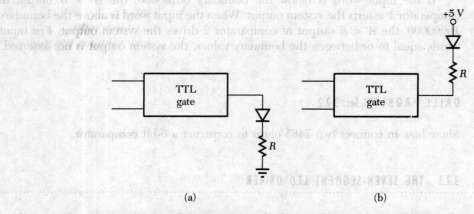

(a)   (b)

FIGURE 3.36 LED circuit: (a) high-assertion input, (b) low-assertion input.

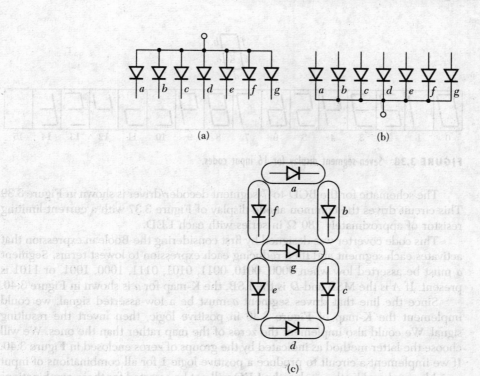

FIGURE 3.37 Seven-segment LED display: (a) common anode, (b) common cathode, (c) actual physical arrangement.

produce at the high voltage level; thus, the common anode configuration is generally used with TTL circuits. If a common cathode circuit is used, current amplifiers or buffers must be inserted between the TTL gates and the LEDs. A light diffuser is placed over each LED to create a uniformly illuminated line for each segment.

In order to create each number, the correct segments must be activated. For example, a 1 is created by activating segments b and c. A 3 requires segments a, b, c, d, and g. An 8 requires that all segments be activated.

A 4-output digital circuit can represent the binary code for any number from 0 to 15. In order to use the LED to indicate the number contained on the 4 lines, a code-converting circuit must be used to drive the correct segments. The 7447A and 7448 are chips that contain a BCD to 7-segment decoder. The 7447A is an open-collector circuit and is the more popular of the two chips because of its low asserted outputs. The 7448 has high asserted outputs. The numbers 10 through 15 do not cause displays of corresponding decimal numbers, but display the symbols indicated in Figure 3.38. There are hex displays available that produce the symbols A, B, C, D, E, and F for the numbers 10 through 15.

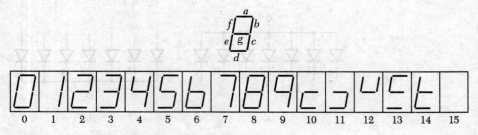

FIGURE 3.38 Seven-segment display for 16 input codes.

The schematic for the BCD-to-7-segment decoder/driver is shown in Figure 3.39. This circuit drives the common anode display of Figure 3.37 with a current limiting resistor of approximately 180 Ω in series with each LED.

This code coverter was designed by first considering the Boolean expression that activates each segment and then reducing each expression to lowest terms. Segment a must be asserted low when 0000, 0010, 0011, 0101, 0111, 1000, 1001, or 1101 is present. If A is the MSB and D is the LSB, the K-map for a is shown in Figure 3.40.

Since the line that drives segment a must be a low-asserted signal, we could implement the K-map of Figure 3.40 in positive logic, then invert the resulting signal. We could also implement the zeros of the map rather than the ones. We will choose the latter method as indicated by the groups of zeros enclosed in Figure 3.40. If we implement a circuit to produce a positive logic 1 for all combinations of input variables enclosed by the circles, the LED will not be asserted for these combinations but will be asserted for all other combinations. The signal to drive line a is then

$$a = B\bar{D} + AC + \bar{A}\bar{B}\bar{C}D$$

A careful examination of Figure 3.39 will indicate that this expression is implemented for segment a on the 7447A chip. The expressions for all other segments were reduced in the same way before implementing with gates.

There are two blanking control pins of special interest in the construction of multidigit LED displays. These are the $\overline{RBI}$, ripple-blanking input, and BI/$\overline{RBO}$, blanking input or ripple-blanking output. If the BI/$\overline{RBO}$ pin is held low, the display is blank; that is, no segment will be illuminated regardless of the input. If this pin is held high or floated, there is another possibility for blanking to occur. This requires that the $\overline{RBI}$ pin be held low and $A = B = C = D = 0$. For this combination of binary zero at the input and $\overline{RBI}$ low, the pin BI/$\overline{RBO}$ is pulled low internally and the display is blanked. If both $\overline{RBI}$ and BI/$\overline{RBO}$ are high, no blanking occurs.

One of the major purposes of these control pins is to blank leading or trailing zeros of a multidigit display. For example, if the number 0070.1600 is to be displayed as 70.16, the two leading zeros and the two trailing zeros must be blanked. This can be done by tying the $\overline{RBI}$ pin of the most significant digit (MSD) low, which will blank the MSD zero. The BI/$\overline{RBO}$ pin of the MSD should be tied to the $\overline{RBI}$ pin of the next MSD. This arrangement should continue down to the LSB integer display.

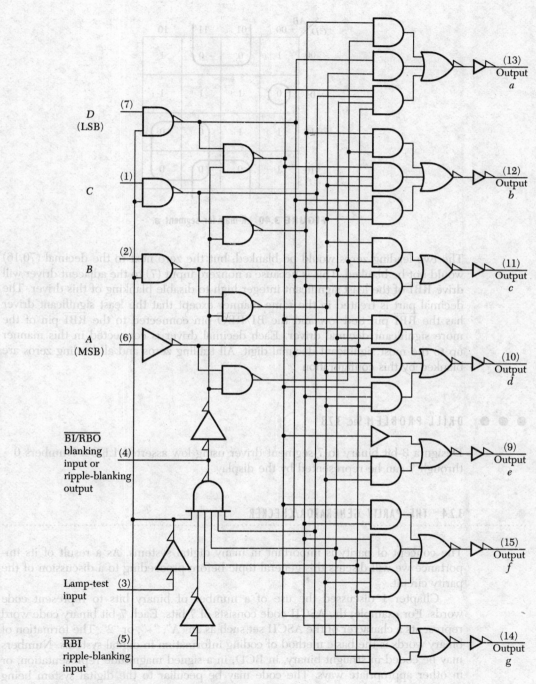

FIGURE 3.39 BCD-to-7-segment decoder driver.

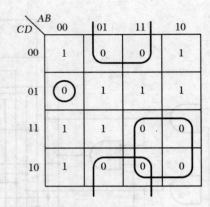

FIGURE 3.40   K-map for segment *a*.

The two leading zeros would be blanked, but the zero next to the decimal (70.16) would not be blanked. This is because a nonzero input (7) to the adjacent driver will drive $\overline{RBI}$ of the least significant integer high to disable blanking of this driver. The decimal part is treated in the same manner except that the least significant driver has the $\overline{RBI}$ pin tied low and the $BI/\overline{RBO}$ pin connected to the $\overline{RBI}$ pin of the more significant decimal driver. Each decimal driver is connected in this manner up to the most significant decimal digit. All trailing zeros and all leading zeros are blanked by this configuration.

● ● ●   **DRILL PROBLEM** Sec. 3.2.3

Design a 3-bit binary to 7-segment driver using low asserted LEDs. Numbers 0 through 7 can be represented by the display.

3.2.4   THE PARITY GENERATOR/CHECKER

The concept of parity is important in many digital systems. As a result of its importance we will discuss this general topic before proceeding to a discussion of the parity circuit.

Chapter 1 discussed the use of a number of binary bits to represent code words. For example, the ASCII code consists of 7 bits. Each 7-bit binary code word represents a character of the ASCII set such as an "A", "+", or "2". The formation of binary words is the basic method of coding information in digital systems. Numbers may be coded in straight binary, in BCD, in a signed magnitude representation, or in other appropriate ways. The code may be peculiar to the digital system being used, or it may be an accepted standard.

As a digital system carries out its function, it moves binary words from one section of the system to another, or it may exchange words with other digital systems. A computer, for example, stores binary words in memory and retrieves them as needed by the overall system. A computer terminal transmits binary words to a digital computer and receives other binary words back from the computer.

As this movement of binary information takes place, an occasional error may be introduced into the information. A bit of a word may be changed due to noise in the system. Extraneous signals or unwanted signals are classified as noise and represent a common, but serious, problem in all digital systems. Although it is beyond the scope of this book to discuss the sources of noise, for our purposes we will state that this problem can change the value of an occasional binary bit from its correct value. When a bit of a word is changed, an incorrect word results.

There are two methods of dealing with this problem. One simply identifies words containing errors while the second actually corrects the errors. These two methods are called error detection and error correction. We will briefly discuss the notion of error detection.

Typically, noise will introduce single errors that are widely separated in time. Perhaps a terminal transmitting information to a computer may have one bit identified incorrectly for each one thousand 8-bit words transmitted. Thus, we assume that no word will ever contain more than one error. Based on this assumption, the set of words are now chosen to have a minimum distance of two. This means that in order to confuse one word of the set with any other word, at least 2 bits must be changed. Figure 3.41 demonstrates the concept of distance.

If the minimum distance of a set of words is two, a single error introduced into a word will result in a new word that is not a member of the set. This word then can be identified to the system as an erroneous word and proper steps can be taken to recover the correct word.

Although it may appear complex to implement the ideas of the preceding paragraphs, in fact it is surprisingly simple. If all words of a set contain an even number of 1 bits, the minimum distance is two. Similarly, the same minimum distance results if all words of a set contain an odd number of 1 bits. When a word contains an even number of 1 bits, the word is said to have even parity. Odd parity requires an odd number of 1 bits in the word.

Transmission of words all having the same parity allows us to detect single errors by checking parity at the receiving end. If correct parity does not exist for the

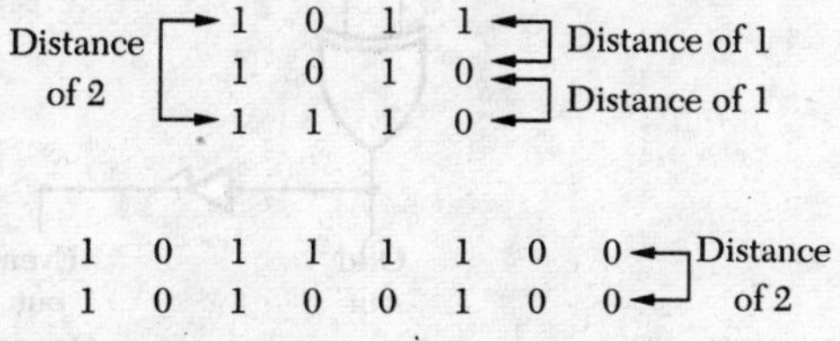

FIGURE 3.41 Distance between code words.

Original 7-bit word   Transmitted 8-bit word
                             with even parity

0	1	1	0	0	1	0		0	1	1	0	0	1	0	1
0	1	0	0	1	0	0		0	1	0	0	1	0	0	0
0	1	1	1	1	0	0		0	1	1	1	1	0	0	0
0	0	0	1	0	0	1		0	0	0	1	0	0	1	0
1	0	0	1	0	1	0		1	0	0	1	0	1	0	1
                                                            ↑
                                                       parity bit

FIGURE 3.42 Creation of even parity words.

received word, it indicates that this is an erroneous word. All words of a given set of information may not exhibit the same parity, but an additional bit position can be used to create the same parity for each word. This extra bit position is called the parity bit. In Chapter 1, we saw that the ASCII code requires 7 bits to represent each character. To use error detection, an eighth bit is added to create parity for each word. Figure 3.42 shows the creation of even parity for several different 7-bit words.

When a word is received, a parity checker circuit is used to determine if an error has occurred. This parity checking circuit can be composed of EXOR gates as shown in Figure 3.43. Each EXOR will check a 2-bit group for parity, generating a 1 if odd parity exists and a 0 for even parity. If an even number of 2-bit groups generate odd parity, the overall word is an even parity word. If all 4 groups exhibit odd parity, the

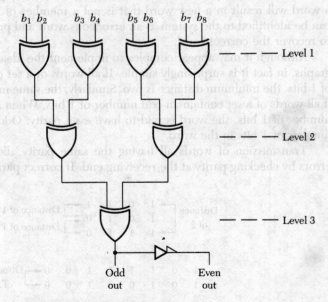

FIGURE 3.43 A parity checker.

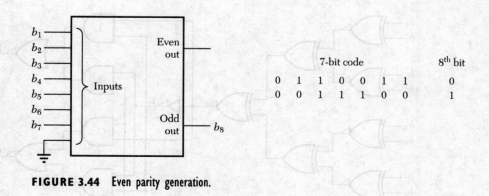

FIGURE 3.44 Even parity generation.

second level EXOR gates will generate 0 outputs. If 2 groups generate odd parity and 1s are applied to different EXOR gates at level 2, the level 3 EXOR will output a 0. The inverter produces a high assertion output for even parity. Odd parity will result in an output of 1 by the third-level EXOR gate.

The parity checker can also be used as a parity generator. Before transmitting a 7-bit ASCII code character, the parity is checked. The output of the parity checker, corresponding to the opposite type parity desired, will generate the correct parity bit. This function is demonstrated in Figure 3.44 for an even parity system.

Because 7- and 8-bit codes such as ASCII or extended ASCII commonly occur in data transmission, a standard MSI chip has been designed to generate or check parity for words up to 9 bits. This chip is the 74180, which is shown in Figure 3.45. This circuit is similar to that of Figure 3.43 with two exceptions. We note that NEXOR gates are used at the first and third levels of gating and additional control gates appear at the output.

Each input gate produces a high output if even parity exists for the pair of bits this gate checks. An even parity input will result in either 0, 2, or 4 of these gates having a high output. Point X will then be at a high level if an even parity word is applied to the inputs. If our system requires that an even parity input result in a high-asserted output on the Even Out line, the control inputs P_E and P_O are high and low, respectively. The high level on P_E opens output gates 2 and 3 resulting in Even Out being high and Odd Out being low for an even parity word. For an odd parity input word, Even Out and Odd Out would be low and high, respectively. Reversing the levels of the inputs P_E and P_O results in a reversal of output levels as indicated by the table of Figure 3.45. This discussion assumes that the 1 bit is asserted high at the input.

Two items of information must be known in order to generate parity for a word. We must know whether the generated parity bit is to be low or high asserted and we must know whether odd or even parity is to be produced. Knowing these specifications allows us to select the correct levels on P_E and P_O. In order to demonstrate, let us assume we want to produce a high assertion parity bit that results in odd parity for a 9-bit word. The 8 code bits would be applied as inputs, and P_E could be driven

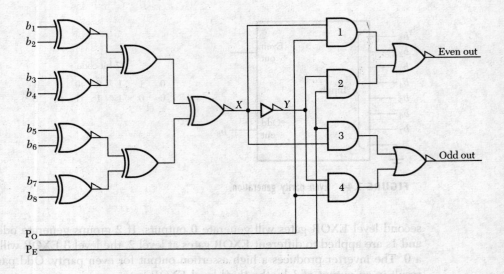

Inputs			Outputs	
Number of high data inputs $b_1 - b_8$	P_E	P_O	Even out	Odd out
Even	H	L	H	L
Odd	H	L	L	H
Even	L	H	L	H
Odd	L	H	H	L
θ	H	H	L	L
θ	L	L	H	H

FIGURE 3.45 The 74180 parity generator/checker and function table.

low while P_O is driven high. This will result in Odd Out being high whenever the 8-bit code word has even parity. Hence, Odd Out can be used as the ninth parity bit of the word.

If parity checking of a 9-bit input word is desired, 8 bits of the word are applied to the normal inputs while the ninth bit connects to P_E or P_O. For true high asserted parity checks, the ninth bit connects to P_O with an inverter connected from P_O to P_E. The Even Out will then be asserted high if even parity exists, while the Odd Out line will be asserted high if odd parity exists.

Two items of information must be known in order to generate parity for a word. We must know whether the generated parity bit is to be low or high asserted and we must know whether odd or even parity is to be produced. For example, suppose we want to produce a high asserted parity bit that relates to odd parity.

EXAMPLE 3.5
. . . .

Show how to connect the 74180 to generate even parity for a 9-bit word that is to be transmitted. The circuit should also check parity on a 9-bit received word. Assume that the words have high-asserted 1 bits and that a control signal called GENCHK is high when generating parity and low when checking.

SOLUTION

This problem requires the production of two signals: an error signal when the received word has odd parity, and a parity bit that must be added to each word that will be transmitted. We will produce these signals on two separate output lines. We will arbitrarily assign the Even Out line to generate the ninth bit of a transmitted word, while Odd Out will produce the error signal when odd parity of the received word is detected.

In checking parity for a 9-bit word, the ninth bit can be connected to P_O with an inverter connected from P_O to P_E. This will result in Odd Out being asserted high if odd parity is detected; thus, this output will become the error indicator.

In generating parity, if P_E is low and P_O high, Even Out will be asserted high when the 8-bit code has odd parity. This output then contains the correct parity bit to be inserted as the ninth bit.

Figure 3.46 shows the implementation of these ideas. For parity checking of a received word, GENCHK is taken low. When the GENCHK line is low, the ninth bit of the received word is gated to P_O and Odd Out is gated to the error output. This assumes that the other 8 bits of the received word appear at the bit inputs of the 74180. An odd parity input word will lead to the high assertion of the ERROR line. When GENCHK is taken high, a parity bit is to be generated. The 8 bits of the word to be transmitted must now appear at the bit inputs of the 74180. The pin P_O is now forced high and P_E is low. If odd parity of the word exists, the Even Out will be asserted high, producing the necessary ninth parity bit. ●

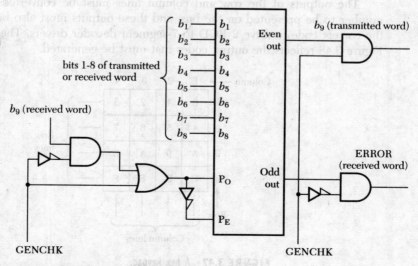

FIGURE 3.46 A parity generator/checker system.

The 74180 is expandable in 8-bit increments. The Even Out and Odd Out pins of the first chip connect to the P_E and P_O input pins, respectively, of the second chip.

● ● ● **DRILL PROBLEMS** Sec. 3.2.4

1. Show how to connect the 74180 to check parity of a 7-bit word. When odd parity is not true, a high asserted output should occur.

2. Repeat Prob. 1 for a 12-bit input word.

3.3
··

DESIGN PROBLEM USING MSI CIRCUITS

It is desired to interface a 16-key keypad with 2 LED displays and also with a bus. When a key is depressed, the decimal number of the key should be displayed on the 2 LED displays and the binary code for the number should be presented on a 4-line bus.

The keypad is arranged in a matrix fashion as shown in Figure 3.47. We will not go into detail relative to the keypad design, but we will assume each time a key is depressed, a low level voltage appears on the row line and the column line corresponding to the location of the depressed key. For example, when the key for number 6 is depressed, the second row line and the third column line are asserted low while all other row and column lines remain at the high voltage level.

The outputs of the row and column lines must be converted into a binary number to be presented on the bus, and these outputs must also be converted to the correct code to drive 2 BCD to 7-segment decoder drivers. The truth table of Figure 3.48 reflects the output codes that must be generated.

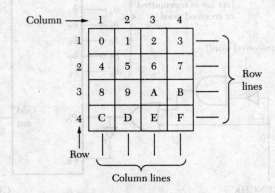

FIGURE 3.47   A hex keypad.

R	C	Bus lines M	L	MS Driver M	L	LS Driver M	L
1	1	0 0	0 0	0 0	0 0	0 0	0 0
1	2	0 0	0 1	0 0	0 0	0 0	0 1
1	3	0 0	1 0	0 0	0 0	0 0	1 0
1	4	0 0	1 1	0 0	0 0	0 0	1 1
2	1	0 1	0 0	0 0	0 0	0 1	0 0
2	2	0 1	0 1	0 0	0 0	0 1	0 1
2	3	0 1	1 0	0 0	0 0	0 1	1 0
2	4	0 1	1 1	0 0	0 0	0 1	1 1
3	1	1 0	0 0	0 0	0 0	1 0	0 0
3	2	1 0	0 1	0 0	0 0	1 0	0 1
3	3	1 0	1 0	0 0	0 1	0 0	0 0
3	4	1 0	1 1	0 0	0 1	0 0	0 1
4	1	1 1	0 0	0 0	0 1	0 0	1 0
4	2	1 1	0 1	0 0	0 1	0 0	1 1
4	3	1 1	1 0	0 0	0 1	0 1	0 0
4	4	1 1	1 1	0 0	0 1	0 1	0 1

FIGURE 3.48 Truth table for keypad example.

The columns MS Driver and LS Driver correspond to the code that must be presented to the most significant LED and least significant LED driver circuits.

The code conversion from row-column assertion to binary output code can be accomplished with two 4-line to 2-line encoders. Figure 3.49 outlines this conversion. The row and column lines drive the encoders to produce a binary code for each of

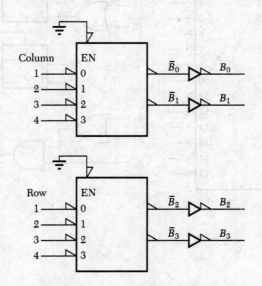

FIGURE 3.49 Conversion of row-column information to binary.

the 16 possible input keys. From a practical standpoint, 8-line to 3-line decoders are readily available and may be used to implement the 4-line to 2-line modules shown. We note that the typical encoder requires low asserted inputs and produces low asserted outputs. Consequently, 4 inverters are used to convert the encoder outputs to high asserted signals.

The next task is the development of a system to drive two 7447A BCD to 7-segment circuits to display the results of each key depression on 2 LEDs. One scheme is shown in block diagram form in Figure 3.50. We note that the decoder is driven by the bus lines produced by the encoders and inverters of Figure 3.49. The four 2:1 MUXs connect the same bus lines to the inputs of the least significant

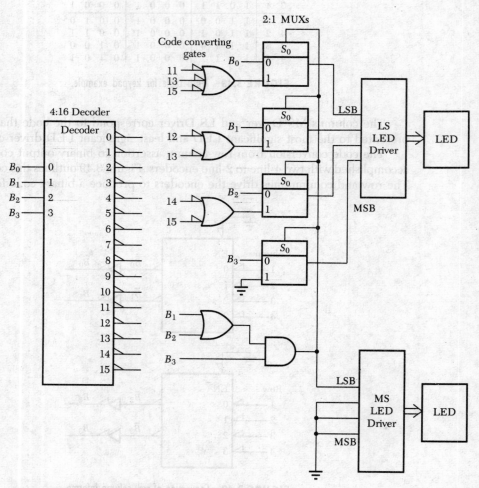

FIGURE 3.50   Keypad display system.

LED driver circuit for all binary numbers between 0 and 9. For the numbers 10 to 15, the select line is driven to 1, connecting the outputs of the code converter gates to the least significant LED driver circuit. The most significant driver should display a 0 for all keys depressed up to the tenth key. For the keys 10 to 15, a 1 should be displayed on the most significant driver. The same line that drives the select line of the MUXs is used to drive the least significant bit of the most significant LED driver circuit.

Typically, MSI circuits require a small number of gates as does the system of Figure 3.50.

SUMMARY
• • • • • • • • • • • • • • • •

1. Logic functions can be realized by MUXs and decoders, which are MSI circuits. Although gate circuits may lead to a lower component cost, MSI circuits can yield a lower chip count, require less wiring, or allow simpler function changes.
2. The number of inputs and outputs of MUX and decoder circuits are limited by the pins available on MSI circuit chips. These smaller elements can be used as building blocks to create larger MUXs and decoders.
3. Several common logic functions have been implemented as MSI circuits. Among these are the encoder, the comparator, the 7-segment LED driver, and the parity generator/checker.
4. It is important to understand the operation of these MSI circuits in order to apply them effectively in digital system design.

CHAPTER 3 PROBLEMS
• • • • • • • • • • • • • •

● Sec. 3.1.1

*3.1 Realize the expression $Y = ABCD + \bar{A}BCD + \bar{A}B\bar{C}D + \bar{A}BC\bar{D} + A\bar{B}\bar{C}\bar{D} + A\bar{B}\bar{C}D + AB\bar{C}\bar{D} + ABC\bar{D}$ using a 16:1 MUX.

3.2 Realize the expression in Prob. 3.1 using an 8:1 MUX.

3.3 Use a 4:1 MUX preceded by a minimal number of gates to realize Y of Prob. 3.1.

*3.4 Realize the expression $X = \bar{A}\bar{B}\bar{C} + \bar{A}B\bar{C} + A\bar{B}\bar{C} + \bar{A}BC$ with three 2:1 MUXs.

3.5 On a quad 2:1 MUX chip, all 4 select lines are connected to a common pin to minimize the number of pins on the chip. Using this type of a chip and a minimal number of gates, realize the function X of Prob. 3.4.

3.6 Use seven 8:1 MUXs to realize the 7-segment LED driver of Prob. 2.40.

3.7 Use seven 4:1 MUXs preceded by gates to realize the 7-segment LED driver of Prob. 2.40.

● **Sec. 3.1.2**

°3.8 Realize the expression X of Prob. 3.4 using a 3-line to 8-line decoder plus a gate.

3.9 Realize the expression Y of Prob. 3.1 using a 4-line to 16-line decoder plus a gate.

3.10 Realize the expression Y of Prob. 3.1 and the expression X of Prob. 3.4 using a 4-line to 16-line decoder plus 2 gates.

3.11 Show how to construct a 5-line to 32-line decoder from standard MSI decoders.

°3.12 Use a 4-line to 16-line decoder and gates to realize the 7-segment driver of Prob. 2.40.

● **Sec. 3.2.1**

3.13 Design a 10-line to 4-line BCD encoder using gates. *Hint:* Assume there are no input lines for 10, 11, 12, 13, 14, and 15.

3.14 Design a 10-line to 4-line BCD priority encoder using gates. *Hint:* Assume there are no input lines for 10, 11, 12, 13, 14, and 15.

°3.15 Using 74148 chips, design a 10-line to 4-line BCD encoder.

3.16 Using a 4-line to 16-line decoder and a 16-line to 4-line encoder, build a code converter to convert binary to Gray code (see Example 3.2).

3.17 Repeat Prob. 3.16 converting Gray code to binary.

● **Sec. 3.2.2**

3.18 Design a 4-bit comparator using gates.

3.19 Design a 5-bit comparator using a 7485 chip and additional gates.

● **Sec. 3.2.3**

3.20 Show how to connect a 7447A chip to a 7-segment, common anode, LED display such as that shown in Figure 3.37(a). The display should be unblanked at all times. Assume that the diode requires 2.2 V and a minimum of 20 mA when active. Also assume that $V_{OLmax} = 0.4$ V. Calculate the resistor value.

3.21 Show the pin diagram of four 7447A chips driving 7-segment displays, used to display numbers from .01 to 99.99. Connect chips to blank leading and trailing zeros.

● **Sec. 3.2.4**

3.22 Show how to connect two 74180 chips to check even parity of a 16-bit word. When even parity is not true, a high asserted output should occur.

REFERENCES AND SUGGESTED READING

1. K. J. Breeding, *Digital Design Fundamentals,* 2nd ed. Englewood Cliffs, NJ: Prentice-Hall, 1992.
2. P. Burger, *Digital Design.* New York: John Wiley and Sons, 1988.
3. Engineering Staff, *TTL Data Manual.* Sunnyvale, CA: Signetics, 1986.
4. W. I. Fletcher, *An Engineering Approach to Digital Design.* Englewood Cliffs, NJ: Prentice-Hall, 1980.
5. M. M. Mano, *Digital Design,* 2nd ed. Englewood Cliffs, NJ: Prentice-Hall, 1991.
6. C. H. Roth, Jr., *Fundamentals of Logic Design,* 4th ed. St. Paul, MN: West, 1993.

Chapter 4

Flip-Flops, Counters, and Registers

The circuits considered in the first three chapters of this text are combinational logic circuits. As explained earlier, these systems have outputs that depend only on the applied input signals. For a given set of inputs, the outputs will always be the same. Combinational circuits are important, but so is another class of digital circuits referred to as sequential circuits.

Sequential circuits behave quite differently from combinational circuits. A sequential circuit depends not only on the present inputs, but also on past history of the inputs and time. This chapter will consider an important subclass of sequential circuits called flip-flops. These are rather simple sequential circuits designed for specific applications. The flip-flop is perhaps the most important single type of circuit in the digital field. Certain types of flip-flops are used as building blocks to construct larger sequential circuits called state machines. In Chapters 5 to 7, we will discuss state machine design following a discussion of the simple but useful flip-flop in this chapter.

After considering several aspects of the flip-flop and the related Schmitt trigger circuit, this chapter will proceed to applications of the flip-flop in counter and register systems. Both the flip-flop and Schmitt trigger belong to a class of circuits called bistable multivibrators.

4.1

THE BISTABLE MULTIVIBRATOR

SECTION OVERVIEW A significant component in logic circuits is the flip-flop, which is used much more than any other type of bistable circuit. This element will be considered in various forms before we proceed to the Schmitt trigger circuit, a bistable of importance in a limited number of applications.

Flip-flops operate in one of two modes: direct or clocked. Direct-mode flip-flops respond directly to applied inputs. The outputs change as a direct result of the inputs. In clocked flip-flops, a change of input has no effect on the output until a clock signal is applied. When a clock transition from one voltage level to another occurs, the output changes to a value dictated by the input levels. The SR flip-flop and gated D flip-flop are examples of direct-mode operation devices. The toggle, JK, and clocked D flip-flops are clocked devices. Some clocked flip-flops also have direct inputs, allowing operation in either mode.

4.1.1 DIRECT INPUT FLIP-FLOPS

The SR Flip-Flop

This form of flip-flop has two input lines and either one or two outputs. The output line that is always present is often labeled Q. Generally, a second output called $\bar{Q}$ will also be present. One input line is used to set the device to the $Q = 1$ state while the other input sets the device to the $Q = 0$ state. These inputs are called the SET or S input and the RESET or R input; hence the name SR flip-flop. The outputs Q and $\bar{Q}$ make up what is called a double-rail output; that is, $\bar{Q}$ always equals the complement of Q except when both S and R are asserted simultaneously. This condition is normally avoided in actual circuit operation. One type of SR flip-flop responds to high-asserted inputs while a second type responds to low-asserted inputs. The symbols for both active high input and active low input flip-flops are shown in Figure 4.1 along with the function and characteristic tables for each.

The function table is similar in purpose to the function table used for combinational circuits. The flip-flop table implies a time dependence, however, while the function table for a gate does not. Outputs Q and $\bar{Q}$ refer to conditions applying after the inputs S and R take on particular values. Q_0 and $\bar{Q}_0$ refer to conditions before the input conditions are applied. For example, row 1 of the high assertion input SR function table tells us that the state of Q and $\bar{Q}$ prior to applying $S = R = L$ will remain after these input conditions are applied.

For the high assertion input SR flip-flop, inputs are normally kept low until a change in output state is desired. To guarantee that a 1 is set into the device $(Q = 1)$, S is taken high while R remains low. The input S can then be returned to a low value and Q will continue at the high level. In order to reset the circuit,

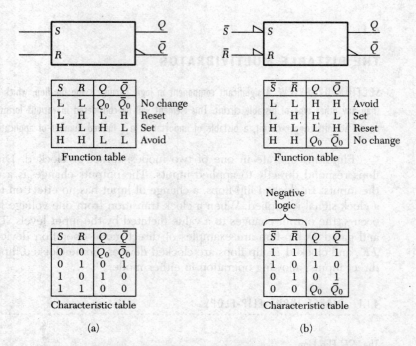

S	R	Q	$\bar{Q}$	
L	L	Q_0	$\bar{Q}_0$	No change
L	H	L	H	Reset
H	L	H	L	Set
H	H	L	L	Avoid

Function table

$\bar{S}$	$\bar{R}$	Q	$\bar{Q}$	
L	L	H	H	Avoid
L	H	H	L	Set
H	L	L	H	Reset
H	H	Q_0	$\bar{Q}_0$	No change

Function table

S	R	Q	$\bar{Q}$
0	0	Q_0	$\bar{Q}_0$
0	1	0	1
1	0	1	0
1	1	0	0

Characteristic table

(a)

Negative logic

$\bar{S}$	$\bar{R}$	Q	$\bar{Q}$
1	1	1	1
1	0	1	0
0	1	0	1
0	0	Q_0	$\bar{Q}_0$

Characteristic table

(b)

FIGURE 4.1 The set-reset flip-flop and function tables: (a) active high inputs, (b) active low inputs.

R is taken high while S remains low. When this occurs, $Q = 0$ and this condition will persist even after R returns to a low level. Because the flip-flop remains in a particular state after the inputs move to the no-change condition, it is also called a latch.

From the characteristic table we see that Q may equal 0 or 1 for the input condition of $S = R = 0$, depending on the input conditions prior to setting both inputs low. This is a distinct difference from combinational circuits, which always produce a unique output for a given set of inputs.

When both S and R are asserted, Q and $\bar{Q}$ are driven low. If both S and R are returned simultaneously to the low level, it is impossible to determine if Q will assume the 1 state or the 0 state. Generally, we avoid the input condition of $S = R = 1$ unless the following input condition is either $S = 1$ and $R = 0$ or $S = 0$ and $R = 1$. In either of these cases, the output will be uniquely determined.

The high assertion input SR flip-flop can be constructed from NOR gates as shown in Figure 4.2(a). For this latch, applying a 1 on S (high level) and a 0 on R results in $\bar{Q}$ being asserted low. Since $\bar{Q}$ drives an input to the upper NOR gate, both inputs are low, resulting in a high value for Q. When S returns to a low value, the high level of Q continues to drive the lower NOR gate to keep $\bar{Q}$ at a low level. If R is now asserted, the upper NOR gate drives Q low, which now allows $\bar{Q}$ to return to a high level.

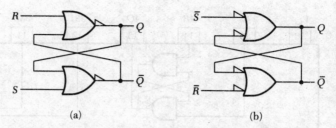

FIGURE 4.2 (a) A high assertion input *SR* latch. (b) A low assertion input *SR* latch.

The low assertion input flip-flop behaves in a similar way except the inputs must be asserted low. Before the use of dependency notation, this circuit was referred to as an $\bar{S}\bar{R}$ flip-flop. The low assertion OR gates or NAND gates of Figure 4.2(b) can implement the low assertion input *SR* flip-flop. An equation describing the *SR* latch is

$$Q = S + \bar{R}Q_0 \qquad (4.1)$$

This tells us that Q will be asserted if we assert S or if R is not asserted and Q has previously been set to 1. A popular SSI latch is the 74279, which is a quad low assertion input *SR* latch. Two of the latches have an extra set input on them. This circuit is shown in Figure 4.3.

The *SR* flip-flop can function as a 1-bit memory. To store a 1 in this device, the S input is asserted. This input can then be deasserted, and the flip-flop remains in the $Q = 1$ state as long as the two inputs remain deasserted. The 0 bit can be stored by asserting R. Several *SR* flip-flops can be combined to form a storage register for several bits as we shall see later.

A very common application of the *SR* flip-flop is that of debouncing switch contacts in digital systems. Toggle or push-button switches do not make sudden, firm, mechanical contact as a switch is closed or opened. If the output of the switch of Figure 4.4(a) were observed on an oscilloscope with proper triggering and time-scale setting, the waveform would typically appear as shown in Figure 4.4(b). This contact bounce lasts 5 to 10 ms for the small switches used in electronic circuits.

If the switch output were connected to a gate or inverter input, the output would change state several times as the electrical signal bounced through the trigger level of the gate. In some systems, this multiple switching at the output of the gate cannot be tolerated. If the signal were applied to the S input of an *SR* latch, the latch would set the first time the switch output exceeded the 1-state level. If the switch output then dropped low again, the latch would remain set in that state until the S signal returned to the low level and the R input went high. Thus, a clean signal with only one transition would result at the latch output. A latch can be used to debounce a switch for both positive and negative transitions as shown in Figure 4.5. In this instance, the low assertion input *SR* latch is used. When the switch is in the reset

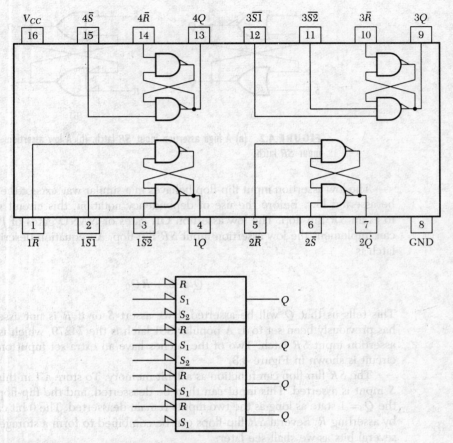

FIGURE 4.3 The 74279 TTL quad *SR* latch and IEEE/IEC symbol.

position, *R* is at the 0 V level and *S* is at +5 V due to the pull-up resistor. As the switch changes to the set position, *R* moves to a voltage of +5 V and *S* is pulled toward ground. Although this voltage on *S* may fluctuate rapidly, once the *S* input

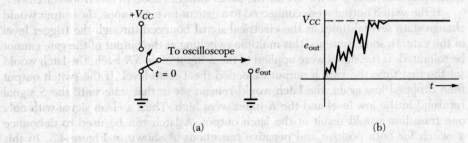

FIGURE 4.4 (a) Switch as state is changed. (b) Typical output.

drops to the low logic level, the latch switches to $Q = 1$, creating a single transition. When the switch moves from the set to the reset position, a single transition to $Q = 0$ also occurs. The waveform is shown in Figure 4.6.

A useful variation of the SR latch is the gated SR latch of Figure 4.7. This circuit will not allow the S or R inputs to affect the output when G is low; when G is high, the circuit behaves like a normal SR latch.

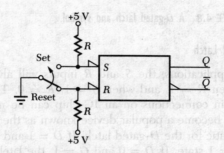

FIGURE 4.5 Switch debouncer.

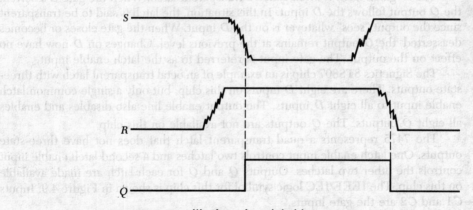

FIGURE 4.6 Waveform of a switch debouncer.

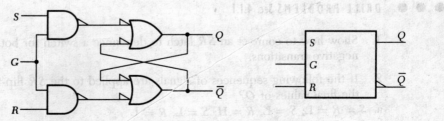

FIGURE 4.7 A gated SR latch and symbol.

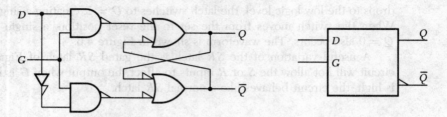

FIGURE 4.8 A D-gated latch and symbol.

The D-Gated Latch

In some applications, the S and R inputs will always be complementary, that is, $S = 0$ when $R = 1$, and when $S = 1$, $R = 0$. This can be expressed as $S = \bar{R}$. Because pin connections on an IC chip can be minimized for this situation, this circuit has become a popular device known as the D-gated latch. Figure 4.8 shows the schematic for the D-gated latch. If $D = 1$ and $G = 1$, then the latch will be set to the $Q = 1$ state. If $D = 0$ and $G = 1$, the latch will return to the $Q = 0$ state. If $G = 0$, the state cannot be changed by the D input.

The D-gated latch is often called a transparent latch. When the gate is asserted, the Q output follows the D input. In this situation, the latch is said to be transparent since the output "sees" whatever is on the D input. When the gate closes or becomes deasserted, the Q output remains at the previous level. Changes on D now have no effect on the output. The gate input is referred to as the latch enable input.

The Signetics 8TS807 chip is an example of an octal transparent latch with three-state outputs. There are eight D inputs on this chip, but only a single common latch enable input to all eight D inputs. The output enable line also disables and enables all eight $\bar{Q}$ outputs. The Q outputs are not available on this chip.

The 7475 represents a quad transparent latch that does not have three-state outputs. One latch enable input controls two latches and a second latch enable input controls the other two latches. Outputs Q and $\bar{Q}$ for each latch are made available on this chip. The IEEE/IEC logic symbol for this chip is shown in Figure 4.9. Inputs $C1$ and $C2$ are the gate inputs.

● ● ● **DRILL PROBLEMS Sec. 4.1.1**

1. Show how to connect an SR latch to debounce a switch for both positive and negative transitions.

2. If the following sequences of signals are applied to the SR flip-flop, what are the final values of Q?
   a. $S = R = L$; $S = L$, $R = H$; $S = L$, $R = L$
   b. $S = R = L$; $S = H$, $R = L$; $S = L$, $R = L$
   c. $S = R = L$; $S = L$, $R = H$; $S = H$, $R = L$; $S = L$, $R = H$

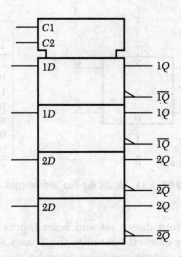

FIGURE 4.9 The 7475 quad transparent latch.

4.1.2 CLOCKED FLIP-FLOPS

The T Flip-Flop

A multivibrator that changes state or toggles with each successive input is called a T or toggle flip-flop. The symbol for this device is shown in Figure 4.10 along with its typical input and output waveforms. If the T input is high, positive transitions of the CK or clock input will cause the output to change state. The wedge-shaped symbol on the CK input indicates that the output change will occur on the positive transition. If T goes low, then the CK input has no effect on the flip-flop output. T flip-flops that toggle on the negative clock transition can also be constructed. This circuit can be used as a frequency divider for rectangular waveforms and is useful in counting applications.

The JK Flip-Flop

This bistable circuit has two gating inputs along with a clock input. The voltage level of the gates determines the output state to which the clock input will shift

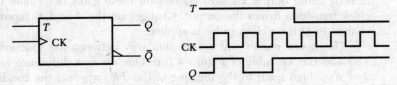

FIGURE 4.10 A T flip-flop and output waveform.

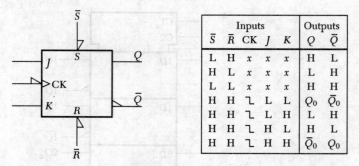

Inputs					Outputs	
$\bar{S}$	$\bar{R}$	CK	J	K	Q	$\bar{Q}$
L	H	x	x	x	H	L
H	L	x	x	x	L	H
L	L	x	x	x	H	H
H	H	⌐	L	L	Q_0	$\bar{Q}_0$
H	H	⌐	L	H	L	H
H	H	⌐	H	L	H	L
H	H	⌐	H	H	$\bar{Q}_0$	Q_0

FIGURE 4.11   A *JK* flip-flop and function table.

the flip-flop. Often direct set and reset inputs are provided that can override the clock input. For clocked operation the R and S inputs are set to the inactive level; but if a certain output condition is to be preset into the flip-flop before clocked operation takes place, these inputs can be used. Figure 4.11 shows the symbol for the JK flip-flop and its corresponding function table. This particular flip-flop has low assertion direct inputs.

The x in the function table means that, regardless of what state this input has, the output will not be affected. When the direct inputs are high, clocked operation results. The notation Q_0 means the state of Q prior to the occurrence of the negative clock transition. If J and K are both high, the clock input toggles the flip-flop indicated by Q taking on a value $\bar{Q}_0$ and by $\bar{Q}$ taking on a value of Q_0. When J and K are both low, the clock input has no effect on the output. To set a 1 into the flip-flop with the clock input, J must be high and K low prior to the negative clock transition. There is generally some minimum time before the clock pulse occurs that the J and K inputs must be stable. This time is referred to as gate set-up time and, for TTL flip-flops, it is typically in the nanosecond range. The 74112 chip contains two JK devices with operation similar to that indicated in Figure 4.11. The small triangle preceding the wedge on this circuit indicates the fact that output changes take place on the negative-going clock transition.

In the clocked mode of operation, the gating conditions are summarized in Figure 4.12. The condition of both J and K low is called the "no change" condition. The toggle condition corresponds to both J and K high. The set condition occurs when $J = 1$ and $K = 0$; while the reset condition occurs when $J = 0$ and $K = 1$. It must be emphasized that the voltage levels applied to the J and K gates do not directly cause output changes. Levels on these gates determine to what level the clock transition drives the output. Changes on the J and K inputs cannot cause a change in Q if no clock pulse is applied.

The casual reader will note a similarity between the function table of Figure 4.12 and the SR table of Figure 4.1(a). One obvious difference is the condition of $J = K =$ high leads to the toggling of the JK, whereas the condition of $S = R =$ high should be avoided for the SR flip-flop. A more subtle difference is that the SR

J	K	Q	$\bar{Q}$	Condition
0	0	Q_0	$\bar{Q}_0$	No change
0	1	0	1	Reset
1	0	1	0	Set
1	1	$\bar{Q}_0$	Q_0	Toggle

FIGURE 4.12 Four possible gating conditions.

flip-flop operates in direct response to the S and R inputs while the levels on J and K determine the state that will exist after a clock transition occurs.

The D Flip-Flop

The clocked D flip-flop or simply D flip-flop is closely related to the JK flip-flop and has become quite useful since the development of the integrated circuit. This device eliminates the K gate external input connection by including an on-chip inverter from the J input to the K. This always forces K to equal $\bar{J}$. The input is then labeled D rather than J. If $D = 1$, a clock transition will result in $Q = 1$. If $D = 0$, then $Q = 0$ after the transition. The symbol and function table for the D flip-flop are shown in Figure 4.13. This table reflects only the clocked operation of the device. The 7474 is a typical example of a D flip-flop. The inputs S and R are direct inputs.

Constructing a T Flip-Flop

Manufacturers do not produce T flip-flops. Instead, they are created from JK or D flip-flops. The JK flip-flop of Figure 4.11 will toggle on each negative clock transition if $J = K = 1$. Tying J and K together creates the T input of the toggle

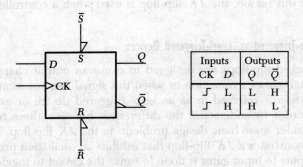

Inputs		Outputs	
CK	D	Q	$\bar{Q}$
⌐	L	L	H
⌐	H	H	L

FIGURE 4.13 A D flip-flop and function table.

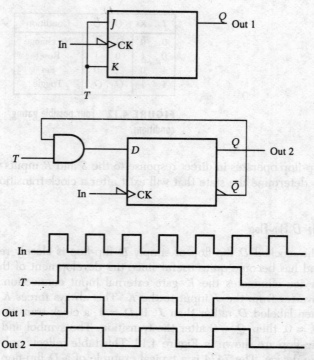

FIGURE 4.14 Two methods of creating a controlled toggle.

flip-flop. When this input is dropped low, the toggle function is disabled. A D flip-flop can be connected to toggle by connecting $\bar{Q}$ to D. Once connected, the toggling function can only be disabled by overriding clocked operation with the direct set or reset input unless an additional AND gate is used as shown in Figure 4.14. The two methods depicted in Figure 4.14 exhibit one difference when the T input is deasserted. The JK output remains at the value present when the T input is deasserted. The D flip-flop always reverts to the $Q = 0$ state after T is deasserted. For this reason, the JK flip-flop is used when a controlled toggle is required.

Edge-Triggered vs. Level-Triggered Devices

Clocked flip-flops are designed to cause an output change either when the clock signal makes a transition or when this signal reaches some particular level. A given flip-flop is referred to as an edge-triggered device or as a level-triggered device. In order to understand the differences between these triggering methods, let us consider some basic design problems in the JK flip-flop. The circuit of Figure 4.15 demonstrates a JK flip-flop that exhibits an oscillation problem. The coupling from outputs to input gates is done to cause the circuit to toggle when both J and K are high and the CK input is applied. If $Q = 1$, the K gate is enabled while the J gate

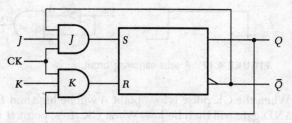

FIGURE 4.15 An impractical *JK* flip-flop.

is disabled. A CK input will cause a high level to reach R with a low level remaining on S. The SR flip-flop will then change to the $Q = 0$ state. At this point, the J gate opens and the K gate closes. Herein lies the problem. This transition from $Q = 1$ to $Q = 0$ takes place very shortly after the CK pulse goes high. Typically, the propagation delay through a TTL gate and SR flip-flop is in the nanosecond range. Since the clock pulse is still high when the J gate opens and the K gate closes, a high level reaches S while a low level is applied to R. The circuit toggles again and instead of toggling just once, it will continue toggling until CK goes low to disable both gates.

There are two popular means of overcoming this problem to create a circuit that toggles only once for each CK pulse applied. The first method uses edge-triggering, while the second method uses a master-slave arrangement. Edge-triggering implies that the flip-flop is sensitive to the rising or falling edge of the CK input. In a positive edge-triggered circuit, the flip-flop can only change state in response to the positive edge of the CK signal. This resolves the continuous toggling problem of the circuit of Figure 4.15. The circuit can only toggle once each time the CK input goes positive. The question remains of how to produce a circuit that responds only to the edge of the CK waveform. The answer is to use a pulse-narrowing circuit for the CK signal before applying it to the gates. Figure 4.16 shows this modification. The pulse-narrowing circuit could be an RC differentiating circuit as it was in the era of discrete flip-flops. For integrated circuits, a delaying circuit is used to generate the narrow pulse width. Although the circuit of Figure 4.17 is not precisely the one applied in IC flip-flops, it demonstrates the basic operating principle of a pulse-narrowing circuit.

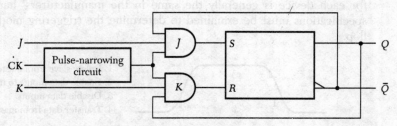

FIGURE 4.16 A practical *JK* flip-flop.

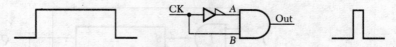

FIGURE 4.17   A pulse-narrowing circuit.

When the CK pulse is low, point A will be high and B will be low. The output of the AND gate will then be low. When CK rises, point B rises to the high level while A remains high. The AND condition will be satisfied until the delaying inverter allows point A to drop to its low level. This time will be approximately equal to the propagation delay time of the inverter. The AND gate output will be high during the short period of time that both A and B are high and drops to the low level as A goes low.

When the pulse-narrowing circuit is used for the JK flip-flop, the pulse width must be less than the total propagation delay through the flip-flop. This allows the narrow pulse to initiate a change of state of the SR flip-flop when either the J or K gate is enabled, but then the pulse disables both gates before the flip-flop output changes. The edge-triggered circuit can be designed to respond to either the positive-going or negative-going transition of CK.

The master-slave flip-flop is a level triggered-circuit that solves the oscillation problem in a different way. The master-slave flip-flop consists of two latches, the first of which accepts the input information on one clock transition and transfers this information to the second latch on the alternate clock transition. The latch that accepts the input information is called the master latch; the output latch is called the slave. Figure 4.18 is a timing chart of a master-slave unit.

The operation of a master-slave JK can be explained in terms of the circuit of Figure 4.19. As the clock signal goes positive, the slave gates are closed by CK. Information is then shifted through the master gates to the master latch. When the clock signal goes negative, the master gates are first closed so that information on the master latch cannot be changed, and then the slave gates open to transfer data from the master latch to the slave latch. The cross-coupling of outputs to gates guarantees that the flip-flop will toggle when $J = K = 1$.

There are some very subtle differences between the edge-triggered and master-slave flip-flops that must be understood by the digital system designer. The symbol for each device is generally the same in the manufacturers' handbooks, and the specifications must be examined to determine the triggering mode of a given flip-flop.

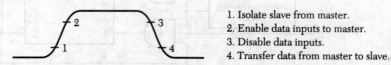

1. Isolate slave from master.
2. Enable data inputs to master.
3. Disable data inputs.
4. Transfer data from master to slave.

FIGURE 4.18   Master-slave timing chart relative to clock input.

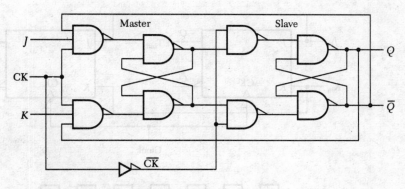

FIGURE 4.19 A master-slave *JK* flip-flop.

Before discussing these differences, we will return to a consideration of the meaning of the wedge-shaped clock symbol used with clocked flip-flops. Figure 4.20 shows this symbol. The symbol in Figure 4.20(a) indicates that the output will only change as the CK input goes positive while the symbol of (b) indicates that output changes can only occur as CK goes negative. It should be emphasized that these symbols are used for both the edge-triggered and the master-slave circuit. The major difference between the two types of flip-flop is in the time that input information must be presented to result in an output change. A negative-transition edge-triggered JK will shift to that output state determined by the data on the J and K inputs at the time of the negative clock transition. Until the next negative transition, the J and K inputs have no effect on the flip-flop output.

For the master-slave device, the two preceding sentences do not apply. When the clock goes positive on a negative-transition device such as that of Figure 4.19, the master flip-flop will accept information. If J and K change while the clock is positive, the master flip-flop state will change also. Let us assume that CK goes positive while $Q = 0$ and $J = K = 0$, and the master flip-flop contains a 0. If J goes positive during positive clock time, the master flip-flop will switch to the 1 state. Even if J goes back to the zero level prior to the negative transition of CK, a 1 remains in the master. When the negative clock transition occurs, the output switches to $Q = 1$, although both J and K equal zero at this point in time. Figure 4.21 demonstrates these differences by showing the results of driving an

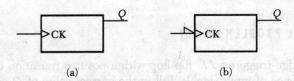

FIGURE 4.20 Symbols for flip-flops that change on: (a) positive-going clock, (b) negative-going clock.

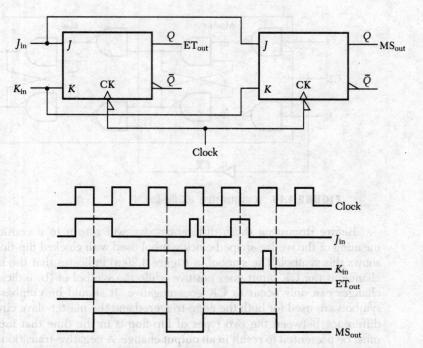

FIGURE 4.21 Circuit and waveforms for edge-triggered and master-slave devices.

edge-triggered and a master-slave flip-flop with the same inputs. We note that the short pulses on J and K during the positive half-cycle of the clock signal have no effect on the edge-triggered flip-flop but determine the output of the master-slave device. For this reason, the master-slave circuit is called a "ones-catching" flip-flop. That is, it will register the fact that a one occurred on an input even if that input has dropped to zero before the clock transition. In some cases, this "ones-catching" feature is useful, but in state machine design we will see that the master-slave device is generally avoided to simplify design. We will also see in a later section that both the master-slave and the edge-triggered flip-flops are useful in counter and register applications.

● ● ●   **DRILL PROBLEM Sec. 4.1.2**

An edge-triggered JK flip-flop with a positive transition clock input has the given sequence of input signals. Fill in the correct value of Q after each clock transition.

$$J = 0, \qquad K = 1, \qquad CK\uparrow, \qquad Q = \ ?$$
$$J = 1, \qquad K = 0, \qquad CK\downarrow, \qquad Q = \ ?$$
$$J = 1, \qquad K = 1, \qquad CK\uparrow, \qquad Q = \ ?$$

$$J = 1, \quad K = 1, \quad CK\downarrow, \quad Q = ?$$
$$J = 1, \quad K = 0, \quad CK\uparrow, \quad Q = ?$$
$$J = 1, \quad K = 1, \quad CK\downarrow, \quad Q = ?$$

4.1.3   THE SCHMITT TRIGGER

The Schmitt trigger is a bistable circuit, but its function is different from that of the flip-flop. This circuit is used to detect levels of an input signal. When a particular level is exceeded by the input signal, the Schmitt trigger asserts the output voltage. This input level that must be exceeded is called the upper trip point (UTP). As the input signal decreases below the UTP, the output remains asserted until the input drops to a value called the lower trip point (LTP). As the signal drops below the LTP, the output is deasserted.

A normal trigger-level detector has a single trip point. When the input exceeds this value, the output is asserted; when the input drops below the trip point, the output is deasserted. The Schmitt trigger is said to have hysteresis, which refers to the two trip points. It is this difference in trigger level that allows the Schmitt trigger to discriminate against noise as a trigger level is detected. The noisy waveform of Figure 4.22 is used to compare a conventional trigger-level detector to the Schmitt trigger. Assuming a trigger level for the conventional circuit equal to the UTP of the Schmitt, the waveform of Figure 4.22(b) results. The noise causes the input signal to

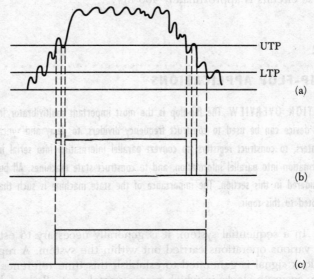

FIGURE 4.22   (a) Input signal with noise. (b) Output of conventional trigger-level detector. (c) Output of Schmitt trigger.

cross the UTP three times. The trigger-level detector exhibits a transition each time the input level is crossed. This rapid switching of the output is called "chattering" and can have an adverse effect on following circuits in many instances. The Schmitt trigger eliminates chattering since the output will become asserted when the UTP is exceeded, but will remain asserted when the noise drops the input below this level. It is not until the input waveform drops below the LTP that the Schmitt trigger output is deasserted. Although the noise may again cause the input to exceed the LTP, the output will not change unless the UTP is exceeded. For the Schmitt trigger to discriminate against noise effectively, the maximum noise amplitude must be less than the difference UTP − LTP.

We may note some similarity in the noise discrimination property of the Schmitt trigger and the switch-debouncing ability of the SR flip-flop. Both circuits output a single transition when a trigger level is crossed, but suppress further transitions when the level is crossed repeatedly. The flip-flop requires a second input signal to resume its original state, while the Schmitt trigger will move to its original state when the single input signal decreases below the LTP.

Although it is possible to construct Schmitt trigger circuits that allow both the UTP and LTP to be determined by external resistors, most logic circuits involving the Schmitt trigger have fixed trigger levels. The 7414 is a hex inverter with a Schmitt trigger input. The UTP is typically 1.7 V while the LTP is 0.9 V. This would effectively discriminate against noise with a maximum amplitude of less than 800 mV. The 7413 is a dual 4-input NAND with Schmitt trigger inputs. The UTP and LTP are equal to those of the inverter. The 74244 chip contains eight 3-state buffers with Schmitt trigger inputs. The difference between UTP and LTP with these circuits is approximately 400 mV.

4.2

FLIP-FLOP APPLICATIONS

SECTION OVERVIEW The flip-flop is the most important multivibrator in digital system applications. This device can be used to construct frequency dividers, to delay and synchronize signals, to construct counters, to construct registers, to convert parallel information into serial information, to convert serial information into parallel information, and to construct state machines. All but the latter application will be considered in this section. The importance of the state machine is such that several chapters are later devoted to this topic.

In a sequential system, it is generally necessary to establish a time reference for various operations carried out within the system. A repetitive waveform called a clock signal is generated to establish this time reference. Some systems require very accurate clocks in which case a crystal-controlled astable circuit may be used. In computers, the master clock will generally consist of such a circuit. Several other digital systems do not require the precision of the master clock and can use simpler astable circuits. The duty cycle of many clock signals used in digital systems is 50%, although some microprocessor chips require a two-phase clock with a lower duty cycle.

In many instances, the rectangular waveform produced by an astable or oscillator circuit may be operated on to produce the final clock signal. Furthermore, some subsystems of a digital system may require a lower frequency clock than the master clock signal. The following paragraphs consider various modifications of a rectangular waveform to produce an appropriate clock signal.

4.2.1 FREQUENCY DIVISION AND CLOCK GENERATION

A rectangular waveform can be reduced in frequency by a factor of 2^n, where n is the number of cascaded flip-flops in the divider. Each flip-flop in the frequency divider is connected so that it will toggle. The signal to be divided is applied to the clock input of the first stage and the Q output of each stage is connected to the clock input of the succeeding flip-flop. A three flip-flop divider is shown in Figure 4.23. The waveform at $Q1$ shows a frequency that has divided the input frequency by 2. The waveform at $Q2$ divides the frequency of $Q1$ by 2 and the input frequency by 4. The frequency at $Q3$ is a factor of 8 lower than the input frequency.

The number of stages used is limited by the total propagation delay time through the divider. Each flip-flop has a clock-to-output propagation delay time that is cumulative through the stages. In low frequency applications, this factor may not be significant, but in high frequency systems even one or two stages may introduce an unacceptable delay.

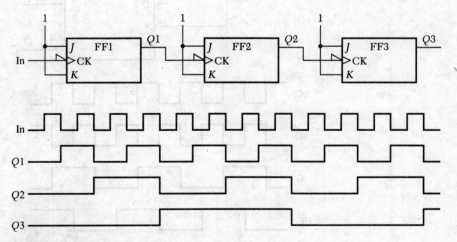

FIGURE 4.23 A frequency divider.

Many digital systems are driven by a clock signal with a 50% duty cycle. The clock transitions are used to initiate actions in various components of the systems. Because a transition occurs only every half-period for a 50% duty cycle clock, it is sometimes useful to produce a second clock signal that is delayed one-quarter of the clock period. The presence of both the clock signal and this delayed clock signal allows more choice of times when transitions occur.

If a system requires a clock signal, C, of a particular frequency, a signal of twice this frequency, $2C$, is useful. The delayed clock signal, DC, which lags the clock signal by one-quarter cycle, is produced by first inverting $2C$ and then applying this signal to the clock input of a T flip-flop. This scheme is shown in Figure 4.24 along with pertinent timing diagrams. The signal DC now provides a clock transition at the one-quarter and three-quarter points of the period while C provides a transition at the beginning and at the one-half point of the period. For very high frequency clocks this method is inappropriate since the oscillator is required to produce an output frequency equal to twice that of the desired clock frequency.

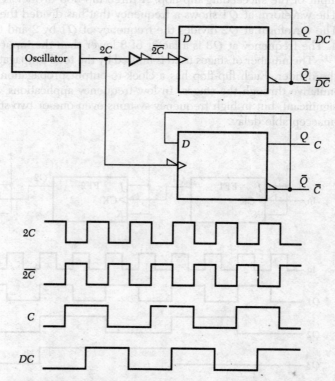

FIGURE 4.24 Clock system with delayed clock signal.

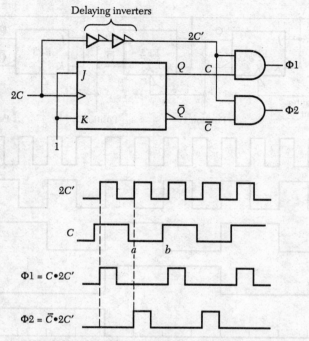

FIGURE 4.25 A two-phase clock circuit.

A second type of clock signal that has become more popular since the advent of the microprocessor is the two-phase clock. This type of clock consists of two separate signals, each of which goes high for a portion of the clock cycle but not simultaneously. The two-phase clock is required by several microprocessor chips. Figure 4.25 shows one method of generating these signals. The delaying inverters are required to eliminate possible glitches at point a on $\Phi1$ and point b on $\Phi2$. The signal C must be deasserted before $2C'$ goes positive to avoid a glitch at point a. Similarly, $\bar{C}$ must drop low before $2C'$ goes high to avoid the glitch at point b on $\Phi2$. The inverters must introduce a total delay that exceeds that of the flip-flop.

4.2.2 BINARY COUNTERS

A binary pulse counter can be constructed with a T flip-flop circuit as its basic element. One flip-flop is required for each column of the maximum binary number to be counted. For example, if the counter must count from 0 to decimal 1000, ten stages are required since nine stages can reach a maximum decimal number of $2^9 - 1 = 511$. Ten stages can reach binary 1111111111, which corresponds to decimal $2^{10} - 1 = 1023$.

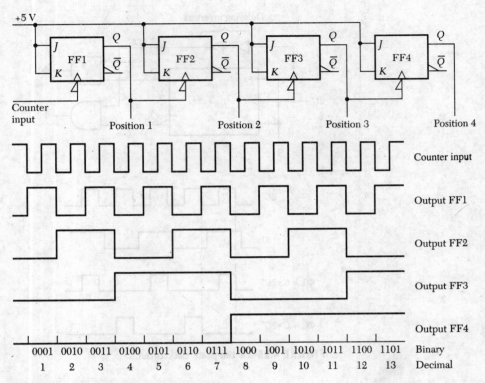

FIGURE 4.26 A binary counter and waveforms.

Figure 4.26 shows a 4-position binary counter that will proceed from 0 up to decimal $2^4 - 1 = 15$. The first negative transition of the counter input line will cause $Q1$ to assume a high level. All other stages are still in the 0 state. As $Q1$ makes this transition from 0 to 1, a positive transition is presented to CK2 with no corresponding change in state of FF2. The counter now reads 0001. A second negative transition causes $Q1$ to switch from 1 back to 0. This negative transition is presented to CK2, resulting in a change in state to $Q2 = 1$. The counter now reads 0010, corresponding to decimal 2. As additional negative input transitions occur, the count advances to consecutive binary numbers. Figure 4.26 shows a timing chart for the counter input and all outputs.

The 7493 is a direct implementation of the 4-bit counter of Figure 4.26 with two minor differences. This circuit, shown in Figure 4.27, incorporates a means of directly resetting the counter to contain all zeros. To do so requires assertion of inputs R_1 and R_2 simultaneously. As long as these inputs are both asserted, all flip-flops will remain in the 0 state regardless of the signal presented to the clocked inputs. Direct sets or resets always override clocked operation in flip-flop circuits.

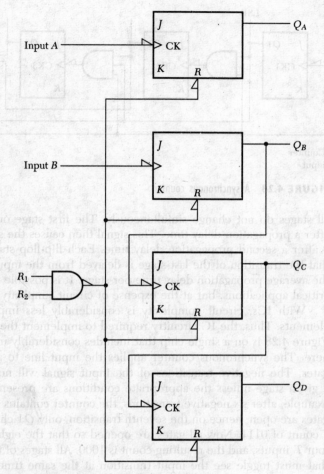

FIGURE 4.27 A 1-, 3-, or 4-bit counter.

We note that the output of FFA is not connected to the input of FFB in the 7493. If a 4-bit counter is desired, Q_A must be externally wired to input B and the counter input becomes input A. A 3-bit counter results if input B is taken as the counter input with Q_B, Q_C, and Q_D as the outputs. A 1-bit counter results if input A is used as the counter input and Q_A is used as the counter output. A 4-bit binary counter is also called a divide-by-16 counter because the output frequency equals the counter input frequency divided by 16. One-bit, 2-bit, and 3-bit counters are called divide-by-2, divide-by-4, and divide-by-8 counters, respectively.

The 7493 presents speed problems in very high frequency applications. If three 4-bit circuits are used to create a 12-bit counter and all stages are in the 1 state, the next negative input transition will cause all stages to revert to 0. Unfortunately,

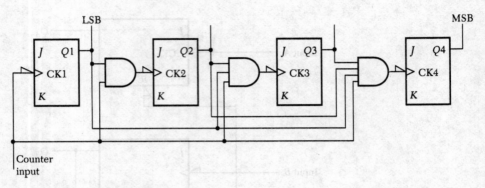

FIGURE 4.28 A synchronous counter.

all stages do not change simultaneously. The first stage output changes to 0 only after a propagation delay time. This signal then causes the second stage to revert to 0 after a second propagation delay time. Each flip-flop stage adds to the delay so that the transition of the last stage is delayed from the input transition by 12 times the average propagation delay time per stage. It is possible to decrease this delay in critical applications, but at the expense of circuit simplicity.

With ICs, circuit complexity is considerably less important than in discrete elements. Thus, the IC circuitry required to implement the synchronous counter of Figure 4.28 is on a single chip that includes considerably more circuits than shown here. The synchronous counter applies the input line to all stages through AND gates. The negative transitions of the input signal will not reach the T input of a given stage unless the appropriate conditions are present to open the gate. For example, after six negative transitions, the counter contains 0110. None of the AND gates are open; hence on the seventh transition, only $Q1$ changes to binary 1, giving a count of 0111. Now all gates are opened so that the eighth transition reaches all four T inputs, and the resulting count is 1000. All stages of the synchronous counter that must toggle see the input transition at the same time, resulting in minimum delay. More will be said about synchronous design in Chapter 6.

There are occasions in digital systems that require a backward or down counter. A binary number is set into the counter that then counts toward zero as input pulses occur. Figure 4.29 shows a simple method of constructing a down counter. The timing chart assumes that the original count was 1011 or decimal 11. After 11 negative transitions of the input signal the counter contains a count of 0000.

An example of an IC up-down counter is the 74193. This is a synchronous, 4-bit, up-down counter with a clear input, and borrow and carry outputs. Data can be loaded into the counter in parallel if a preset count is desired. Figure 4.30 shows the block diagram for this circuit, which has a complexity equal to approximately 55 gates. This counter can be cascaded with other stages to increase the bit capacity by 4 per chip. The carry and borrow outputs connect directly to the up-count and down-count inputs, respectively, of the adjacent chip. The clear input resets all bits to 0 at anytime this input is activated.

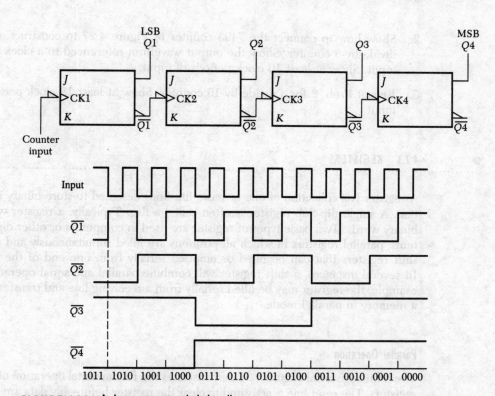

FIGURE 4.29 A down counter and timing diagram.

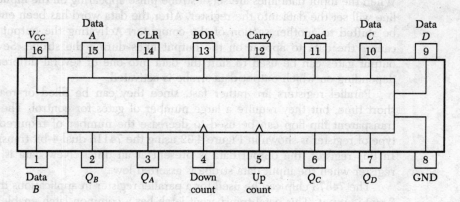

FIGURE 4.30 The 74193 synchronous up-down counter.

● ● ● **DRILL PROBLEMS** Sec. 4.2.1 and 4.2.2

1. Design a clock circuit having a signal with the same frequency but delayed from a reference clock by one-eighth of the clock period.

2. Show how to connect the 7493 counter of Figure 4.27 to construct a divide-by-8 counter. Show the output waveform referenced to a clock signal input. Show at least 10 clock periods of input.

3. Repeat Prob. 2 for a divide-by-16 counter. Show at least 18 clock periods of input.

4.2.3   REGISTERS

A register is a collection of one or more flip-flops designed to store binary information. A single flip-flop register is often called a flag. Typically, a register will store binary words. Two basic types of register are used in computers or other digital systems: parallel registers in which all positions are filled simultaneously and dynamic shift registers that can be filled or emptied serially from one end of the register. In several instances, a shift register will combine parallel and serial operation. For example, the register may be filled serially from a receiving line and transferred into a memory in parallel mode.

Parallel Operation

The 8-bit register of Figure 4.31 demonstrates the fundamental operation of parallel registers. The reset line is activated to clear the register before any data are entered. When the input data lines are set, a strobe pulse appearing on the input data strobe line will set the data into the register. After the data word has been entered, it can be shifted to another section of the computer. Activating the output data strobe causes the data to appear on the output lines during the strobe. Several sets of output gates can be used to shift the data into one of several different locations, depending on which output data strobe is activated.

Parallel registers are rather fast, since they can be filled or read in a very short time, but they require a large number of gates for control. The gated D or transparent flip-flop can be used to decrease the number of required gates. This type of register is shown in Figure 4.32 using the 74116 dual 4-bit transparent latch. In this register, the output data is present at all times. New data is set into the register when the input data strobe is asserted low.

The 74373 chip can be used for a parallel register in applications that require a 3-state output. This octal transparent latch has a common latch enable input and a common output enable line. Typically, the 3-state register is used to interface with a microprocessor bus. The output is enabled only when this register is to drive the bus to input data to the microprocessor. The 74LS373 is a positive edge-triggered octal register without 3-state outputs. The clock input is applied in parallel to all 8 flip-flops to set the 8 bits of data into the register. A low asserted direct reset can force all flip-flops to the zero state.

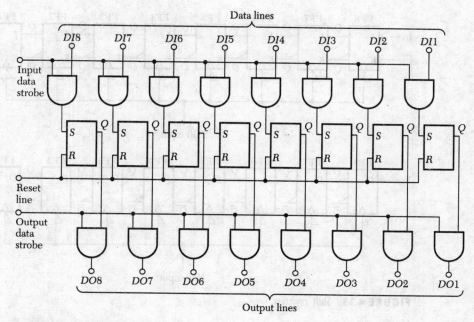

FIGURE 4.31 Parallel register.

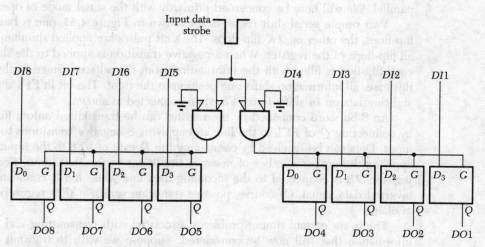

FIGURE 4.32 A parallel register based on the 74116 transparent latch.

Serial Operation

A shift register is used to receive or transmit serial information. Often it is designed also to receive or transmit parallel information. Some applications require the reg-

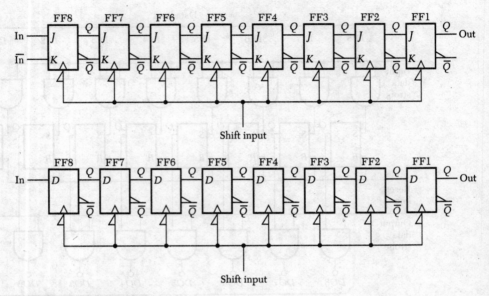

FIGURE 4.33 Shift registers.

ister to receive in parallel and transmit serially or to receive serially and transmit in parallel. We will here be concerned primarily with the serial mode of operation.

Two simple serial shift registers are shown in Figure 4.33; one is based on D flip-flops, the other on JK flip-flops. The shift pulses are applied simultaneously to all flip-flops of the register. When a negative transition is applied to the shift input, each flip-flop is filled with the information from the adjacent stage to the left. In this case, all information shifts one position to the right. The bit in FF1 disappears, and the data on *In* shifts into FF8 when connected as shown.

An 8-bit word contained by the register can be transmitted onto a line simply by connecting Q of FF1 to the line and applying 8 negative transitions to the shift input. Data can be received by connecting the D gate of FF8 to the input line and applying the proper number of negative transitions to the shift input. For the JK register, $J8$ is connected to the incoming data line while $K8$ is connected to the inverted data signal. Of course, positive transition sensitive shift registers are also available.

There are certain timing problems associated with transmitting and receiving information that will now be considered. Suppose we wish to transmit the 7-bit ASCII code for the character "J" which is 1001010. We will proceed from the LSB to the MSB as we transmit, but we must precede the LSB with a start bit of 1 to signal the receiver that a digital word follows.

One method of transmitting this information onto a line uses 9 flip-flops. The register is loaded with 100101010. The rightmost 0 is not part of the data word but is added to produce equal spacing of the bits. The timing chart for the transmission

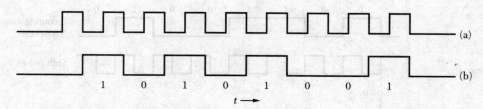

FIGURE 4.34 (a) Gated clock applied to shift input. (b) Serial code appearing on Q of FF1.

of this word is shown in Figure 4.34. The first negative transition sets the start bit onto the line. Succeeding transitions shift the remaining 7 bits onto the line, while the last transition clears the line. If the start bit had been in the first flip-flop rather than the second, the line would have been high prior to the application of the first negative transition.

A word or series of bits to be transmitted sequentially over a pair of lines may use one of two common formats. These are the return to zero or RZ code and the nonreturn to zero or NRZ code as shown in Figure 4.35.

The RZ code represents the bit value of the code during the first half of the clock period and always returns to the zero level during the second half of the period. A 1 is indicated by a high value during the initial half of the clock period, while a 0 is indicated by a low level during this time. The NRZ code remains high for the entire clock period when a 1 is present and remains low for the entire clock period to represent a 0.

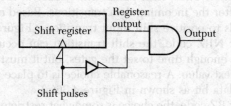

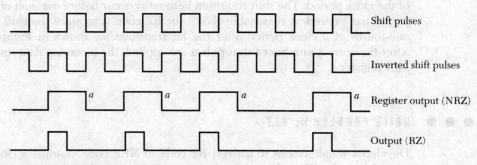

FIGURE 4.35 Conversion of NRZ to RZ code.

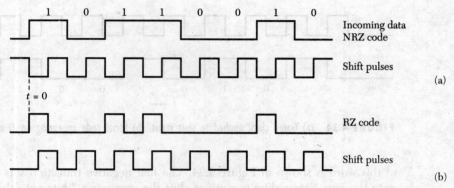

FIGURE 4.36 Relationship between shift pulses and incoming data: (a) NRZ code, (b) RZ code.

We note that the code produced by the shift register is NRZ code. If RZ code is desired, the circuit of Figure 4.35 can be used. The inverter must delay the inverted clock signal applied to the gate enough to allow the shift register to change before the gate is opened. This prevents possible slivers from occurring in the output RZ code at the points marked a. In order to use the shift register as a serial receiver, an input data line is connected to the J input of the leftmost flip-flop, and an inverter is connected to K. Note that D flip-flops could be used instead of JK flip-flops to construct this register. If an 8-bit word is to be received, exactly 8 shift pulses must be applied to the shift input to fill the register. Furthermore the proper time relationship must exist between the incoming data and the shift pulses. The gates $J8$ and $K8$ must be set prior to the application of a shift pulse to satisfy specified gate setup times. Thus, shift pulses must occur at intervals of 1 bit time, but must also occur after the incoming data transitions. Based on the assumption that the register responds to positive shift input transitions, Figure 4.36 demonstrates this point.

For NRZ code, the shift transition can occur at any point after each data bit has had enough time to set the gates, but it must occur before the data bit changes to the next value. A reasonable choice is to place the shift transition at the midpoint of the data bit as shown in Figure 4.36(a).

For RZ code, the choice is somewhat restricted, since the 1 bit lasts only one-half of the clock period. The shift transition here must occur before one-half of the clock period has expired. A reasonable choice for RZ code is to place the shift transition one-fourth of a clock period after the bit transition, as shown in Figure 4.36(b). After the correct number of shift pulses are applied, the incoming data reside in the shift register.

● ● ●   **DRILL PROBLEM Sec. 4.2.3**

Develop a simple circuit to convert RZ code to NRZ code. Assume a clock signal synced with the RZ code.

Integrated Circuit Registers

IC shift registers are available from several manufacturers. Several hundred bit registers can be fabricated on a single chip for serial-in serial-out registers. Obviously, there cannot be lines available for direct sets to each stage. There are various types of 8-bit registers available in monolithic circuits. The 74164 is an 8-bit, parallel-out, serial shift register; the 74165 is a parallel-load shift register. These two units can be used in a digital communication system, with the 74164 as the sending register and the 74165 as the receiving register. The sending register might be filled in parallel, then transmit serially to the receiver. When this register is filled, a parallel transfer can store its contents into the appropriate unit.

The 74194 is a 4-bit bidirectional universal shift register that can be used to construct larger registers. It features parallel input, parallel output, or serial input and output (right or left shift). Figure 4.37 includes the pin diagram and function table for the 74194. The mode controls, $S0$ and $S1$, should be changed only when the clock input is high.

4.2.4 GENERATION OF GATED CLOCK SIGNALS

In transmitting or receiving data by means of a serial shift register, a certain number of transitions must be presented to the shift input. These pulses shift the serial data from the register to the line for a transmitting register and fill the register from data on the line for a receiving register. When all data has been shifted into or out of a register, no more pulses should be applied to the shift input. For reception of serial data, the shift pulses must start after a suitable delay from the appearance of the first bit on the incoming data line. This requirement is due to the gate setup time of the flip-flops.

A common method of applying the correct number of pulses is to gate the clock signal. An AND gate with one input connected to the clock signal is opened and closed by another signal, referred to as the gating signal. The gating signal must begin and end at the appropriate times to produce a clean gated clock signal at the output of the AND gate. The gating signal is asserted for the correct number of clock cycles to produce the prescribed number of shift pulses.

The next paragraphs will first consider the generation of gated clock pulses for transmission of data and then proceed to the receiving register.

Clock Pulses for Serial Data Transmission

In a system that is to transmit serial data from a shift register, the controlling circuit will generate a signal to initiate the gated clock. We will refer to this as the Generate signal, which may or may not be synchronized to the continuous clock signal. We will assume that the assertion of Generate lasts longer than the gated clock pulses to be produced. If the signal is not synchronized, we may use the simple synchronizing circuit of Figure 4.38. The unsynchronized Generate signal is applied to the D input

MODE SELECT—FUNCTION TABLE

OPERATING MODE	INPUTS							OUTPUTS			
	CP	MR	S1	S0	DSR	DSL	DN	Q0	Q1	Q2	Q3
Reset (clear)	X	L	X	X	X	X	X	L	L	L	L
Hold (do nothing)	X	H	L	L	X	X	X	Q0	Q1	Q2	Q3
Shift left	↑	H	H	L	X	L	X	Q1	Q2	Q3	L
	↑	H	H	L	X	H	X	Q1	Q2	Q3	H
Shift right	↑	H	L	H	L	X	X	L	Q0	Q1	Q2
	↑	H	L	H	H	X	X	H	Q0	Q1	Q2
Parallel load	↑	H	H	H	X	X	DN	D0	D1	D2	D3

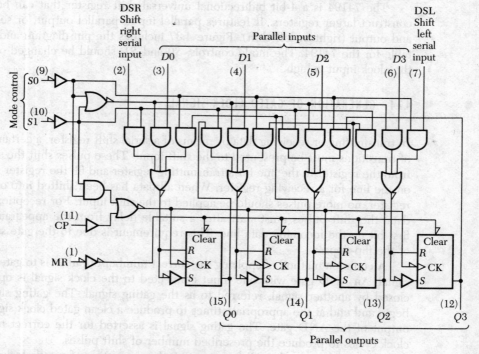

FIGURE 4.37 A 4-bit bidirectional shift register.

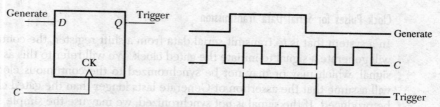

FIGURE 4.38 Trigger synchronization using a D flip-flop.

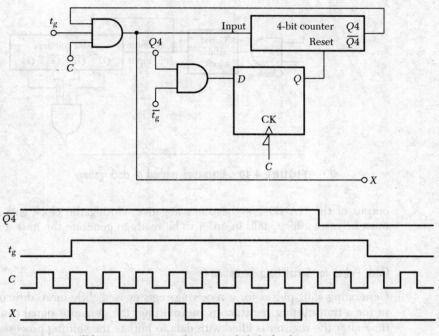

FIGURE 4.39 Using a counter to produce a gated clock signal.

of the flip-flop and the clock signal is applied to the clock input. The output signal will now be asserted in synchronism with the positive transition of the clock signal.

Using the synchronized trigger signal, a simple method of generating eight clock pulses is shown in Figure 4.39. The trigger that opens the gate is designated t_g. The counter is at 0 count prior to this time; thus $\overline{Q4} = 1$. The gated clock pulses are applied to the counter, which responds to the negative transitions. The eighth negative transition switches the fourth flip-flop of the counter to the 1 state. This closes the gate since $\overline{Q4} = 0$ after the eighth negative transition. The counter must be reset before it can be used again to gate the clock. The D-input of the flip-flop is gated to binary 1 when the trigger strobe falls, since $Q4 = 1$ at this point. The first negative clock transition after t_g has returned to a 0 value raises the reset line to reset the counter to 0. Because $Q4$ and the D-input now equal 0, the following negative clock transition returns the flip-flop to the 0 state, deactivating the reset line. The circuit is now ready to generate another series of clock pulses when t_g is reasserted.

Figure 4.40 shows an alternate method that uses a counter to generate nine negative transitions of a subclock (the reset circuitry is not shown for the sake of simplicity). Assuming a zero count on the counter, the NAND gate output will be high. When t_g goes positive, the astable clock is gated on. After nine negative clock transitions the NAND gate input will consist of $Q1 = \overline{Q2} = \overline{Q3} = Q4 = 1$. The

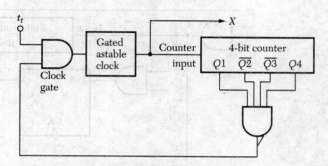

FIGURE 4.40 Alternative method of clock gating.

output of the NAND now becomes negative, closing the clock gate. The counter must be reset after t_g falls in order to be ready to generate the next series of pulses.

Clock Pulses for Serial Data Reception

Generating shift pulses for a receiving register is slightly more difficult than doing so for a transmitting register. In transmitting, the generate signal can occur at any time after the register is filled with data to initiate the shifting process. In receiving, some other system or subsystem puts data on the line, and the receiving register must accept this data whenever it occurs. In receiving, we must detect the start bit or first bit, then generate the shift pulses so that after each individual bit appears on the incoming data line, a shift pulse occurs. When receiving NRZ code, the logical point for the shifting transition to occur is at the middle of the bit time. Since it is impossible to generate a receiving clock with the exact frequency as the transmitting clock, it is necessary to synchronize the receiving clock with the start bit, and then place the shifting transitions at the midpoint of the bit times. If the receiving clock drifts from the correct frequency, an error of ± 0.5 bit times must accumulate to cause a bit error.

One approach to generating the correct number of clock transitions at a receiver is shown in Figure 4.41. Prior to reception of a word the counter must contain a count of 1000, leading to $Q4 = 1$ and $\overline{Q4} = 0$. The flip-flop must also be reset. The gated clock is now off. When the start bit occurs, the flip-flop is set and, after a slight delay, the counter is reset. This drives $\overline{Q4}$ to 1, which initiates the output of the gated clock. The assertion of $\overline{Q4}$ also resets the flip-flop, which now removes the reset signal from the counter. The gate to the FF is closed to prevent the assertion of the S input during the time the word is being received. The output pulses of the gated clock are applied to the receiving register and are also counted by the counter. When a count of 8 is reached, $\overline{Q4}$ goes low, terminating the gated clock output. The system is now ready to receive another word. Two inverters are used to delay the signal applied to the counter reset. This delay is necessary to stretch the width of the reset pulse to an acceptable level. If the FF Q signal were connected directly to the counter reset, it would disappear immediately upon the counter's beginning to reset. The signal $\overline{Q4}$ would start positive and $Q4$ would start negative, which would reset the flip-flop. With the reset signal removed, the counter might not reset properly—which brings up an important point about logic design that deserves emphasis. A signal that drives a circuit reset should not be removed as the circuit is reset because oscillations or improper reset may occur. Using the delaying inverters allows the reset pulse width to stretch to a value slightly greater than the total delay, ensuring proper reset.

There is a more reliable method of designing the transmitting and receiving clock circuitry discussed in this section. This method is state machine design, which will be covered in the following three chapters. We have not attempted to propose design principles here; we have simply proposed systems to produce the gated clock signals. In Chapters 5 to 7 we will develop principles that allow the orderly design of control circuits that can be used for many different purposes.

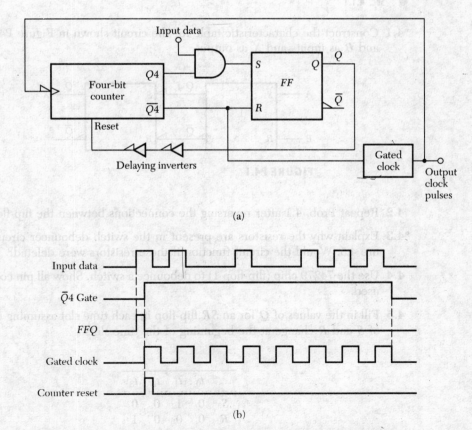

FIGURE 4.41 (a) Receiving clock signal generator using counter. (b) Timing chart.

SUMMARY
. . . .

1. The bistable flip-flop is perhaps the most important logic element in use today. This circuit is used in counter and register systems and forms the basis of the state machine to be considered in the next three chapters. The parallel and serial registers are important components of several digital systems.

2. Serial transmission and reception of binary words can be accomplished with shift registers and properly timed gates. The gated clock is necessary to transmit or receive information with shift registers.

CHAPTER 4 PROBLEMS
. . . .

● **Sec. 4.1.1**

°**4.1** Construct the characteristic table for the circuit shown in Figure P4.1 with A and B as inputs and X as output.

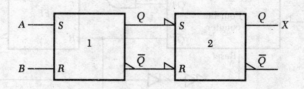

FIGURE P4.1

4.2 Repeat Prob. 4.1 after reversing the connections between the flip-flops.

°**4.3** Explain why the resistors are present in the switch debouncer circuit of Figure 4.5. Would the circuit function if these resistors were deleted?

4.4 Use the 74279 chip (flip-flop 1) to debounce a switch. Show all pin connections used.

4.5 Fill in the values of Q for an SR flip-flop in each time slot assuming the values of S and R change at the beginning of the time slots.

	t_0	t_1	t_2	t_3
S	0	1	0	0
R	0	0	0	1
Q	0			

4.6 Fill in values for S and R to cause the Q values of the SR flip-flop.

	t_0	t_1	t_2	t_3
S	0			
R	0			
Q	1	0	0	1

*4.7 Fill in the values of Q for a gated D flip-flop in each time slot assuming the values of D and G change at the beginning of the time slots.

	t_0	t_1	t_2	t_3
D	0	1	0	0
G	0	0	0	1
Q	1			

4.8 Plot the output waveform for the inputs shown in Figure P4.8 assuming the initial contents of the flip-flop is $Q = 0$.

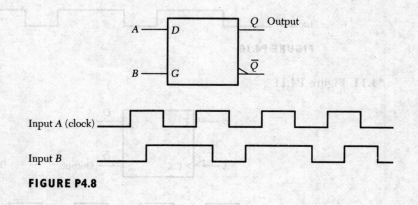

FIGURE P4.8

● Sec. 4.1.2

In Probs. 4.9 to 4.17, plot the output waveform referenced to the clock signal assuming the initial contents of all flip-flops is $Q = 0$. Assume all flip-flops are edge triggered.

4.9 Figure P4.9

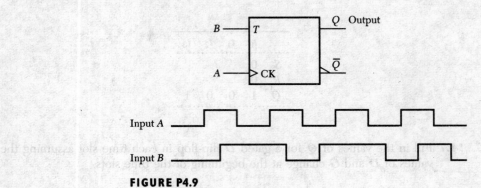

FIGURE P4.9

4.10 Figure P4.10

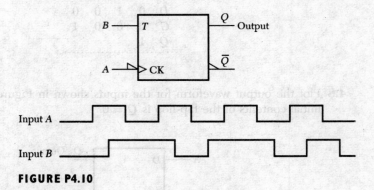

FIGURE P4.10

°4.11 Figure P4.11

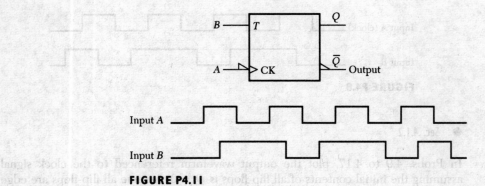

FIGURE P4.11

4.12 Figure P4.12

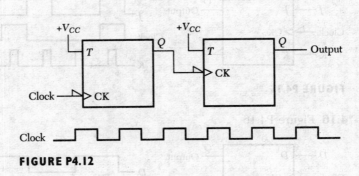

FIGURE P4.12

4.13 Figure P4.13

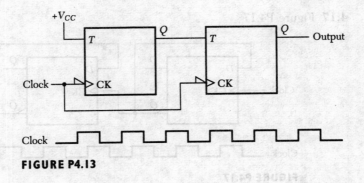

FIGURE P4.13

4.14 Figure P4.14

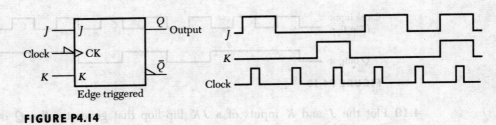

FIGURE P4.14

4.15 Figure P4.15

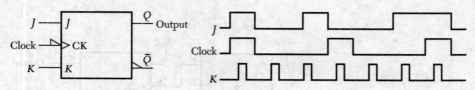

FIGURE P4.15

°4.16 Figure P4.16

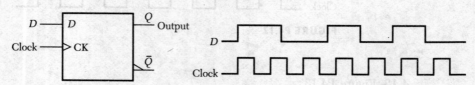

FIGURE P4.16

4.17 Figure P4.17

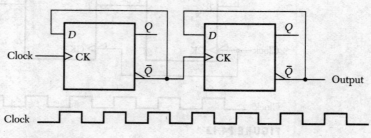

FIGURE P4.17

For Probs. 4.18 to 4.20, assume that the flip-flops are positive edge-triggered devices.

4.18 Plot the T input of a T flip-flop that generates the Q output of Figure P4.18.

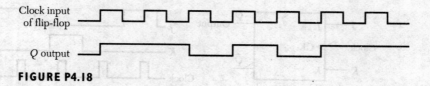

FIGURE P4.18

4.19 Plot the J and K inputs of a JK flip-flop that generates the Q output of Figure P4.18.

°4.20 Plot the D input of a D flip-flop that generates the Q output of Figure P4.18.

4.21 If the flip-flop is edge triggered and the initial value of Q is 0, plot the waveform for Q in Figure P4.21.

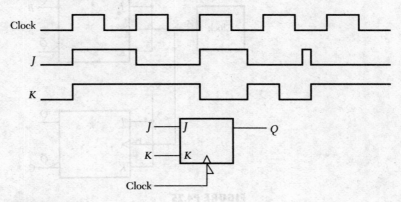

FIGURE P4.21

4.22 Repeat Prob. 4.21 for a master-slave flip-flop.

● **Sec. 4.1.3**

4.23 Assume the UTP of the 7414 is 1.7 V while the LTP is 0.9 V. The high output of the 7414 is 3.4 V while the low output is 0.2 V. Plot the output for the input shown in Figure P4.23.

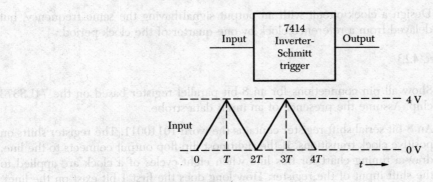

FIGURE P4.23

4.24 Repeat Prob. 4.23 if the input is attenuated by a factor of 2. The peak voltage now only reaches 2 V.

● **Sec. 4.2.1**

°4.25 Sketch the waveforms appearing at points A, B, and C of Figure P4.25. Assume that $B = C = 0$ initially.

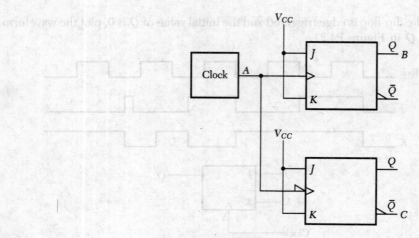

FIGURE P4.25

● Sec. 4.2.2

4.26 Show how to connect the 7490 counter to construct a divide-by-5 counter. Show the output waveform referenced to a clock signal input. Show at least eleven clock periods of input.

4.27 Repeat Prob. 4.26 for a divide-by-10 counter. Show at least 21 clock periods of input.

4.28 Design a clock circuit with an output signal having the same frequency, but delayed from a reference clock by one-quarter of the clock period.

● Sec. 4.2.3

°**4.29** Show all pin connections for an 8-bit parallel register based on the 74LS373 chip. Assume the presence of an input data strobe.

4.30 An 8-bit serial shift register contains the word 10110011. The register shifts on positive clock transitions. If the rightmost flip-flop output connects to the line, draw a timing chart for this line when eight cycles of a clock are applied to the shift input of the register. How long does the first 1 bit exist on the line?

4.31 Repeat Prob. 4.30 if a ninth flip-flop, containing a 0, is added between the register and the line. Assume nine cycles of a clock are now applied. What is the purpose of this ninth flip-flop?

● Sec. 4.2.4

4.32 Design a system that produces 14 negative clock transitions on a line after which the line remains at the 0 level. Assume that a start transmission strobe is available that is synced with the positive half-cycle of the clock to initiate the 14 transitions.

4.33 Repeat Prob. 4.32 assuming 17 transitions are required and the start transmission strobe is asynchronous.

4.34 Design a system to generate 7 positive clock transitions to load a serial register with an incoming start bit plus 6 data bits. The appearance of the start bit on the incoming line should initiate the clock transitions.

REFERENCES AND SUGGESTED READING

1. K. J. Breeding, *Digital Design Fundamentals*, 2nd ed. Englewood Cliffs, NJ: Prentice-Hall, 1992.

2. P. Burger, *Digital Design*. New York: John Wiley and Sons, 1988.

3. Engineering Staff, *TTL Data Manual*. Sunnyvale, CA: Signetics, 1986.

4. W. I. Fletcher, *An Engineering Approach to Digital Design*. Englewood Cliffs, NJ: Prentice-Hall, 1980.

5. C. H. Roth, Jr., *Fundamentals of Logic Design*, 4th ed. St. Paul, MN: West, 1993.

REFERENCES AND SUGGESTED READING

1. J. F. Wakerly, Digital Design Fundamentals, 2nd ed. Englewood Cliffs, NJ: Prentice-Hall, 1992.

4. J. P. Hayes, An Engineering Approach to Digital Design. Englewood Cliffs, NJ: Prentice-Hall, 1990.

5. C. H. Roth, Jr., Fundamentals of Logic Design. St. Paul, MN: West, 1995.

Chapter 5

Introduction to State Machines

5.1

THE NEED FOR STATE MACHINES

Before we begin developing state machine theory, we should establish the fact that this is an important area of digital system design. We will do this by emphasizing three advantages of the state machine approach.

The first advantage is that many electronic systems require the type of sequential operation exhibited by state machines. Therefore, state machine design can be applied to the solution of a wide variety of practical circuit problems.

The second advantage is that state machine design methods lead to minimal design. In combinational circuits, we found that the Karnaugh map was useful in minimizing the number of gates required to implement a logic function. Although the importance of this tool has perhaps diminished as a result of the availability of MUXs, PLAs, PALs, and decoders, it remains a significant method in combinational design. The state machine design procedure relates to sequential circuit design in the same way the K-map relates to combinational circuit design. This method results in the minimum number of required flip-flops and can minimize other circuitry in the system as well.

The third advantage of the design method is that it is a well-developed, orderly procedure that anticipates and solves commonly occurring problems of sequential

208

circuits. Trial-and-error design procedures often result in the appearance of very narrow unwanted pulses or glitches on output lines or occasional oscillation problems. State machine methods eliminate these problems and reduce the time taken to debug the implemented hardware.

The preceding three advantages of state machine design over other types of design make it the predominate method used in sequential logic system design.

5.2
THE STATE MACHINE

SECTION OVERVIEW The state machine is defined and classified in this section. Three applications of the state machine as a system controller are considered.

5.2.1 STATE MACHINE DEFINITION

A state machine, also called a sequential machine, is a system that can be described in terms of a set of states that the system may enter [1]. Once in a particular state, the system must be capable of remaining in that state for some finite period of time even if the system inputs change. This requirement dictates memory capability for the state machine. Furthermore, the state machine must have a set of inputs and a set of outputs. As the system progresses from one state to another (from present state to next state), the next state reached depends on the inputs and the present state. The outputs also depend on inputs and the present state.

Before leaving this description of state machines, let us consider the practical meaning of the existence of states. A state may be a set of values measured at various points in a circuit. A simple flip-flop has two states in which it can exist: the output Q can equal either 1 or 0. A set of n flip-flops can produce 2^n possible unique output codes and thus could have 2^n possible unique states of existence. We note here that n lines driven by combinational variables can also take on 2^n different unique codes, but these are not states since they cannot exist independently of the input variables. These n lines change immediately as the input variables change and cannot remain at any code different from the input values. A set of n flip-flops can exist in a particular state different from the set of input values, and this is a requirement of the state machine.

5.2.2 CLASSIFICATION OF STATE MACHINES

It was earlier stated that a sequential machine or state machine must possess memory capability. A large majority of practical state machines use clocked flip-flops as the

Inputs

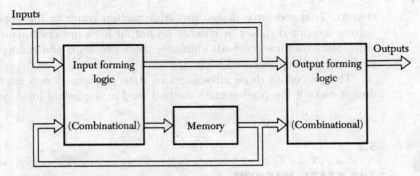

Outputs

FIGURE 5.1 General model of a sequential machine.

storage elements. The code that defines each state then corresponds directly to the code contained by the flip-flops.

The general model of the sequential machine is shown in Figure 5.1. This model is also called the Mealy machine after the man who first proposed the model [1, 2]. The input forming logic (IFL) and the output forming logic (OFL) sections are made up of combinational logic circuits. The memory section contains the state of the system. A path is provided from memory output to the IFL. Both input signals and present state signals drive the IFL to determine the next state of the system. The outputs are determined by the present state and the system inputs. A slight variation of the Mealy machine is the Moore machine [1, 2], which uses only the memory to drive the OFL. In this case, the output is a function only of the state of the system.

Another important characteristic of the state machine depends on whether the system is clock-driven or not. In many digital systems, a timing reference signal is required. Some type of astable multivibrator is generally used to produce a continuous clock signal. We refer to a variable as clock-driven if that variable changes value only at the time of a clock transition. It need not change at every clock transition to be clock-driven, but when it does change it must do so as the clock is changing values. When a variable is clock-driven, it is considered a synchronous variable.

If a variable can change at a time not related to transitions of the reference clock, it is called an asynchronous variable. This variable might change as a human operator closes a switch, as an analog voltage reaches a trigger level, or as a vehicle passes a checkpoint. There is no fixed relationship between an asynchronous variable change and the reference clock change. We should also note that a variable might be clock-driven by the clock in one system, but still be considered asynchronous with respect to the reference clock of a second system.

State machines are normally classified as synchronous or asynchronous; however, we will expand the number of categories to three. If a machine changes state in response to the clock and all inputs are synchronous, we will classify that circuit as a synchronous system. If the state changes occur in response to the clock, but one or more inputs are not clock-driven, the machine will be called a mainly synchronous system. If the state changes are input-driven rather than clock-driven, the system is an asynchronous one.

Mainly synchronous systems are encountered more than any other type in the digital field. Synchronous systems are the easiest to design and asynchronous systems are the most difficult. We will consider synchronous systems first and then proceed to mainly synchronous machines in Chapter 7. Chapter 10 considers asynchronous design.

5.2.3 THE STATE MACHINE AS A SEQUENTIAL CONTROLLER

Before we proceed to the fundamental ideas of state machine design, we wish to provide some motivation for this study by considering a few applications of the state machine.

In the control of certain devices or processes, it is necessary to generate unique sets of controlling signals during specified time periods. Control of machinery may require time slots ranging from milliseconds to seconds, whereas control of a computer or electronic device may use time slots as short as nanoseconds.

A given process may involve controlling the speed or position of electric motors. For example, a computer-controlled vehicle may use electric motors in the guidance system. The access mechanism in a magnetic disk drive might be powered by an electric motor of some sort. Another process could involve timing of events. The control of stop signals at an intersection to maximize traffic flow on both streets is a typical application of a sequential controller. Here timing is a variable depending on the density of vehicles at various points near the intersection. Control signals must be developed to assert the proper lights at appropriate times.

The process being controlled is referred to, quite reasonably, as the controlled process. The device generating the controlling signals is called a sequential controller, or simply a controller. Generally, a controller will produce output values depending on certain events or inputs that have occurred earlier or are now present. In these applications, the input signals are generated by commands or by sensors that monitor the process. These inputs help determine the sequence of states through which the system proceeds. The time that the system resides in each state provides the time slots required for driving the controlled system. A different set of outputs can be produced at each state.

A significant requirement of a controlled process is that it must be repeated over and over again with only slight variations. The process is then cyclic in nature. So also is a state machine. Although it may run through a different sequence of states before returning to the starting state, the system has a finite number of such paths and repetition is inevitable.

We will demonstrate these ideas by considering three practical problems in the digital field. The first problem can occur in the digital communication area. Suppose some sort of measurement data is being gathered at a location far away from a central computer. This information is transmitted to the central computer to be processed and stored. There may be several sets of variable length data transmitted. For example, one set of data may correspond to pressure measurements, another set may correspond to temperature values, another may correspond to radiation measurements. The data may only be meaningful when the measured value exceeds some threshold; thus, each set of data is a variable length, depending on the process being monitored.

In order to separate the sets of data, a frame separator word can be transmitted. This particular word will be unique to avoid confusion with any transmitted data word. It may be transmitted first to signal the computer that data follow. The first set of data is then transmitted followed by the frame separator word. When the receiver identifies this word, the computer recognizes that the next data set corresponds to another measurement. The frame separator word precedes each data set and terminates the final one. Knowing the order of transmission of data sets allows the computer to then process each set appropriately.

The identification of the frame separator word can be accomplished in various ways. One method would be to run the incoming words continually through a shift register and check the register contents at each word boundary for the frame separator word. Conceptually, this is a simple method. If the word length is 16 bits, a 16-bit shift register is required by this method. A 4-bit counter that keeps track of the word boundaries is also required along with a 16-input decoding circuit.

Another method that can be used is the state machine approach. In this instance, the design procedure might be more time consuming, but the resulting circuit will require fewer components. The basic goal here would be to create 16 time slots in which the value of each bit of each incoming word is checked. As bits are checked, if they correspond to the bits of the frame separator word, a particular sequence of states is entered. If all 16 states of this sequence are entered, an output is generated that signifies the occurrence of the frame separator word. At any point in the sequence, the detection of a bit that does not correspond to that of the frame separator word sends the system into a sequence of states that will not generate an output. This sequence is designed only for timing purposes. Although the occurrence of an incorrect bit immediately tells the system that the word being checked is not the frame separator word, the system must not start a new check until the beginning of a new word boundary. The use of states for timing purposes eliminates the need for counting circuits.

This particular application would require only five flip-flops and some gating circuits to implement with a state machine. This can be compared to 20 flip-flops plus gating in the shift register scheme. Such a comparison clearly demonstrates the advantage of component minimization in state machine design.

A second problem that can be considered is that of performing serial addition with a single full adder without using a shift register. The organization of a digital system might be primarily a parallel register structure. In order to conserve components, a single full adder rather than a parallel adder may be used. Sixteen-bit words require a 16-bit adder for parallel addition. Gating circuitry to store the sum must also be used. A single-bit full adder and a state machine can accomplish the same result with fewer components. Obviously, a serial addition will require more time than a parallel addition.

Let us consider the sequence of operations needed to accomplish a 4-bit addition using a single-bit full-adder. The following steps should accomplish this procedure. Here we assume that one word is stored as bits $A3$ $A2$ $A1$ $A0$ while the other is stored as $B3$ $B2$ $B1$ $B0$. The full-adder has inputs $I1$ and $I2$ for the two operands and CIN, the carry input. The outputs are SUM and COUT.

1. Present $A0$ and $B0$ to inputs $I1$ and $I2$. CIN must be 0.
2. Store SUM as $S0$.
3. Present $A1$ and $B1$ to inputs $I1$ and $I2$. If COUT in step 1 was 0, set CIN = 0; if COUT was 1, set CIN = 1.
4. Store SUM as $S1$.
5. Present $A2$ and $B2$ to inputs $I1$ and $I2$. If COUT in step 3 was 0, set CIN = 0; if COUT was 1, set CIN = 1.
6. Store SUM as $S2$.
7. Present $A3$ and $B3$ to inputs $I1$ and $I2$. If COUT in step 5 was 0, set CIN = 0; if COUT was 1, set CIN = 1.
8. Store SUM as $S3$.
9. Store COUT as $S4$.

The final result is $S4$ $S3$ $S2$ $S1$ $S0$.

To implement this set of procedures with a state machine requires that we choose states to accomplish the various steps listed. We would require a state or time slot in which to present $A0$, $B0$ and CIN = 0 to the adder. Still another would present $A1$, $B1$, and CIN to the adder. As the process begins, the first state would create and store $S0$ and carry COUT. Based on the value of COUT in this state, the next state would be one that causes CIN = 0 or one that causes CIN = 1. As the system proceeds through the sequence of states, the path taken is determined by the value of COUT. Thus, we see the capability of the state machine to create time slots in which outputs are generated. These outputs depend on previous inputs that have occurred in the past and also on present inputs.

A final example of an application that can be satisfied using a state machine is that of a traffic light controller. At a busy intersection the movement of vehicles may be monitored by pressure-sensitive sensors buried in the roadways. The output of these sensors can be used to drive counters to determine the traffic density in each lane for some fixed time interval. Figure 5.2 shows the layout of the intersection and turn lanes.

The traffic lights are to operate in one of several modes. When there are no vehicles in the turn lanes (even-numbered lanes) and no significant traffic density

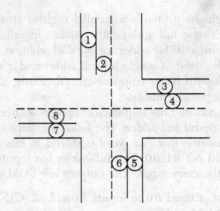

FIGURE 5.2   An intersection of streets.

imbalance, the odd-numbered lanes should see a 50% duty cycle for the lights. This means the lights controlling lanes 1 and 5 will be green 50% of the time as also will those lights controlling lanes 3 and 7. A second mode occurs if the total traffic density in lanes 1 and 5 differs from the density of lanes 3 and 7 by more than 50%. When this occurs, the lights controlling the higher density lanes should remain green for 60% of the time while the less dense lane lights are green for only 40% of the time.

If vehicles are detected in a single turn lane, the turn light for this lane should be green when the corresponding through lane light turns green. That is, if cars are detected in lane 6, but in no other even-numbered lane, this light will turn green when the lane 5 light turns green. Of course, the lane 1 light must remain red at this time. This condition will prevail for 5% of the total period, and then the turn signal will change to red while the lane 1 light turns green.

When cars are in facing turn lanes, either 2 and 6 or 4 and 8, the turn lights of one of these pairs change to green while all other lights are red. After a time duration of 5% of the total period, this pair of lights changes to red while the corresponding through lane lights change to green.

It is obvious that a state machine could serve as the controller for this system. Certain time slots must be created, the length of which will vary according to input information from the counters or sensors. The sequence of output signals required to drive the lights will also change in response to traffic patterns. The state machine can generate the correct sequence of outputs by allowing the inputs to influence the sequence of states assumed by the system.

From these three examples we can begin to see that the state machine is a very versatile and important device for digital control systems. Although other methods can be used to create controllers, the state machine is generally accepted as the most popular controller with the best design procedures available.

5.3

. . . .

BASIC CONCEPTS IN STATE MACHINE ANALYSIS

SECTION OVERVIEW This section introduces the excitation table for the flip-flop. With this table, the signals necessary to produce desired flip-flop outputs can be found. The next state table is considered as an aid in analyzing a state machine. Next, state maps are also discussed. The state diagram is developed to describe graphically the possible state movement of a system.

5.3.1 EXCITATION TABLES FOR FLIP-FLOPS
. . . .

Chapter 4 introduced the characteristic table for various flip-flops. These tables indicate the outputs resulting from each input combination and allow us to analyze circuits containing flip-flops. In designing circuits, we are given the output signals required and must find the inputs necessary to create these outputs. For flip-flops, the excitation tables yield this information. These tables list the input conditions necessary to cause all possible output transitions of the flip-flop. Figure 5.3 shows the excitation tables for the D and JK flip-flops. The variable Q_n represents the Q output prior to the assertion of the CK input and Q_{n+1} is the output after CK has been asserted. The conditions on D or J and K are assumed to be stable prior to the assertion of CK.

The excitation table for the D flip-flop is rather straightforward. Regardless of the initial state Q_n, the state after the clock transition will equal the value of D. The JK table is somewhat more complex. If the initial state of the flip-flop is $Q_n = 0$, the value of $Q_{n+1} = 0$ can be obtained with two different sets of input conditions. If $J = 0$ and $K = 0$, the flip-flop will not change states, and thus the zero value for Q remains after the clock transition. A second set of inputs that results in this output condition is $J = 0$ and $K = 1$. This always forces the flip-flop to the $Q_{n+1} = 0$ state. Row 1 of the excitation table combines these input conditions to yield $J = 0$ and $K = \Theta$ ("don't care") for the 0 to 0 output transition.

The 0 to 1 transition can be caused by setting the flip-flop with $J = 1$ and $K = 0$ or by toggling with $J = 1$ and $K = 1$. These conditions can be specified

Q_n	Q_{n+1}	D		Q_n	Q_{n+1}	J	K
$0 \to 0$		0		$0 \to 0$		0	Θ
$0 \to 1$		1		$0 \to 1$		1	Θ
$1 \to 0$		0		$1 \to 0$		Θ	1
$1 \to 1$		1		$1 \to 1$		Θ	0

FIGURE 5.3 Excitation tables.

by $J = 1$ and $K = \Theta$. The same type of reasoning leads to the conditions $J = \Theta$ and $K = 1$ for the 1 to 0 transition and $J = \Theta$ with $K = 0$ for the 1 to 1 output transition.

● ● ●   **DRILL PROBLEMS** Sec. 5.3.1

1. Assuming the state changing clock transitions define the time slots, and $Q(t_0) = 1$, specify $D(t_0)$, $D(t_1)$, $D(t_2)$, and $D(t_3)$ to result in $Q(t_1) = 1$, $Q(t_2) = 1$, $Q(t_3) = 0$, and $Q(t_4) = 1$.

2. If the flip-flop of Prob. 1 is replaced by a JK device, specify the values of J and K required at t_0, t_1, t_2, and t_3.

3. Develop an excitation table for a T flip-flop.

5.3.2   USING THE EXCITATION TABLE

A major use of the excitation table occurs in designing a set of flip-flops to generate a desired output waveform. For example, the output waveforms, Q_1 and Q_2, of Figure 5.4 may be required in a particular application. The outputs change only on

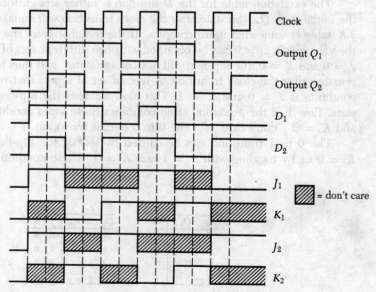

FIGURE 5.4   Generation of output waveforms using flip-flops.

negative clock transitions, thus these signals can be generated using flip-flops. If D flip-flops are used, the required inputs can be found from the D excitation table. A change in Q from 0 to 1 requires a 1 on D, prior to the negative-going transition of the clock. A change in Q to a zero value requires a zero on D prior to the negative clock transition. In Figure 5.4, it is assumed that the inputs are allowed to change on the positive-going clock transition. If JK flip-flops are used, two inputs are required for each flip-flop. The JK excitation table allows the construction of the J_1, K_1, J_2, and K_2 input waveforms.

A common problem in sequential design consists of a set of specified inputs that must produce certain outputs. Figure 5.5 shows an example of two required outputs and the given inputs that should cause these outputs. The D flip-flop input required to produce Q_1 and Q_2 are also shown. The design problem is solved by designing the combinational circuitry that produces the proper values of D_1 and D_2 when inputs X and Y are applied. These flip-flop inputs will then drive Q_1 and Q_2 to the desired values.

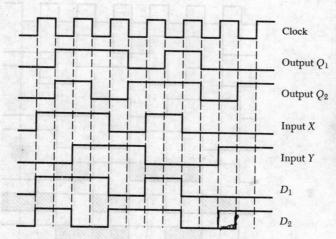

FIGURE 5.5 Input and output waveforms for D flip-flops.

$$D_1 = X$$

$$D_2 = X \oplus Y$$

FIGURE 5.6 K-maps for D flip-flops.

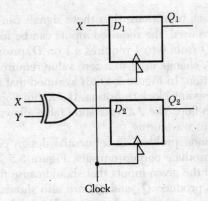

FIGURE 5.7 *D* flip-flop circuit that generates outputs of Figure 5.5.

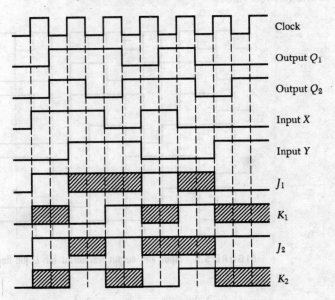

FIGURE 5.8 Input and output waveforms for *JK* flip-flops.

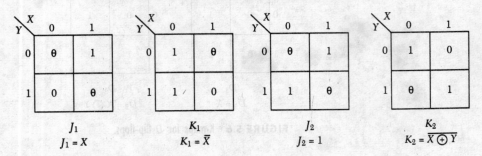

J_1
$J_1 = X$

K_1
$K_1 = \overline{X}$

J_2
$J_2 = 1$

K_2
$K_2 = \overline{X \oplus Y}$

FIGURE 5.9 K-maps for *JK* flip-flops.

The K-maps for D_1 and D_2 as functions of inputs X and Y are derived from Figure 5.5 and are given in Figure 5.6. The circuit that produces outputs Q_1 and Q_2 is shown in Figure 5.7.

The circuit can also be realized using JK flip-flops. Figure 5.8 shows the same input and output waveforms along with the necessary J and K values to produce the output waveforms. Information from Figure 5.8 is used to develop the K-maps for J_1, K_1, J_2, and K_2 as indicated in Figure 5.9. We note in constructing these K-maps that certain combinations of X and Y occur more than once over the seven clock periods. The combination $X = Y = 0$ occurs twice, but each occurrence leads to no change in Q_1 or Q_2. On the other hand, the input combination $X = 0$ and $Y = 1$ also occurs twice, but the first occurrence results in Q_1 changing from 1 to 0 while the second results in Q_1 remaining at 0. A change in Q_1 from 1 to 0 dictates that $J_1 = \Theta$ and $K_1 = 1$. If $Q_1 = 0$ and must remain at 0 after the clock transition, then $J_1 = 0$ and $K_1 = \Theta$. Thus, when $X = 0$ and $Y = 1$ both sets of gating requirements must be met. This can only be accomplished by selecting $J_1 = 0$ and $K_1 = 1$ as indicated in the K-maps.

Output Q_2 also has two sets of gating conditions to satisfy for $X = 1$ and $Y = 0$. This results in $J_2 = 1$ and $K_2 = 0$ in the K-maps for this input combination. The circuit that implements these expressions is shown in Figure 5.10.

From these two circuit examples we conclude that the flip-flop excitation tables are valuable in design situations. When a given set of inputs must generate a set of specified outputs, the excitation table is helpful in designing the needed circuit.

The characteristic tables introduced in Chapter 4 are useful in the analysis of a circuit while the excitation tables are more important in design.

● ● ●　**DRILL PROBLEM Sec 5.3.2**

Can you design a T flip-flop circuit to produce the outputs of Figure 5.5 from the given inputs X and Y?

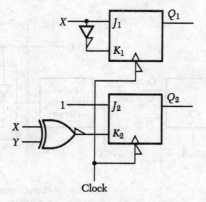

FIGURE 5.10 *JK* flip-flop circuit that generates outputs of Figure 5.8.

5.3.3   THE FLIP-FLOP AS A STATE MACHINE

One of the most basic state machines is a flip-flop circuit. A D or JK flip-flop has a set of inputs (D or J and K), a set of outputs (Q), and a set of states ($Q = 0$ and $Q = 1$). Figure 5.11 shows one version of the clocked D flip-flop drawn to conform to the general model of a state machine given in Figure 5.1.

The SR flip-flop is the memory element and the two NAND gates and inverter comprise the input forming logic, or IFL. The feedback from memory to IFL is derived from the Q and $\bar{Q}$ outputs. There is no output forming logic, or OFL, for this state machine. Since the output Q is a direct function of the memory only, this is a Moore machine. The clock signal that drives the state changes is not considered an input signal, although it is a necessary signal in the operation of synchronous and mainly synchronous state machines. It is simply understood that these types of systems will be driven by a reference clock signal.

While it is true that a single flip-flop is a state machine, its two possible states are too few to allow this device to be a very practical system. Two or more devices can be used to create a reasonable number of states to result in quite useful state machines.

5.3.4   ANALYZING STATE MACHINES

Although our ultimate objective is state machine design, we can learn much about the operation of these systems through analysis of some rather simple versions. We can also see the value of certain design aids such as the next state table and next

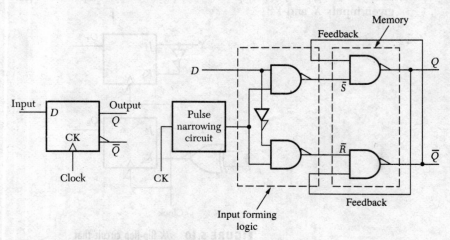

FIGURE 5.11   The D flip-flop as a state machine.

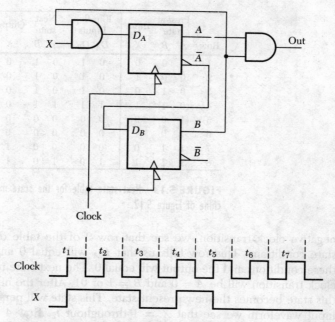

FIGURE 5.12 A state machine using D flip-flops.

state map. We will consider the synchronous system of Figure 5.12 consisting of two D flip-flops to introduce some basic ideas relating to state machine analysis.

The IFL consists of the AND gate that drives flip-flop A with the signal $D_A = XB$ and the wire that drives flip-flop B with $D_B = \bar{A}$. The OFL is the AND gate that results in Out $= AB$. We recognize in this circuit the basic characteristic of the state machine, that is, the next states are functions of the present state and the input X since X, A, and B determine the values of D_A and D_B.

It is worth noting that the input changes on the opposite clock transition to that which drives the state changes. The state changes take place on the negative clock transition while the value of X is updated on the positive transition. This allows the signals driving the flip-flop inputs to be stable prior to the state change.

In order to find the states through which the system moves, the next state table of Figure 5.13 is constructed. This table first lists all possible present states and input values. A consideration of the input equations, $D_A = XB$ and $D_B = \bar{A}$, allows the flip-flop inputs to be listed for each present state and input combination. The next state for each combination can be found from the D flip-flop excitation table of Figure 5.3. A negative clock transition causes the next state to take on the values presented to the D inputs prior to the transition. The output is determined from the equation Out $= AB$ using present state values.

Once the next state table has been constructed, we can use it to trace the state movement of the system. Assuming that the initial state during t_1 of the waveform in Figure 5.12 is $A = 0$ and $B = 0$, and noting that $X = 1$ prior to the first

Row	Present state		Input	Flip-flop Inputs		Next state		Output
	A	B	X	D_A	D_B	A	B	X
1	0	0	0	0	1	0	1	0
2	0	0	1	0	1	0	1	0
3	0	1	0	0	1	0	1	0
4	0	1	1	1	1	1	1	0
5	1	0	0	0	0	0	0	0
6	1	0	1	0	0	0	0	0
7	1	1	0	0	0	0	0	1
8	1	1	1	1	0	1	0	1

FIGURE 5.13 Next state table for the state machine of Figure 5.12.

negative clock transition, we see that row 2 of the table describes these present state conditions. This row tells us that D_A will equal 0 and D_B will equal 1 for these conditions and the output will equal 0. The next state reached after a negative clock transition will be $A = 0$ and $B = 1$ or 01. After the negative clock transition, this state becomes the new present state. This state will persist during t_2. From the input waveform we see that $X = 1$ throughout t_2. Row 4 of the next state table applies to this case. This row indicates an output of Out = 0 and a next state of 11. After the negative clock transition, the present state will become 11 and will remain at this value during t_3. We note that X drops to 0 prior to the next negative clock transition, thus row 7 applies to this time period. The output is Out = 1 and the next state will be 00. During t_4, the present state is 00 and X returns to 1 before the beginning of t_5. Row 2 is now used to predict an output of 0 and a next state of 01. This state of 01 lasts through t_5 with X dropping to 0 prior to t_6. Row 3 describes these conditions and dictates an output of 0 and a next state of 01. Since X remains low, the state 01 remains through t_6 and t_7. The timing chart of Figure 5.14 reflects the information of the preceding discussion.

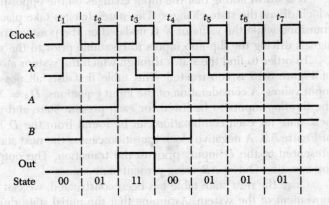

FIGURE 5.14 Timing chart for the circuit of Figure 5.12.

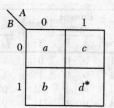

FIGURE 5.15 Present state map.

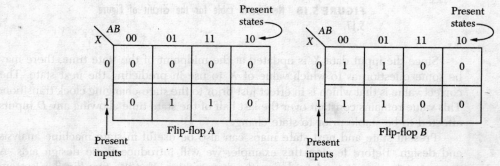

FIGURE 5.16 Next state maps.

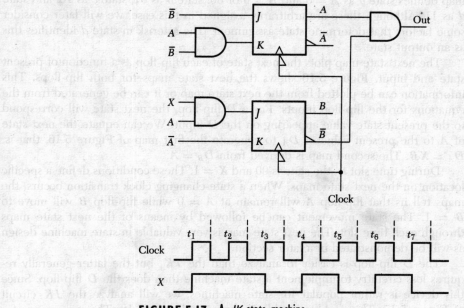

FIGURE 5.17 A *JK* state machine.

Row	Present state A	Present state B	Input X	J_A	K_A	J_B	K_B	Next state A	Next state B	Output Out
1	0	0	0	1	0	0	1	1	0	0
2	0	0	1	1	0	0	1	1	0	0
3	0	1	0	0	0	0	1	0	0	0
4	0	1	1	0	0	0	1	0	0	0
5	1	0	0	1	1	0	0	0	0	1
6	1	0	1	1	1	1	0	0	1	1
7	1	1	0	0	0	0	0	1	1	0
8	1	1	1	0	0	1	0	1	1	0

FIGURE 5.18 Next state table for the circuit of Figure 5.17.

Since the input data X is updated at the midpoint of the state time, there may be some question as to which value of X to use in predicting the next state. The correct value is that which is in effect just prior to the state-changing clock transition. This value remains constant over the last half of the state time allowing any D inputs affected to be stable prior to state change.

Present state and next state maps can also be useful in state machine analysis and design. Before leaving this example, we will introduce these design aids. A present state map is used to define the state names. For the two flip-flop system of Figure 5.12, there are four possible states. The present state map shows these possible states in Figure 5.15 as functions of the flip-flop outputs A and B. This map defines state a as $A = 0$ and $B = 0$ or 00, state b as 01, state c as 10, and state d as 11. Although these are arbitrarily assigned in this case, we will later consider some factors that determine state assignment. The asterisk in state d identifies this as an output state.

The next state map plots the next state of each flip-flop as a function of present state and input. Figure 5.16 shows the next state maps for both flip-flops. This information can be plotted from the next state map or it can be generated from the equations for the flip-flop inputs. For a D flip-flop, the next state will correspond to the present state value appearing on the D input. We can equate the next state of A to the present value of D_A to generate the first map of Figure 5.16, that is, $D_A = XB$. The second map is derived from $D_B = \bar{A}$.

During time slot t_1, the state is 00 and $X = 1$. These conditions define a specific location on the next state maps. When a state-changing clock transition occurs, the maps tell us that flip-flop A will remain at $A = 0$ while flip-flop B will move to $B = 1$. The state movement can be followed by means of the next state maps through each time slot. The next state map is very valuable in state machine design as will be demonstrated in a later section.

The D flip-flop is easier to analyze than the JK, but the latter generally requires less circuitry to implement a state machine than does the D flip-flop. Since this device is rather popular for state machines, we will analyze the JK circuit of Figure 5.17. The next state table is shown in Figure 5.18. The expressions for

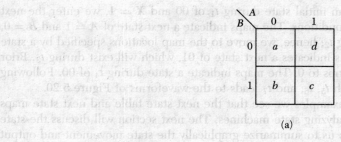

(a)

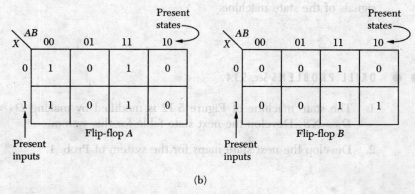

(b)

FIGURE 5.19 (a) Present state map. (b) Next state map.

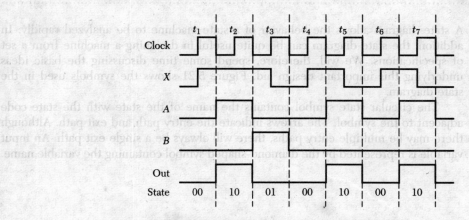

FIGURE 5.20 Waveforms for the circuit of Figure 5.17.

flip-flop inputs are found by examining the IFL of the circuit. This gives $J_A = \bar{B}$, $K_A = A\bar{B}$, $J_B = AX$, and $K_B = \bar{A}$. The equation for the output signal is seen to be Out $= A\bar{B}$. The excitation table for the JK flip-flop of Figure 5.3 is used to determine the next state from a knowledge of the J and K inputs.

The present and next state maps are shown in Figure 5.19. We will apply the next state maps to analyze this state machine, although the next state table could be used also. Assuming an initial state during t_1 of 00 and $X = 1$, we enter the next state maps for these conditions. The maps indicate a next state of $A = 1$ and $B = 0$, which will exist during t_2; hence, we move to the map locations specified by a state of 10 with $X = 1$. This indicates a next state of 01, which will exist during t_3. Prior to the end of t_3, X drops to 0. The maps indicate a state during t_4 of 00. Following this procedure through t_5, t_6, and t_7 leads to the waveforms of Figure 5.20.

From these two examples we see that the next state table and next state maps are quite useful in analyzing state machines. The next section will discuss the state diagram, which allows us to summarize graphically the state movement and output signals of the state machine.

● ● ● **DRILL PROBLEMS Sec. 5.3.4**

1. The state machine of Figure 5.12 is modified by making $D_A = \bar{A}B$ and $D_B = X\bar{B}$. Develop the next state table for this system.

2. Develop the next state maps for the system of Prob. 1.

● **5.3.5 STATE DIAGRAMS**

A state diagram allows the behavior of a state machine to be analyzed rapidly. In addition, the state diagram can be quite useful in designing a machine from a set of specifications. We will, therefore, spend some time discussing the basic ideas underlying this important design aid. Figure 5.21 shows the symbols used in the state diagram.

The circular state symbol contains the name of the state with the state code adjacent to the symbol. The arrows indicate the entry path and exit path. Although there may be multiple entry paths, there will always be a single exit path. An input variable is represented by the diamond-shaped symbol containing the variable name.

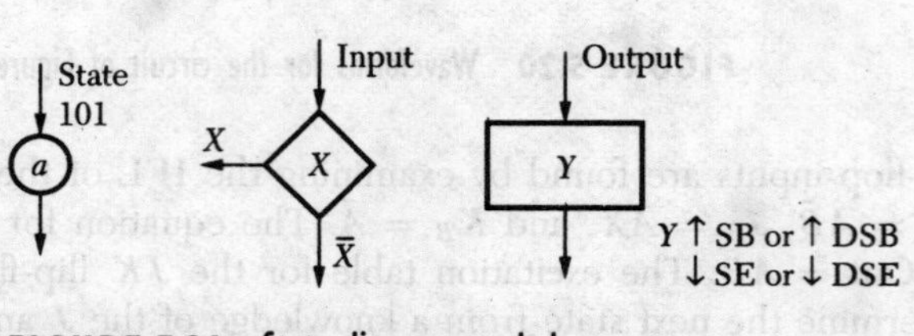

FIGURE 5.21   State diagram symbols.

This symbol will always have a single entry path, but will have two exit paths determined by the value of the variable. When an output is to be generated, a rectangular block is used containing the name of the output variable. The upward arrow indicates that the output is asserted, while the downward arrow indicates deassertion. If the output is asserted at the beginning of the state, the notation $Y \uparrow$ SB is used. If Y is asserted at mid-state, the notation $Y \uparrow$ DSB is used indicating assertion is delayed from the beginning of state time by one-half clock period. The notations $Y \downarrow$ SE and $Y \downarrow$ DSE mean Y is deasserted at the end of state time and on the next clock transition after the state time end, respectively.

To illustrate its use, we will construct the state diagram for the circuit of Figure 5.12. Using the next state table or next state maps and assuming the circuit can be reset to the a state, the state diagram of Figure 5.22 results. From state a (00), the diagram indicates that the circuit will move to state b regardless of the value of X. After one negative clock transition, the state of the system is b (01). If X is zero during the time the circuit is in state b, the negative clock transition would not change the state. Instead, state b would continue until X goes to 1. When this input condition occurs, the following negative clock transition will move the system to state d (11). From state d, the circuit will either move to state a, if $X = 0$, or to state c (10), if $X = 1$. When state d is entered, the output is asserted for one clock period. At the end of state c, the circuit returns to state a, regardless of the input

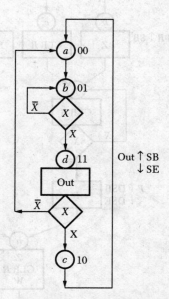

FIGURE 5.22 State diagram for the circuit of Figure 5.12.

condition. This same information is presented by the next state table and maps, but it is represented more concisely by the state diagram.

Returning to the X input waveform of Figure 5.12, we note that the system will move from state a to state b as a result of the first negative clock transition. Since $X = 1$ during t_2, the system then moves to state d at the beginning of t_3. The input X drops to 0 prior to the next negative clock transition, moving the system back to state a during t_4. The system then moves to state b and remains there since $X = 0$ from that point on.

Not only is the state diagram useful in system analysis, it is indispensable in designing a controlling state machine from a set of specifications. A state diagram can consist of many states involving several inputs and outputs. In some cases, an input diamond must be repeated to provide the several possible output paths that may result in a state diagram. Figure 5.23 demonstrates such an example.

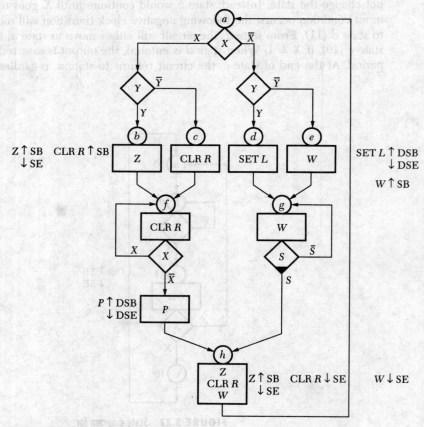

FIGURE 5.23 State diagram.

The four possible combinations of X and Y lead to four different next states following state a; consequently, the duplicate input symbol for Y is required. If possible, all states that can be entered at a given time should be placed on the same horizontal level. States b, c, d, and e are on the same level since one of these states will be entered when state a is exited. We note that input X also controls entry into state h later in the diagram. This situation is not uncommon in state machines. The symbol $W \downarrow$ SE appears indicating that W is deasserted at the end of state h. Output P follows input X and will become asserted when in state f only if $X = 0$.

The symbol representing input S contains a shaded portion at the bottom of the diamond. This indicates that input S is not a synchronous variable. S is therefore an asynchronous input that will change at some time that is not related to a clock transition. All inputs without this shaded region represent synchronous inputs. If state f is reached, the system will remain in this state as long as X is asserted. When X is deasserted, the system can move to state h.

Although the synchronous inputs change as the clock makes a transition, it is important to recognize that state changes should take place on the alternate clock transition. If this is true, the input conditions have one-half of a clock period in which to settle prior to a change in state.

Generally, the state diagram will be obtainable from the specifications of the design problem. The circuit is then developed from the state diagram to satisfy the specifications using the techniques discussed in the following chapters.

The type of state diagram used in this text is also called an algorithmic state machine (ASM) diagram [2]. It resembles the programming flowchart that is used in algorithm development. The use of these symbols in a state diagram implies similar operations to those of the programming flow chart. The diamond represents a decision, and the rectangle represents operations to be performed. In the state machine, however, decisions to be made are based on voltage levels or states of the system. Operations to be performed are represented by the generation of output voltage levels, which will then drive some number of following circuits. In a computer, decisions are based on the comparison of two numerical values, and operations to be performed are carried out by sequences of instructions. Thus, while the symbols and operations of state machines are similar to those of programming flowcharts, the mechanics of implementation are quite different.

Many textbooks use different symbols than those of Figure 5.21 for the state diagram. This textbook will equate the ASM diagram to the state diagram and use this tool in all succeeding synchronous state machine analysis and design. Chapter 10, which deals with asynchronous state machines, will introduce a second type of state machine diagram.

● ● ● **DRILL PROBLEM Sec. 5.3.5**

Draw the state diagram for the system described in drill Prob. 5.3.4-1.

SUMMARY

• • • • • • • • • • • • • • •

1. The state machine is a sequential controller that produces control signals that depend on inputs and input history.
2. The flip-flop excitation table is more important than the function table in the design of flip-flop circuits.
3. Present and next state tables and maps allow a state machine to be analyzed.
4. The algorithmic state machine chart (ASM) or state diagram is useful in understanding the overall behavior of a state machine.

CHAPTER 5 PROBLEMS

• • • • • • • • • • • • • •

● **Sec. 5.2**

*5.1 Completely define the meaning of a synchronous state machine.

5.2 What is a mainly synchronous system?

● **Sec. 5.3.1**

5.3 Develop the excitation table for the circuit shown in Figure P5.3 assuming positive logic for inputs A and B.

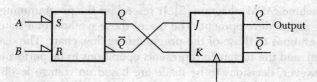

FIGURE P5.3

5.4 Repeat Prob. 5.3 if the connections between flip-flops are reversed.

*5.5 An output sequence of

	t_1	t_2	t_3	t_4	t_5
$Q1$ —	0	1	1	0
$Q2$ —	1	0	1	1

is to be produced with two D flip-flops. What is the sequence of values that $D1$ and $D2$ must take on?

5.6 Repeat Prob. 5.5 if JK flip-flops are used.

5.7 Repeat Prob. 5.5 if T flip-flops are used.

5.8 The characteristic table for a fictitious flip-flop is shown. Construct the excitation table.

Inputs		Output
0	0	Q_n
0	1	$\bar{Q}_n$
1	0	0
1	1	1

● Sec. 5.3.2

°5.9 If the input signals to a gating circuit are as shown, design the circuit to drive two D flip-flops to create the sequence of outputs, Q_1 and Q_2.

	t_1	t_2	t_3	t_4	t_5
$I1$	0	0	1	1	—
$I2$	0	1	0	1	—
$Q1$	—	0	1	1	0
$Q2$	—	1	0	1	1

5.10 Repeat Prob. 5.9 for JK flip-flops.

5.11 Repeat Prob. 5.9 for T flip-flops.

● Sec. 5.3.4

°5.12 Plot $Q1$ and $Q2$ for the sequence of inputs shown in Figure P5.12. Tell why you would or would not consider this circuit a state machine.

5.13 Design the gates of Figure P5.13 using minimum circuitry to cause $Q1$ and $Q2$ to behave as shown, given the inputs $I1$ and $I2$ as shown.

5.14 Repeat Prob. 5.13 if D flip-flops are used.

5.15 Sketch the waveforms X and Y necessary to cause the $Q1$, $Q2$, and $Q3$ waveforms of Figure P5.15.

5.16 Analyze the circuit of Figure 5.12 if $\bar{X}$ is applied to the IFL rather than X.

5.17 Analyze the circuit of Figure 5.17 if $\bar{X}$ is applied to the IFL rather than X.

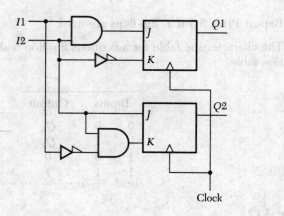

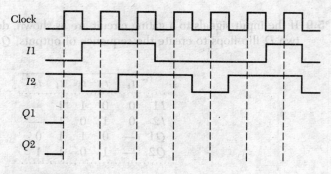

FIGURE P5.12

°5.18 Plot the next state maps of Figure 5.19(b) as variable entered maps with X as the MEV.

5.19 Draw a next state table for the circuit of Figure P5.19. Show present state, gating conditions, and next state. Indicate the sequence of states taken on by the circuit for the inputs in the timing chart.

°5.20 In Figure P5.20, sketch the values taken on by A, B, and R assuming that A and B are initially zero.

● **Sec. 5.3.5**

5.21 Draw the state diagram for the circuit of Figure 5.12.

5.22 Draw the state diagram for the circuit of Figure 5.17.

5.23 Draw the state diagram for a system that checks an incoming data line X that changes on the positive clock. This system is reset to start the process of

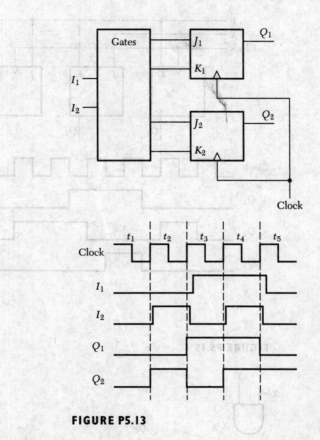

FIGURE P5.13

checking of 3-bit words. If exactly two 1 bits are detected in a word, an output Y is to be generated and the system resumes checking the next word. You may assume that 1 bit time separates each word.

5.24 Repeat Prob. 5.23 if there is no separation between words, that is, the first bit of a word immediately follows the last bit of the previous word.

5.25 Draw the state diagram for the circuit of Figure P5.19.

5.26 Figure P5.26 shows a transmitting shift register along with a state diagram for the control circuit that generates signals LDRG, TRAN, and CNTRST. Plot the waveforms for the signals LDRG, TRAN, CNTRST, and LINE. Assume the register initially contains 00000 and the state machine controller changes states on the positive transitions of the clock.

5.27 Show on the timing diagram in Figure P5.27 what states the system reaches for the X waveform indicated. Also sketch outputs M and K.

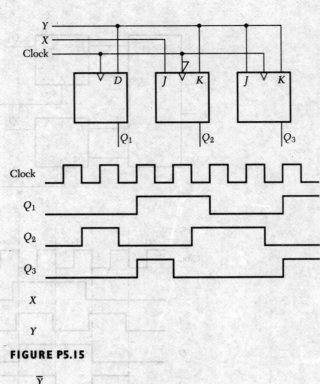

FIGURE P5.15

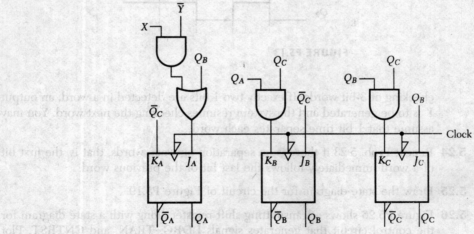

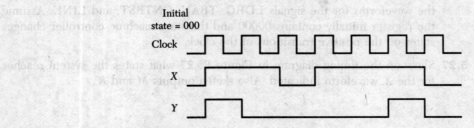

FIGURE P5.19

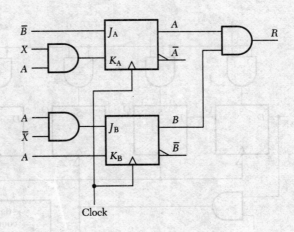

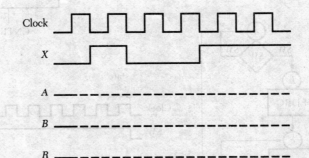

FIGURE P5.20

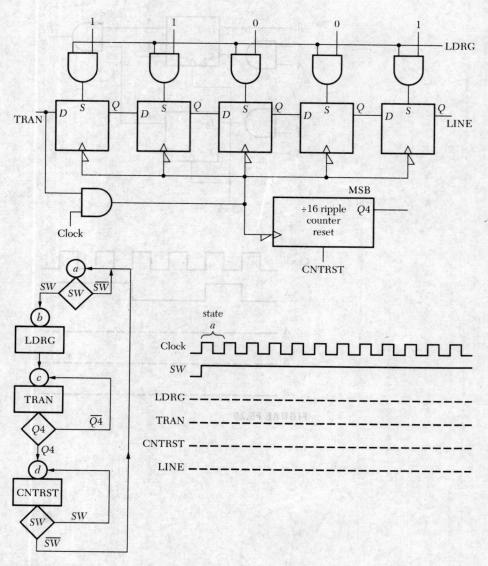

FIGURE P5.26

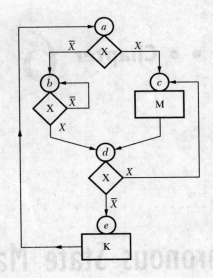

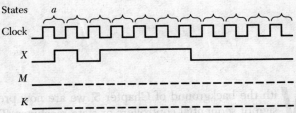

FIGURE P5.27

REFERENCES AND SUGGESTED READING

1. T. L. Booth, *Sequential Machines and Automata Theory*. New York: John Wiley and Sons, 1967.

2. C. R. Clare, *Designing Logic Systems Using State Machines*. New York: McGraw-Hill, 1973.

3. W. I. Fletcher, *An Engineering Approach to Digital Design*. Englewood Cliffs, NJ: Prentice-Hall, 1980.

4. M. M. Mano, *Digital Design*, 2nd ed. Englewood Cliffs, NJ: Prentice-Hall, 1991.

5. C. H. Roth, Jr., *Fundamentals of Logic Design*, 4th ed. St. Paul, MN: West, 1993.

Chapter 6

Synchronous State Machine Design

With the background of Chapter 5, we are now prepared to consider the design of sequential controllers or state machines. The basic principles of synchronous state machine design are covered in this chapter. Input-forming logic design, output-forming logic design, and criteria for state assignment are considered. Much of the succeeding material in later chapters is based on concepts introduced in this chapter.

Each concept will be demonstrated by example to clarify its application. The first system considered is the sequential or synchronous counter. This system has no external inputs, other than direct sets or resets, and is therefore simple enough to serve as a good introductory sequential circuit.

6.1

SEQUENTIAL COUNTERS

SECTION OVERVIEW The simplest form of sequential counter is one with states that progress from zero to some maximum number with increments of one. In general, a sequential counter may have a prescribed sequence of states that does not proceed in increments of one, for example, the sequence may be 1, 3, 0, 2, 1, 3, 0, 2, ... In the sequential counter, the progression of states is repeated on a periodic basis.

The design of sequential counters using both the D and JK flip-flop is discussed.

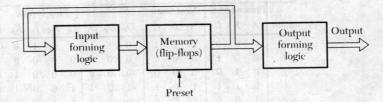

FIGURE 6.1 Block diagram of the sequential counter.

The sequential counter is used to generate some fixed sequence of states in a cyclic manner. For example, the sequence of states desired might be 000, 011, 101, 111, 100, 000, 011, 101, 111, 100, 000, ... Sequential counters are state machines with no inputs other than the preset inputs that initialize the system. The general block diagram for this system is shown in Figure 6.1. This particular type of state machine has a next state that is determined by the present state only. Quite often, the output will be taken directly from the flip-flops, eliminating the need for the OFL. The state diagram for the sequence previously listed is shown in Figure 6.2. The next state table and appropriate maps are shown in Figure 6.3. Gating information for both D flip-flops and JK flip-flops is also given.

The excitation tables for the D and JK flip-flops are used to generate this table. When we are in state a, we see that flip-flop B must change from 0 to 1 on the clock transition. This requires that $J_B = 1$ and $K_B = \Theta$. The next three state maps for the flip-flops have been combined into a single map. The states marked with an asterisk were not defined from the specs in terms of the next state to follow. We would be tempted to call these "don't care" states since they would never be entered if the counter performed in the cyclic manner indicated by the state diagram. However, the counter might be functioning in the prescribed manner when an extraneous noise pulse occurs to send the counter to one of these undefined states. It would be possible for the counter to hang up in some repetitive sequence other than the desired one if these states are not designed to return the system to

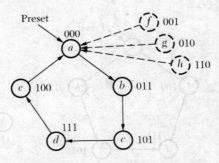

FIGURE 6.2 State diagram for the sequential counter.

State	Present State A	B	C	Next State A	B	C	Flip-Flop Inputs D_A	D_B	D_C	J_A	K_A	J_B	K_B	J_C	K_C
a	0	0	0	0	1	1	0	1	1	0	θ	1	θ	1	θ
f	0	0	1°	0	0	0	0	0	0	0	θ	0	θ	θ	1
g	0	1	0°	0	0	0	0	0	0	0	θ	θ	1	0	θ
b	0	1	1	1	0	1	1	0	1	1	θ	θ	1	θ	0
e	1	0	0	0	0	0	0	0	0	θ	1	0	θ	0	θ
c	1	0	1	1	1	1	1	1	1	θ	0	1	θ	θ	0
h	1	1	0°	0	0	0	0	0	0	θ	1	θ	1	0	θ
d	1	1	1	1	0	0	1	0	0	θ	0	θ	1	θ	1

° Unused States

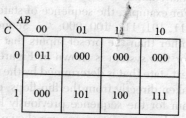

Next state ABC

FIGURE 6.3　Next state table and maps for the sequence counter of Figure 6.2.

the prescribed sequence. For example, suppose that the next state after 001 is 010. Further, assume that the next state after 010 is 110 and the next state after 110 is 001. If this were true, the state diagram of Figure 6.4 would apply. A noise pulse could send the counter to states f, g, or h, and the system would never return to the desired sequence unless the preset signal is applied or another random noise pulse accidentally returns the state to a, b, c, d, or e.

In order to avoid this problem, all unused states can be designed to have a next state of 000 or some other state contained in the desired sequence. If one of these states is inadvertently entered, the system will recover to the desired sequence

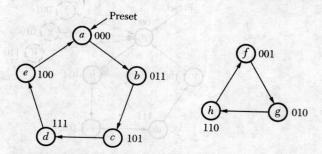

FIGURE 6.4　State diagram for a poorly designed sequence counter.

when the next clock pulse is applied. This type of system is also referred to as a "self-starting" circuit. When the dc power is first turned on, the circuit will come up in an arbitrary state. Regardless of which state comes up, the desired sequence will shortly be entered.

6.1.1 *D* FLIP-FLOP DESIGN

For a *D* flip-flop design, we must produce the logic functions to drive D_A, D_B, and D_C. We begin the design process by plotting the maps of Figure 6.5. The inputs to the IFL are the flip-flop outputs *A*, *B*, and *C*. These variables are used to plot the next state maps of Figure 6.5, which are also the maps for D_A, D_B, and D_C. This counter can be constructed as shown in Figure 6.6. The preset input can be taken low to start the sequence at state 000. Deassertion of this input will allow the counter to cycle through the prescribed sequence. The number of gates used for the IFL can be reduced further by using EXOR type circuits. We will leave these considerations to the student.

6.1.2 *JK* FLIP-FLOP DESIGN

The *JK* realization is more difficult to find, requiring six maps as shown in Figure 6.7 rather than three. These maps are based on the next state or *D* maps of Figure 6.5. The *JK* excitation table aids in this construction. Although more maps are required

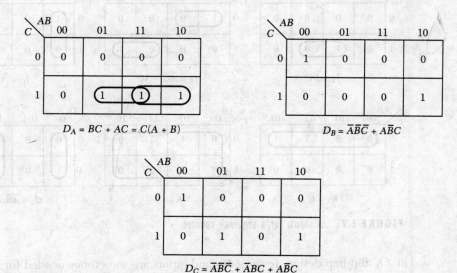

$$D_A = BC + AC = C(A + B)$$

$$D_B = \overline{A}\,\overline{B}\,\overline{C} + A\overline{B}C$$

$$D_C = \overline{A}\,\overline{B}\,\overline{C} + \overline{A}BC + A\overline{B}C$$

FIGURE 6.5 *D* input maps for a sequence counter.

when the next clock pulse is applied. This feedback loop is also referred to as a "tristabilizing" circuit. When the circuit is turned on, the circuit will come on in an arbitrary state. Regardless of which state comes up, the desired sequence will shortly be entered.

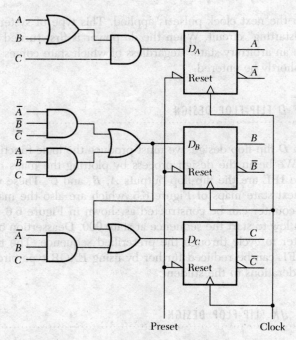

FIGURE 6.6 Sequence counter with D flip-flops.

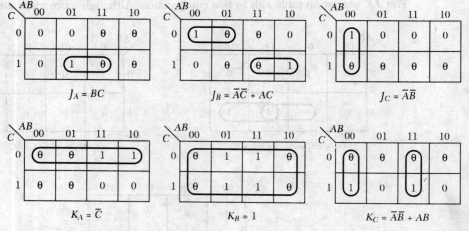

FIGURE 6.7 JK inputs for a sequence counter.

in JK flip-flop design, fewer gates and inputs are sometimes needed for the input-forming logic as demonstrated by the counter of Figure 6.8. Again, if EXOR or NEXOR circuits are used, the IFL can be reduced still further.

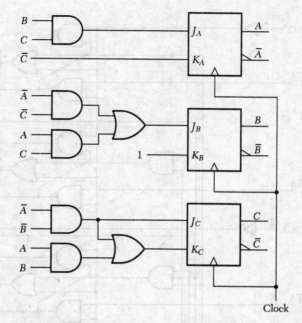

FIGURE 6.8 Sequence counter with JK flip-flops.

Constructing JK Maps from Next State Maps

There is a shortcut method that can be used to construct the JK maps from the next state or D maps. Returning to the excitation tables of Figure 6.3, we note that for a present state of $Q_n = 0$, K is always equal to Θ, while $J = D$ for both values of next state. For a present state of $Q_n = 1$, J always equals Θ, while $K = \bar{D}$ for both values of next state. From these relations, we can develop an algorithm to convert the next state or D map to J and K maps. This algorithm follows.

1. Place a Θ in all locations of the map for K corresponding to a present state of 0. Place $K = \bar{D}$ values in the remaining locations.
2. Place a Θ in all locations of the map for J corresponding to a present state of 1. Place $J = D$ values in the remaining locations.

Since the next state maps will always be known prior to the development of the flip-flop maps, this method can decrease the time required to plot the J and K maps.

In Chapter 4, the ripple counter was discussed. Although this counter is useful in lower frequency counting applications, the ripple effect limits its use in higher frequency work. The sequential counter solves this problem, allowing use at higher operating speeds. In this situation, the states are chosen to be consecutive, starting at the zero state. The resulting counter is called a synchronous counter, and the standard 74160 decade counter is shown in Figure 6.9. This counter can be preset

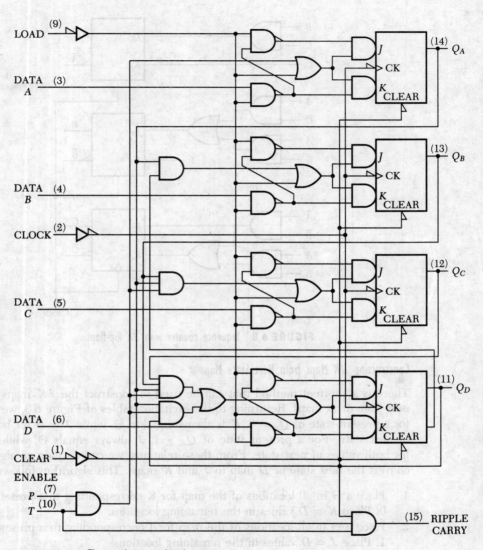

FIGURE 6.9 The synchronous decade counter.

to any desired initial state and includes provision for cascading chips to create larger counters.

The synchronous counter is a sequential system, but it is a special case that involves no external inputs other than the preset input. Only the state flip-flop outputs are used to drive the IFL. We will now consider the design of systems having external inputs that influence the next states of the circuit.

● ● ●   **DRILL PROBLEMS** Sec. 6.1

1. Design a synchronous counter using D flip-flops to count in the sequence 01, 11, 10, 01, 11, 10, ...

2. Repeat Prob. 1 using JK flip-flops.

6.2

. . . .

STATE CHANGES REFERENCED TO CLOCK

SECTION OVERVIEW The inputs to the state flip-flops must be stable just before the state-changing clock transition occurs to avoid incorrect information being set into the flip-flops. This section considers methods of ensuring stable inputs at this critical time.

The flip-flops of the state machine change to a new value as a result of a clock transition. At this transition, the information represented by the flip-flop inputs is set into the flip-flops. There is a requirement called data-to-clock setup time that tells us that the data applied to the D or JK inputs must be stable prior to the occurrence of the state-changing clock transition. A typical minimum setup time for a conventional TTL flip-flop is 10 ns. For low power Schottky, this time is 20 ns, while a value of 3 ns might be typical of Schottky TTL. Figure 6.10 demonstrates the problem that can arise if we do not observe this specification.

If X changes on the positive clock transition, the D input will change slightly before the inversion of this transition $(\overline{CK})$ reaches the C input. The signal X may reach D soon enough to satisfy setup time requirements for the flip-flop, depending on the propagation delay time of the inverter. If so, the Q output will switch to 1

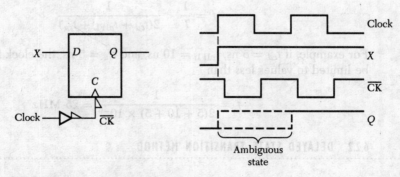

FIGURE 6.10 D flip-flop with synchronous input.

at this time. If the propagation delay of the inverter is not long enough to allow an appropriate setup time, the flip-flop will not change state until the next positive clock transition.

The ambiguous situation depicted in Figure 6.10 must be avoided in order to achieve reliable state machine operation. The input signals to the state machine are applied to the IFL. The output signals from the IFL drive the flip-flop inputs. When an input signal changes, the flip-flop inputs respond only after a delay time introduced by the IFL. For synchronous input signals, driven by a clock transition, the state change must take place long enough after the clock transition to allow the flip-flop inputs to settle and the data-to-clock setup time specification to be satisfied. There are two methods used to delay the state changes sufficiently.

6.2.1   ALTERNATE STATE TRANSITION METHOD

The first is called the alternate state transition method (AST) and the second is called the delayed state transition method (DST). In the AST scheme, the state changes are caused to occur on the alternate clock transition to input data changes. If data changes are driven on the positive clock transition, state changes must take place on the negative clock transition. Data changes taking place on the negative clock transition dictate positive clock transition state changes. For a 50% duty cycle clock, the AST method allows one-half of the clock period between input change and state change. One-half of the clock period must then exceed the delay time between clock transition and input change, t_{cd}, plus the maximum propagation delay of the IFL, t_{pIFL}, plus the data-to-clock setup time, t_{su}, of the state flip-flops. In equation form, this gives

$$\frac{T}{2} > t_{cd} + t_{pIFL} + t_{su} \tag{6.1}$$

The maximum usable clock frequency can then be found to be

$$f = \frac{1}{T} < \frac{1}{2(t_{cd} + t_{pIFL} + t_{su})} \tag{6.2}$$

For example, if $t_{cd} = 5$ ns, $t_{pIFL} = 10$ ns, and $t_{su} = 5$ ns, the clock frequency should be limited to values less than

$$f = \frac{1}{2(5 + 10 + 5) \times 10^{-9}} = 25 \text{ MHz}$$

6.2.2   DELAYED STATE TRANSITION METHOD

In the DST system, the same clock transition drives both the data changes and the state changes, but the clock signal to the state flip-flops is delayed. The number of delaying inverters used between the clock and the flip-flops is determined by two considerations. The total clock delay, t_t, is related to the other delays by

$$t_t > t_{cd} + t_{pIFL} + t_{su} \qquad (6.3)$$

and the correct number of inversions must be present to cause state changes on the same clock transition that drives the data changes.

Unless otherwise stated, we will assume that the AST method is used for all synchronous state machines. This method is more popular in state machine design except for situations involving very critical timing problems. Figure 6.11 demonstrates the AST and DST methods of creating transitions for state machines. For the AST method, Equation (6.1) or (6.2) must be satisfied, while Equation (6.3) must be satisfied by the DST method.

Before we leave the topic of state changes it is important to realize that the state machine is more reliable than many other digital systems as a result of the fact that output changes occur only during clock transitions. Input data changes have a half-period to settle, and thus any hazards produced by inputs are ignored by the state flip-flops. When a state change takes place, the IFL may generate hazards since hazard covers are not used in this section of the system. These hazards will appear after the clock transition that drives the state change take place and will not affect the operation of the state machine. This discussion assumes that the flip-flops used are edge-triggered devices rather than "ones-catching" master-slave circuits.

Even in the presence of noise the state machine operates more reliably than other systems. Not only must a noise pulse be large enough to cause problems of level interpretation, but also it must occur slightly before the state changing clock

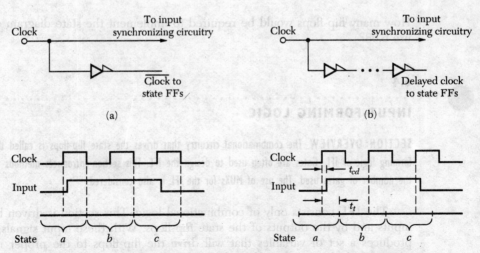

FIGURE 6.11 Two methods of clock generation: (a) the AST method, (b) the DST method.

transition to cause an erroneous state to be reached. The list of advantages of the state machine over other types of digital systems must include that of highly reliable operation.

6.3

NUMBER OF STATE FLIP-FLOPS

SECTION OVERVIEW The number of state flip-flops required in a state machine is related to the number of unique state codes needed.

Another matter that must be considered prior to IFL design is that of the number of flip-flops required for a state machine. This is easily found from the state diagram, which shows the number of states needed for the system. Given n flip-flops, the number of unique codes or states generated is 2^n.

In order to choose n, the number of states of the system is first determined and the smallest value of n that satisfies the relation

$$\text{Number of states} \leq 2^n \qquad (6.4)$$

is selected. For example, if 12 states are needed, four flip-flops are required since $2^4 = 16$. If we used three flip-flops, we see that only $2^3 = 8$ is the maximum number of states possible.

DRILL PROBLEM Sec. 6.3

How many flip-flops would be required to implement the state diagram of Figure 6.68?

6.4

INPUT-FORMING LOGIC

SECTION OVERVIEW The combinational circuitry that drives the state flip-flops is called the input forming logic or IFL. Gates are often used to design the IFL. This section introduces methods to minimize the number of gates used. The use of MUXs for the IFL is also considered.

The IFL consists only of combinational logic. This section is driven by system inputs and by the outputs of the state flip-flops. With these input signals, the IFL produces a set of variables that will drive the flip-flops to the proper next state when the clock transition takes place. If D flip-flops are used in the state machine, one input per flip-flop is required. JK devices require two inputs per flip-flop. Regardless of the type of flip-flop used, the IFL must generate combinational logic functions and can therefore be constructed from gates, MUXs, or decoders.

6.4.1   IFL USING GATES

The traditional method of IFL realization uses gates. This method is still significant in applications using standard IC chips since component cost can be minimized with gates. When IFL is constructed from gates, this circuitry can be minimized by observing the following two principles:

Principle 1. *States having the same next state for a given input condition should have logically adjacent state assignments.*

Principle 2. *States that are next states of a single state should have logically adjacent assignments.*

We will emphasize the point that hazard covers are unnecessary for the IFL since state changes take place after the IFL outputs have stabilized. Two examples will demonstrate the use of these principles before we consider why they lead to a minimal realization of IFL.

A problem that occasionally occurs in digital communications is that of identifying certain bit patterns within a serial data stream. We will simplify this problem to the point of impracticality to introduce the basic ideas of IFL design.

EXAMPLE 6.1

A serial data line X is allowed to change on the falling edge of the clock signal (NRZ code). A state machine is to be designed to detect every next value of X. If $X = 1$, output Y is to be asserted. If $X = 0$, output Z is to be asserted. The output value should be asserted for one bit time, ignoring the value of X during this time, and then the next value of X should be checked.

SOLUTION

The state diagram reflecting the specified behavior is shown in Figure 6.12. Since data changes take place on the negative clock transition, state changes must occur on the positive transition. Depending on the value of X at the first positive clock transition, the system moves to a specific state. If $X = 1$, the system moves to state b; if $X = 0$, it moves to state c. The output in either of these states exists for one

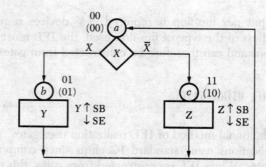

FIGURE 6.12 State diagram for Example 6.1.

clock time after which the system returns to state *a* to check the alternate data bit. A timing diagram is useful to indicate the relationship between data input, clock, state times, and outputs. This chart is shown in Figure 6.13.

Three states are required to satisfy this problem, and thus two state flip-flops are used. We will label these flip-flops *A* and *B*, where *A* represents the MSB of the state code. Prior to designing the IFL, the state assignment must be made. This selection is made in accordance with the two principles stated at the beginning of the section. States *b* and *c* both go to state *a* regardless of the value of *X*. Therefore, these states should be adjacent according to Principle 1. Principle 2 also dictates the adjacency of these two states since both follow state *a*.

In order to demonstrate the results of a correct state assignment, we will assign the states two different ways as shown in Figure 6.14. The first map selects *b* and *c* as adjacent states, while the second map chooses these states to be nonadjacent. The next step of the procedure is to develop the next state maps (or tables) for flip-flops *A* and *B*. This is simplified by adding the assigned codes to the state diagram as shown in Figure 6.12. The improperly chosen states are included in parentheses. Figure 6.15 includes the next state maps for both cases of state assignment.

The next state map for properly chosen states is created by considering the conditions prevailing at each state. At state *a* if $X = 0$, the next state code will be 11; if $X = 1$, this code will be 01. Flip-flop *B* will change to 1 regardless of the

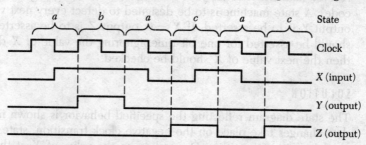

FIGURE 6.13 Timing diagram for Example 6.1.

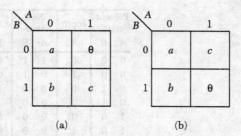

FIGURE 6.14 Present state maps for Example 6.1: (a) for correctly chosen states, (b) for incorrectly chosen states.

value of X. This fact is reflected by placing a 1 in the map for flip-flop B at the location corresponding to state a or 00. Flip-flop A will change to 0 if $X = 1$ and 1 if $X = 0$. This information is noted by placing an $\bar{X}$ in the map for flip-flop A at location 00. The other map locations are filled in a similar manner.

In location 00 of the incorrectly chosen state maps, we see an $\bar{X}$ for flip-flop A and an X for flip-flop B. These values result because flip-flop A must be 0 if $X = 1$ and must change to 1 if $X = 0$. Flip-flop B must change to 1 if $X = 1$ and remain at 0 for $X = 0$. The unused state leads to "don't care" conditions in the maps. Rather than use next state maps, we could use next state tables instead. The next state table for correctly chosen states is shown in Figure 6.16.

After developing the next state maps or table, we must decide which type of flip-flop to select for the state machine. The use of D flip-flops leads to a simpler design procedure since only one input per flip-flop is needed. JK flip-flops require a longer procedure but may result in less complex gating circuitry for the IFL. We will demonstrate the use of both types of flip-flops designing the system first with JKs and then with Ds.

We must consider the JK excitation table of Figure 5.3 along with the next state maps of Figure 6.15 to produce the gate input maps for the system. Figure 6.17

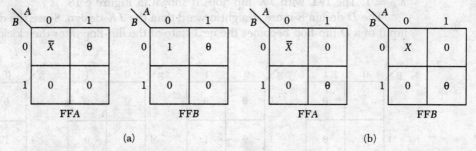

FIGURE 6.15 Next state maps for Example 6.1: (a) for correctly chosen states, (b) for incorrectly chosen states.

Present state		Input	Next state		Outputs	
A	B	X	A	B	Y	Z
0	0	0	1	1	0	0
0	0	1	0	1	0	0
0	1	0	0	0	1	0
0	1	1	0	0	1	0
1	0	0	θ	θ	θ	θ
1	0	1	θ	θ	θ	θ
1	1	0	0	0	0	1
1	1	1	0	0	0	1

FIGURE 6.16 Next state table for Example 6.1 with correctly chosen states.

shows these maps. From the next state map for flip-flop A, we see that when the system is in state a, flip-flop A changes from 0 to 1 if $\bar{X} = 1$. Relating these conditions to the JK excitation conditions leads to $J = 1$, and $K = \Theta$ if $\bar{X} = 1$ and $J = 0$, $K = \Theta$ if $\bar{X} = 0$. This can be expressed as $J_A = \bar{X}$ and $K_A = \Theta$ in the map locations corresponding to state a. This same information could be generated more easily by using the algorithm of Section 6.1.2 for converting from next state map to J and K maps.

When in state a, the next state of flip-flop B is always 1. The excitation table indicates that a transition from 0 to 1 requires $J = 1$ and $K = \Theta$, which is entered in the maps for J_B and K_B. Moving to the proper next state from state b requires that flip-flop A remain at 0 and flip-flop B change from 1 to 0. The excitation table tells us that flip-flop A must have $J_A = 0$ and $K_A = \Theta$, while flip-flop B must have $J_B = \Theta$ and $K_B = 1$. To move from state c to state a requires that both flip-flops change from 1 to 0 resulting in $J_A = J_B = \Theta$ and $K_A = K_B = 1$. Of course, the unused state will contain "don't care" conditions for all flip-flop inputs. The maps for the flip-flop inputs can now be reduced to give $J_A = \bar{X}\bar{B}$, $K_A = 1$, $J_B = 1$, and $K_B = 1$. The IFL with JK flip-flops is shown in Figure 6.18.

The D design is more straightforward than the JK design. Since the data on the input of a D flip-flop becomes the next state of the flip-flop after the clock transition

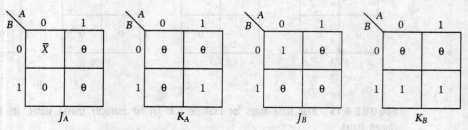

FIGURE 6.17 Gating maps for JK flip-flops.

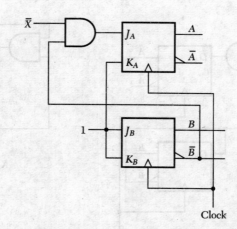

FIGURE 6.18 IFL using *JK* flip-flops for Example 6.1.

occurs, the next state maps of the state machine flip-flops are also the maps for the D inputs. Figure 6.19 indicates these maps for flip-flops A and B. Maps for both correctly chosen and incorrectly chosen states are included. The D input maps for the correctly chosen states lead to the equations $D_A = \bar{X}\bar{B}$ and $D_B = \bar{B}$. For the incorrectly chosen states we get $D_A = \bar{X}\bar{A}\bar{B}$ and $D_B = X\bar{A}\bar{B}$. The resulting IFL for the two cases is shown in Figure 6.20. We note that the system of Figure 6.20(a), which is based on correctly chosen states, uses only a single gate with two inputs, whereas the other system requires three gates and six inputs for the IFL.

Some designers prefer to work with a next state table instead of or in addition to next state maps. The flip-flop input information can be presented with a next state table as indicated in Figure 6.21. Again, we will emphasize that the D input information simply duplicates the next state information. We will defer the design of the OFL to a later section. ●

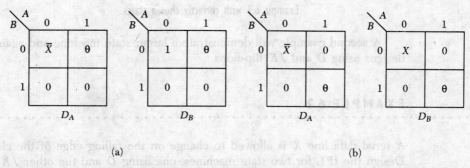

(a)   (b)

FIGURE 6.19 *D* input maps for Example 6.1: (a) for correctly chosen states, (b) for incorrectly chosen states.

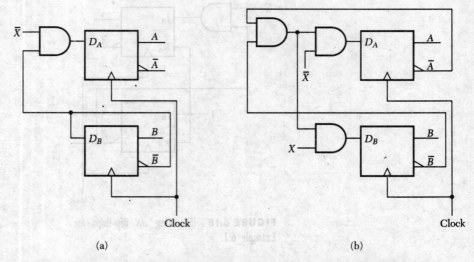

FIGURE 6.20 Two designs for Example 6.1: (a) based on correctly chosen states, (b) based on incorrectly chosen states.

Present state		Input	Next state		Flip-Flop Inputs					
A	B	X	A	B	D_A	D_B	J_A K_A		J_B K_B	
0	0	0	1	1	1	1	1	θ	1	θ
0	0	1	0	1	0	1	0	θ	1	θ
0	1	0	0	0	0	0	0	θ	θ	1
0	1	1	0	0	0	0	0	θ	θ	1
1	0	0	θ	θ	θ	θ	θ	θ	θ	θ
1	0	1	θ	θ	θ	θ	θ	θ	θ	θ
1	1	0	0	0	0	0	θ	1	θ	1
1	1	1	0	0	0	0	θ	1	θ	1

FIGURE 6.21 Next state and flip-flop input table for Example 6.1 with correctly chosen states.

A second example will demonstrate a larger state machine and again compare designs using D and JK flip-flops.

EXAMPLE 6.2

. . . .

A serial data line X is allowed to change on the falling edge of the clock signal. Design the IFL for two state machines, one using D and the other JK flip-flops, to detect the sequence 10 or 01 on the data line. If 10 occurs, output Y should be asserted for one state time before the next check is started. If 01 occurs, output Z

should be asserted for one state time. If 00 or 11 appear as the two sequential bits on X, no output should be generated and the next check should begin after a delay of one state time.

SOLUTION

The state diagram for this system is constructed as shown in Figure 6.22. With six states, three flip-flops are required. This yields eight possible state codes. In order to make a proper state assignment, we note that states d, e, and f all proceed to state a for any value of X. Thus, these states should be logically adjacent to satisfy Principle 1. States b and c both have e as a next state, but for different input conditions. States b and c are next states of a and should be logically adjacent as a result of Principle 2. States d and e are next states of b, and f and e are next states of c. These two pairs should also be logically adjacent. Figure 6.23 shows the timing diagram. An appropriate state assignment is given in Figure 6.24.

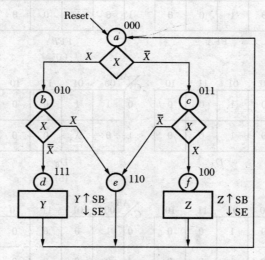

FIGURE 6.22 State diagram for Example 6.2.

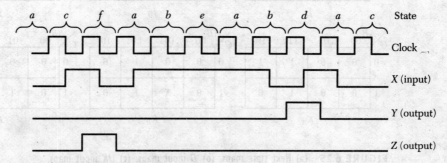

FIGURE 6.23 Timing chart for Example 6.2.

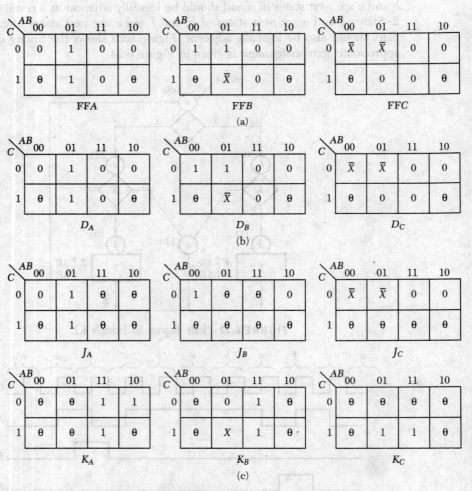

FIGURE 6.24 Present state map for Example 6.2.

FIGURE 6.25 (a) Next state maps. (b) *D* input maps. (c) *JK* input maps.

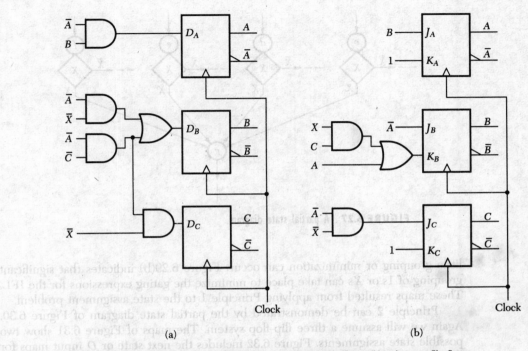

Clock

(a)   (b)

FIGURE 6.26   IFL design for Example 6.2: (a) using D flip-flops, (b) using JK flip-flops.

The next state, D and JK input maps are then developed as shown in Figure 6.25. The equations for the D design are $D_A = \bar{A}B$, $D_B = \bar{X}\bar{A} + \bar{A}\bar{C}$, and $D_C = \bar{X}\bar{A}\bar{C}$. The equations for the JK design are $J_A = B$, $K_A = 1$, $J_B = \bar{A}$, $K_B = A + XC$, $J_C = \bar{X}\bar{A}$, and $K_C = 1$. Both implementations are shown in Figure 6.26. The D implementation uses five gates with 10 inputs while the JK implementation uses three gates with six inputs.   •

Before we leave the subject of IFL design using gates we will consider why Principles 1 and 2 lead to a minimized system. Suppose we have a portion of a state diagram that appears as shown in Figure 6.27. Let us further assume that this system has three D-type state flip-flops, A, B, and C. Figure 6.28 shows two of the many possible assignments for states c through g.

The next state maps for these assignments, which are the same as the D input maps, are included in Figure 6.29. From each of the states c, d, e, and f the system moves to state g if $X = 1$. Thus, all flip-flops change to the 1 state if $X = 1$ from these four states. Since we do not know where the system moves from these states if $X = 0$, we cannot determine if an X or 1 belongs in the map locations corresponding to these states. We do see from Figure 6.29(a) that if these states are not logically adjacent,

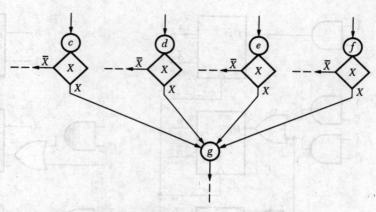

FIGURE 6.27 A partial state diagram.

little grouping or minimization can occur. Figure 6.29(b) indicates that significant grouping of 1s or Xs can take place to minimize the gating expressions for the IFL. These maps resulted from applying Principle 1 to the state assignment problem.

Principle 2 can be demonstrated by the partial state diagram of Figure 6.30. Again we will assume a three flip-flop system. The maps of Figure 6.31 show two possible state assignments. Figure 6.32 includes the next state or D input maps for the two state assignments.

It may appear from a comparison of the maps in Figure 6.31 that little minimization has been accomplished by applying Principle 2. We must recognize that to express a term involving a map-entered variable requires one more variable than is required to express those terms involving a 1. To implement such an expression requires one more input to each gate. The maps of Figure 6.32(b) then represent a slightly simpler logic system than the maps of Figure 6.32(a) represent, since one less MEV is included.

Intuitively, we might expect that observance of Principle 1 does more to minimize the IFL than does observance of Principle 2. This has been shown to be

C	AB 00	01	11	10
0	c	?	e	?
1	?	d	g	f

C	AB 00	01	11	10
0	c	d	?	?
1	e	f	g	?

FIGURE 6.28 State assignments: (a) nonminimizing assignment, (b) minimizing assignment.

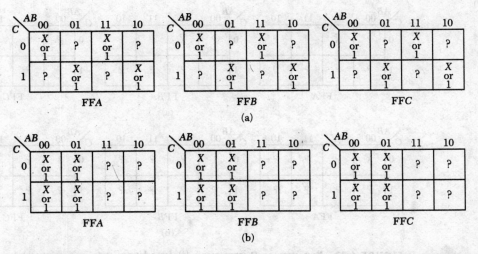

FIGURE 6.29 Next state or *D* input maps: (a) for nonminimizing assignment, (b) for minimizing assignment.

true [2] and dictates how conflicts between the two principles should be resolved. Not all desired adjacencies can be obtained in a given state assignment, so that Principle is given higher priority in system minimization. Once this principle has been satisfied, Principle 2 is applied as far as possible without disturbing the assignments resulting from Principle 1.

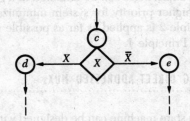

FIGURE 6.30 Partial state diagram to demonstrate the validity of Principle 2.

The IFL of the state machine can be designed with MUXs instead of gates. Although component count is generally higher with MUXs, the flip-flop chip count can be minimized and design time shortened. Thus, the MUX implementation can be placed on a par as an optimal (or "most" optimal) form for overall flexibility in the overall design. We will see in the following sections that state assignment may affect the output forming logic design and state reusability. When IFL is designed with MUXs, state assignment can then be used for these purposes rather than for IFL minimization. Once state assignment is done in an IFL design with MUXs, reusing the state may either be very easy or very difficult, depending on the type of state.

The implementation of a state machine MUX system is shown in Figure 6.31. When the IFL is driven with MUXs, the high asserted outputs of the state flip-flops are connected to the select lines of all MUXs in parallel. Each state address is a different set of MUX input.

C \ AB	00	01	11	10
0	?	c	?	?
1	d		e	

(a)

C \ AB	00	01	11	10
0	c	?	?	?
1	?	d	e	?

(b)

FIGURE 6.31 State assignments: (a) arbitrary, (b) chosen according to Principle 2.

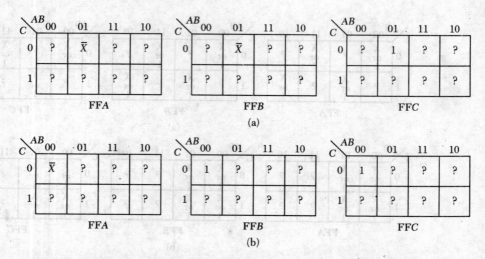

FIGURE 6.32 Next state or D input maps: (a) for arbitrary state assignment, (b) for assignments based on Principle 2.

true [2] and dictates how conflicts between the two principles should be resolved. Not all desired adjacencies can be obtained in a given state assignment; thus, Principle 1 is given higher priority for system minimization. Once this principle has been satisfied, Principle 2 is applied as far as possible without disturbing the assignments resulting from Principle 1.

6.4.2 IFL USING DIRECT-ADDRESSED MUXs

The IFL of the state machine can be designed with MUXs instead of gates. Although component cost may be higher for MUX design, the chip count can be minimized and design time is decreased. Furthermore, no restrictions are placed on state assignment for circuit minimization, allowing more flexibility in the overall design. We will see in the following sections that state assignment may affect the output forming logic design and system reliability. When IFL is designed with MUXs, state assignment can then be used for these purposes rather than for IFL minimization. One last advantage of MUX design lies in the ease of troubleshooting the state machine. It is very easy to isolate problems of faulty wiring or chips with this type of system.

The general arrangement of the direct-addressed MUX system is shown in Figure 6.33. Although the IFL shown consists of gates and MUXs, the gates often are either very simple or not required at all. In this scheme, the high asserted outputs of the state flip-flops are connected to the select lines of all MUXs in parallel. Each state then addresses a different set of MUX inputs.

One MUX is required for each state flip-flop and each MUX must have n select lines, where n is the number of state flip-flops. A direct-addressed MUX design

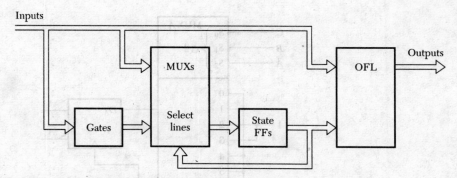

FIGURE 6.33 Architecture of the direct-addressed MUX system.

requires $n - 2^n$:1 MUXs to design the IFL. For example, a six state system utilizes three flip-flops and would use three 8:1 MUXs. This type of design always uses D flip-flops.

In order to demonstrate the use of direct-addressed MUXs for IFL, we will re-design the systems of Examples 6.1 and 6.2. For the first system the D input maps of Figure 6.19(b) will be used. We emphasize that these maps while labeled as corresponding to incorrectly chosen states, are quite appropriate for MUX design since no minimization is involved. The present state of the system selects the input lines of

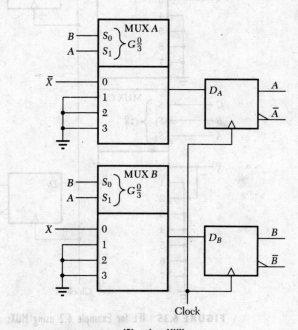

FIGURE 6.34 IFL using MUXs.

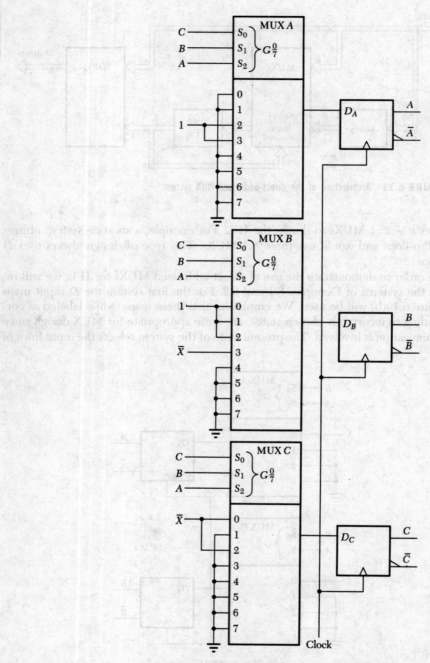

FIGURE 6.35   IFL for Example 6.2 using MUXs.

the MUXs that connect to the flip-flop inputs. If we number the minterm locations in the maps for D_A and D_B, we then connect to the corresponding MUX input lines the values contained in the locations. Figure 6.34 shows the MUX realization of IFL for Example 6.1.

The system of Example 6.2 is redesigned using the D input maps of Figure 6.25(b). Again the minterm numbers determine which values to connect to the MUX input lines. The redesigned system is shown in Figure 6.35.

We see the ease with which the design is carried out with MUXs for IFL. It might be noted, however, that the 8:1 MUXs of Figure 6.35 are not used very efficiently. Several of the inputs connect to the same value. It is possible to use smaller MUXs, but additional circuitry must be added. When the MUX size is reduced, the system is an indirect-addressed MUX arrangement.

It might be instructive to consider the component cost for equivalent systems using gates and MUXs for IFL. Figure 6.26(b) shows the minimized gate realization of IFL for the system of Example 6.2, which uses three 2-input gates. This IFL could be realized with one chip. In small quantities, these chips cost approximately 29 cents each, leading to a 29 cent component cost for IFL. The IFL for the MUX system would require three chips. Each of these 8:1 MUX chips would cost about 39 cents, bringing the total IFL cost to $1.17.

● ● ●   **DRILL PROBLEMS** Sec. 6.4

1. Design the IFL for the diagram of Figure 5.23 using D flip-flops and gates, minimizing the number of gates used.

2. Repeat Prob. 1 using JK flip-flops.

3. Repeat Prob. 1 using D flip-flops and MUXs for the IFL.

6.5

OUTPUT-FORMING LOGIC

SECTION OVERVIEW The combinational circuitry that the state flip-flops drive to produce the output signals is called the output-forming logic or OFL. A common problem in generating the output signals is that of the occurrence of glitches. This section introduces methods to create clean logic signals with no glitches.

When an output is to be generated by a state machine, there are several choices of assertion time. Figure 6.36 demonstrates six different possibilities. The first output is asserted during state c and deasserted at all other times. The up arrow and letters SB indicate that this output is asserted at the state beginning. The down arrow and

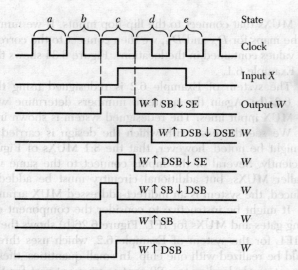

FIGURE 6.36 Six possible output assertion choices.

the letters SE indicate the deassertion of W at the state ending. The second output persists for one state time also, but both assertion and deassertion are delayed for one-half clock period compared with the first case. DSB is a mnemonic for delayed from state beginning while DSE is a mnemonic for delayed from state ending. The delay in each case is one-half clock cycle.

The assertion of the third output is delayed one-half clock period while de-assertion corresponds to the ending of state c. The fourth output is not as common as the others. It is asserted at the state beginning and deasserted one-half clock period later. Outputs 5 and 6 are similar, both becoming asserted during state c and presumably deasserted during some later state. The particular output selected will often depend on output specifications, and we must design the circuit to achieve this signal. At other times, there may be enough flexibility in output signal that the circuit configuration will be chosen with little regard to the resulting output signal shape.

In some cases, the system required may dictate which type of output can be used. It is not always possible to arbitrarily choose an output. The following paragraphs will indicate the considerations that limit this choice.

6.5.1 UNCONDITIONAL AND CONDITIONAL OUTPUTS

An unconditional output (sometimes called an immediate output) is one that depends only on the state of the system. When a particular state is entered, an AND gate or IC decoder reflects this fact by asserting an output line. We will later see that this output information may be delayed by as much as one-half clock period, but its existence is a function only of the state of the system.

A conditional output not only depends on the system state, but also requires certain input conditions before the output is asserted. For example, in state c the system may be required to assert output W only if the input X is asserted.

Returning to Figure 5.23, we can see both conditional and unconditional outputs represented by this state diagram. Output P is conditional while all other outputs are unconditional. P will be asserted only when the system is in state f and the input signal X is not asserted. If X is asserted while the system is in state f, output P will not be asserted. It is important to note here that data changes take place on alternate clock transitions to state changes in AST systems. When a state is entered, we must wait one-half clock period to allow data to be updated before generating the conditional output. If this is not done, an ambiguous situation results as shown in Figure 6.36. State c is entered while input $X = 0$, but the value changes to $X = 1$ at the midpoint of state c. Both possible values of X occur during state c. By convention, only the value of X after the midpoint of the state is allowed to influence a conditional output in the AST state machine. The output cannot be asserted at the state beginning and will typically be asserted at the delayed state beginning and deasserted at the state ending.

This problem can be demonstrated by considering a system that is to sample a data line X to locate the 3-bit word 101. When this word has been located, the system is to generate a signal R and resume checking the next 3-bit word. We will assume data changes occur on the positive-going clock transition, which dictates that state changes take place on the negative-going transition. The state diagram of Figure 6.37 is appropriate for this system. States c and e are timing states inserted to result in a three-clock time repetition period. If the first bit of the word is zero, we know the word cannot be 101, but we must delay two more clock periods before beginning the next word check. The timing chart of Figure 6.38 corresponds to the word 101 appearing on X. When the system is in state a, a value of $X = 1$ causes a state change to state b. The value of X persists for one-half clock time before being updated. If X changes to zero during state b, the system then moves to state d. An output cannot be generated in this third state until the third data value occurs. X is again updated halfway through the state time. The output for this system is conditional and must not occur until the second half of the state time.

In order to cause the conditional output to occur at DSB, the clock signal or inverted clock signal can enable the output gate. If the state change takes place on the positive clock transition, the inverted clock signal would enable the output gate only during the second half of the state time. If the state changes on the negative clock transition, the direct clock would enable the output gate during the second half of state time.

Figure 6.39 shows a partial state diagram having an unconditional output W and a conditional output R. The unconditional output could be generated by the 2-input gate of Figure 6.39(a). This gate produces output W for a state code of 10, which corresponds to state b. Assuming that state changes take place on the positive clock transition, the 4-input gate of Figure 6.39(b) generates the conditional output R. This output will occur only if the state code is 01 (state c), $Z = 0$, and the inverted

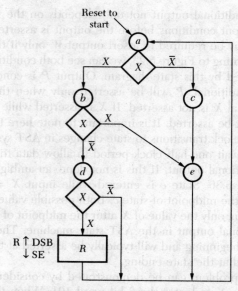

FIGURE 6.37 State diagram with conditional output.

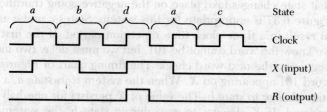

FIGURE 6.38 Timing diagram.

clock is positive. This gate will be enabled only at DSB and disabled at SE. The circuits that produce the output signals are called the output forming logic or OFL.

While the methods of generating outputs shown in Figure 6.39 represent the basic output circuits, these methods may lead to the presence of spurious output glitches. We will consider this very practical glitch problem after we discuss the minimization of OFL.

6.5.2 STATE ASSIGNMENT FOR MINIMIZATION OF OUTPUT-FORMING LOGIC

In OFL design, there are two bases upon which state assignment may be made. The first relates to minimization of gates required to implement the design. Of course, if an IC decoder is used, this consideration is unnecessary. The second basis for state

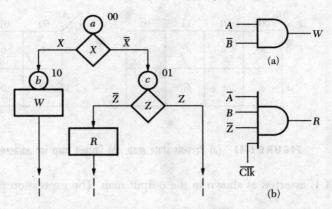

FIGURE 6.39 Partial state diagram and output decoders for (a) unconditional output, (b) conditional output.

assignment is elimination of output glitches that might be troublesome. The relative importance of these considerations depends on the particular application at hand. As we discuss OFL design, we will mention the constraints that will influence the designer's choice of states.

Principle 3. *To minimize OFL design when using gates, those states with identical outputs should be assigned logically adjacent states on the present state map.*

This will allow reduction of the expression for the output equation. If possible, we should use "don't care" conditions of unused states to minimize the output equation further. Let us consider the map shown in Figure 6.40. The *W* in a map location indicates that output *W* is to occur for this state.

The present state map shows that states *d* and *f* are the only ones in which output *W* may occur. This output may not occur even in these states unless input

C	AB 00	01	11	10
0	a	b	e	θ
1	d^W	c	f^W	θ

(a)

C	AB 00	01	11	10
0	0	0	0	θ
1	X	0	X	θ

(b)

FIGURE 6.40 (a) Present state map. (b) Output map.

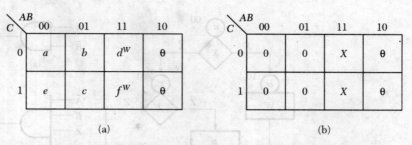

FIGURE 6.41 (a) Present state map. (b) Output map for adjacent states.

X is asserted as shown in the output map. The expression for the output is

$$W = \bar{B}CX + ACX$$

If we rearrange the states to make d and f adjacent and also adjacent to the "don't care" conditions, as shown in Figure 6.41, the output expression becomes

$$W = AX$$

We will emphasize that the state assignment may have been chosen to minimize input forming logic, and a state rearrangement may not be possible. Generally, IFL minimization results in more significant savings than does OFL minimization.

Although it is often more costly, a decoder chip can be used to generate state machine outputs. This approach may minimize chip count especially if several output signals must be generated. The flip-flop outputs are connected directly to the decoder inputs leading to the assertion of a particular line in each state. The decoder output lines corresponding to all states producing the same output signal are ORed to generate the output.

6.5.3 THE OUTPUT GLITCH PROBLEM

Static and dynamic hazards were considered in Section 2.2. These glitches or slivers are very narrow pulses, often having a width of a few nanoseconds, that occur during the switching of variables. They are also referred to as "runt" pulses. The combinational logic expressions implemented by various gates in the circuit may not predict the occurrence of a glitch, but due to differences in switching times of the gates or flip-flops, unwanted glitches may appear.

An earlier section discussed the problem and solution of glitches appearing at the outputs of the IFL gates. These gate outputs must settle to a stable level prior to clocking the state flip-flops. The simplest method of accomplishing this task is the AST approach. In this scheme, one clock transition changes synchronous inputs and the opposite transition is used to drive the state flip-flops. This allows the IFL gate outputs to settle for half the clock period before state changes take place. Runt pulses occurring on the IFL outputs are then ignored.

The output forming logic is driven by the state flip-flops. Although the state-changing clock transition is applied simultaneously to all state flip-flops, due to differences in clock-to-output delay times, the flip-flop outputs will not change at exactly the same time. Since these outputs drive gates or decoders, the possibility of hazards or runt pulses must be considered.

It must be emphasized that a state machine driven by a clock, with synchronous inputs driven by the alternate half-cycle of the clock, will often exhibit output glitch problems. Thus, even in synchronous state machine design, timing problems can cause extraneous pulses on the output lines.

There are applications in which glitches can be tolerated. For example, an output might be generated in a given state for the sole purpose of driving a light-emitting diode or some other indicator device. A short runt pulse does not allow the LED to be active long enough for the human eye to register the assertion. When an output drives a slow-responding device, glitches may be allowable and further design to eliminate these hazards is unnecessary.

Many digital controllers require outputs to drive digital circuits such as counters or flip-flops. In such a case, the output must not occur except when indicated by the specifications. Glitch-free design is essential to produce a reliable system in this type of application.

There are two common sources of output glitches in state machine design. The unequal switching times of the state flip-flops that drive the OFL have been identified as one source of possible glitches. The second source arises from an attempt to enable or disable an output gate coincidentally with other changes of gate inputs. Both sources will be considered in more detail in the following paragraphs.

Glitches Due to Flip-Flop Switching Speed

The problem of unequal flip-flop switching times is demonstrated by the circuit of Figure 6.42. This circuit shows two flip-flops whose outputs drive an AND gate to create an unconditional output. The output X will occur whenever $A = B = 1$. If we happen to be in the 01 state and then a transition is made to the 10 state, a glitch may result if flip-flop A has a shorter propagation delay time than does flip-flop B. In this instance, A changes to 1 before B drops to 0. Actually, the AND gate levels, V_{IHmin} and V_{ILmax}, also affect this situation. For a very short time, both AND gate inputs appear to have 1s applied. A short sliver or glitch may be produced at the gate output. If B drops below V_{IHmin} before A rises to V_{ILmax}, then no glitch could occur. Unfortunately, all flip-flops and gates possess different propagation delays, and these sometimes troublesome glitches tend to be rather prevalent in this type of circuit.

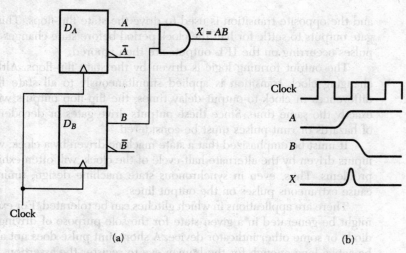

FIGURE 6.42   (a) A circuit that may produce a glitch. (b) Timing diagram.

State Assignment to Eliminate Output Glitches

It is often possible to avoid output glitches in state machines by proper assignment of states. In this method, state transition paths must not cross a state that produces an output on the way to the next state. The possible transition paths can be drawn on a present state map. Those states that should produce an output are identified by an asterisk. If a path crosses a state marked with an asterisk, a glitch may result. For example, consider the diagram and maps of Figure 6.43. Since four states are present, two flip-flops are required. From this state diagram, we see that a transition from state a to c is possible. If the map of Figure 6.43(b) is chosen, the transition from a to c can follow one of two paths. If A is faster than B, the path 00-10-11 will be followed. If B switches faster than A, the path 00-01-11 will be traversed. This latter possibility crosses state d and may produce a glitch on the output. Choosing the present state map of Figure 6.43(c) eliminates any possibility of glitches. As the system moves from state a to b or c, the paths do not cross state d. This output state is reached only when the system conditions dictate entry.

We will note here that if output glitches must be avoided, the selection of state assignment may be based entirely on this criterion. If so, neither the IFL nor the OFL will be minimal (if gates are used), but this is the price paid for a glitch-free output.

A more complex situation is represented in Figure 6.44. From the map of Figure 6.44(b), we see that a transition from state b to e may take a path that crosses state d, that is, 001-011-010-110. The map of Figure 6.44(c) solves this problem. As shown in Figure 6.45, all possible allowed state transitions indicated by the state diagram have possible paths that do not cross output states. From this table, we see that no output state is crossed by any possible path; thus, an output glitch will never

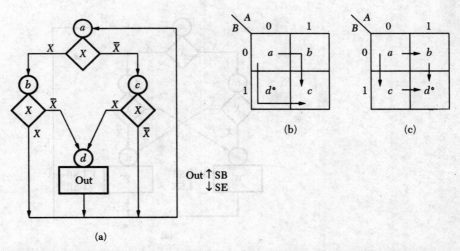

(a)

(b)

(c)

FIGURE 6.43 (a) State diagram. (b) Map leading to possible glitch. (c) Map that guarantees no output glitch.

occur. Although trial-and-error methods are required to select proper states, it is generally appropriate to start by selecting output states as far removed from each other as possible. In this case, d is 001, while each flip-flop must change states to reach state e, which is 110.

Before leaving the discussion of state assignment to avoid output glitches, a word of caution is appropriate. We noted previously that unused or "don't care" states can be placed adjacent to output states to minimize OFL. This implies that when the unused state is entered, an output will be generated. If our states are chosen to avoid output glitches, we must be certain that any "don't care" state used to reduce OFL is not crossed during a state transition.

There are times when it is impossible to avoid glitches by state assignment. Furthermore, conditional outputs are produced by circuits that can lead to glitches without crossing output states; therefore, other methods must be considered.

Glitches Due to Conditional Outputs

A conditional output is not generated when a particular output state is entered. Any output must be delayed one-half state time in the AST state machine until the input data is updated. Figure 6.38 shows a timing chart for such a conditional output. In order to produce a conditional output, the gate of Figure 6.39(b) might be used. This particular configuration can lead to glitches unless certain precautions are taken. The problem here is that as the gate is enabled by the inverted clock, the data line Z may also be changing. Furthermore, as the inverted clock disables the gate, the state could be changing to cause other gate inputs to be in transition as the gate closes. Whenever several inputs of a gate may change at the same time,

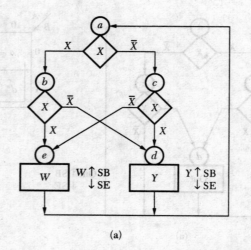

(a)

C \ AB	00	01	11	10
0	a	c	e°	θ
1	b	d°	θ	θ

(b)

C \ AB	00	01	11	10
0	a	b	e°	c
1	d°	θ	θ	θ

(c)

FIGURE 6.44 (a) State diagram. (b) State assignment with possible glitch. (c) Glitch-free assignment.

Transition		Possible
From	To	paths
a	b	000 → 010
a	c	000 → 100
b	d	010 → 011 → 001°
		010 → 000 → 001°
b	e	010 → 110°
c	d	100 → 000 → 001°
		100 → 101 → 001°
c	e	100 → 110°
d	a	001 → 000
e	a	110 → 100 → 000
		110 → 010 → 000

FIGURE 6.45 Transition paths for map of Figure 6.44(c).

the possibility of a glitch exists. Figure 6.46 shows two possible problems that may cause glitching on the output line R. An output should occur only if the system is in state c and $Z = 0$ after the midstate time.

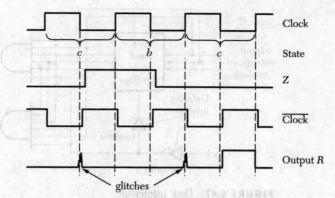

FIGURE 6.46 Possible output glitches for gate of Figure 6.39(b).

We have assumed that the inverted clock is slightly delayed from the clock signal and that input Z is delayed even more than the inverted clock. When the system first enters state c, no output should occur until Z is updated at midstate time or at DSB. The inverted clock signal that enables the output gate at DSB occurs before the value of Z changes from 0 to 1. No output should occur at this time, but a glitch may actually appear on line R just before Z changes. A second unwanted glitch may occur when state c is again entered. During the first half of state time, Z is zero. No output should occur since a conditional output is based on the value of Z after the midstate time. However, when state c is entered, the inverted clock has not yet disabled the output gate. Thus, a glitch may appear on line R at this time.

Clock Suppression to Avoid Glitches

In order to avoid the glitches due to conditional outputs, we should allow the input data to settle before the AND gate is enabled and we should disable the gate before the state change takes place. We can do this by delaying the inverted clock signal with respect to the data changes and delaying the state changes even more than we delay the inverted clock. Figure 6.47 shows a means of accomplishing these delays.

The outputs are formed by ANDing the appropriate state flip-flops and input signals with a delayed, inverted clock signal. State changes are delayed by an additional delaying inverter resulting in the timing chart of Figure 6.48.

In Figure 6.48, input data changes on the negative-going clock transition. Delaying inverter 1 allows any changing inputs to become stable before the output AND gates are enabled. Delaying inverter 2 delays the state changes even further, and so the output AND gates are disabled just prior to state changes. With this arrangement, during the time the output gates are enabled, no other gate inputs will be changing.

If the data changes on the positive clock transition and the state changes take place on the negative transition, a different arrangement of inverters is required. In this case, the gate again must be opened after one-half state time. The positive clock will then enable the output gate, but this signal must be delayed to avoid data

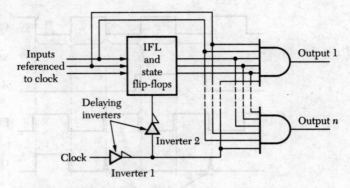

FIGURE 6.47 Clock suppression.

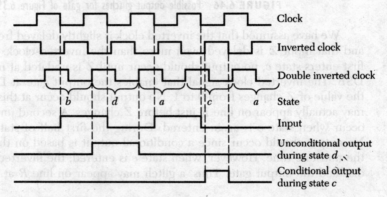

FIGURE 6.48 Timing diagram for clock suppression.

change conflicts. Consequently, two inverters are used to delay the clock signal to the gate. Two more inverters are then inserted that connect to the clock inputs of the flip-flops. Figure 6.47 would be modified by using negative transition flip-flops and replacing each of the inverters by a pair of inverters.

While the clock suppression method is relatively simple, the output lasts for one-half of the state time rather than the full state time. This may be a problem in a small number of applications, but will generally not be a significant factor.

EXAMPLE 6.3
. . . .

The state machine of Figure 6.49 should produce output M near the DSB of state b, which has a code of 01. The maximum clock-to-output delay time for the flip-flops is 10 ns. The change in input X trails the positive clock transition by a maximum of 4 ns. The inverters to be used have a minimum propagation delay time of 6 ns. Design the output circuit to generate M using clock suppression.

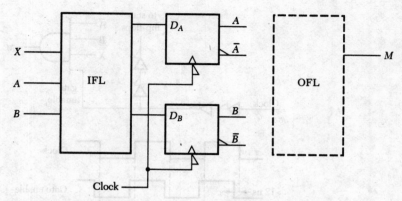

FIGURE 6.49 A state machine with a conditional output.

SOLUTION

Since the state transition takes place on the negative clock transition, the output gate must be enabled when the clock goes positive. This will ensure that the output M will not appear when a new state is entered until X has had a chance to be updated at midstate time. Figure 6.50 shows an output gate that will generate M.

The key to this design is the generation of the enable signal and the signal to drive the state changes. The clock signal must be delayed by at least 4 ns to allow X to change at midstate time. Because the gate must be enabled with the positive half of the clock signal, an even number of inverters must be used to create the enable signal. Two inverters will delay the gate enable by a minimum of 12 ns. Unless the state transition is delayed more than the gate enable signal, a state change will occur before the output gate is disabled. Two additional inverters should then drive the state flip-flops, resulting in the circuit of Figure 6.51. The state change now occurs after the gate is disabled as shown in the timing chart. ●

While this example is seen to consist of tedious, detailed considerations, much attention must be given to timing details in order to avoid the problem of glitches in practical design.

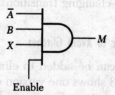

FIGURE 6.50 Output-forming logic using clock suppression.

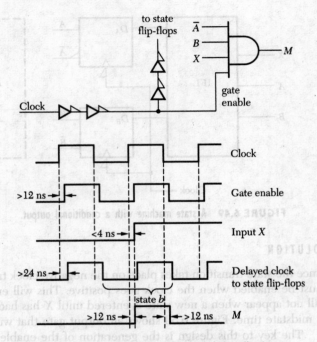

FIGURE 6.51 Output-forming logic and timing diagram.

Clock suppression can also be used to generate unconditional outputs when state assignment cannot be used to eliminate all glitches. This can be done in two different ways. If the output is to be generated during the first half of the state time, the signal that enables the output gate is delayed relative to the state-changing transition. This allows all state flip-flops that drive the output gate to settle before the gate is enabled. If the output is to be generated during the second half of the state time, the timing problem occurs at the end of state time. To avoid the changing of gate inputs as the output gate is disabled, the state changes are delayed relative to the signal that enables this gate. The gate is then disabled prior to the occurrence of the state-changing transition. Both methods are shown in Figure 6.52, assuming R should occur in state b. In neither case does the gate enable assertion of the signal to overlap the state-changing transitions.

Output Holding Register to Avoid Glitches

A component that can be added to eliminate output glitches is an output holding register. Figure 6.53 shows one version of this scheme. One flip-flop is required for each output of the system. For unconditional outputs the decoder gates generate output signals at the beginning of any output-producing state. These signals are presented to the corresponding D input of the holding register. This information can be set into the holding register near the state beginning or at the delayed state

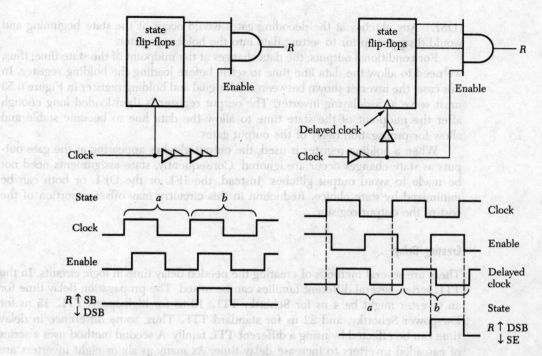

FIGURE 6.52 Clock suppression for unconditional outputs.

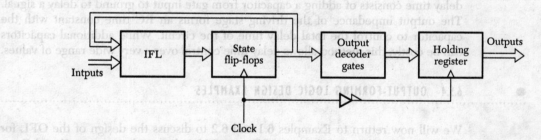

FIGURE 6.53 Use of an output-holding register.

beginning. If it is done at the state beginning, the transition that loads the holding register must be delayed relative to the state-changing transition. This delay must exceed the clock-to-output delay of the state flip-flop plus the propagation delay of the output gates. If the information is loaded into the holding register at the delayed state beginning, the half-period of the clock must exceed the clock-to-output delay of the output gates. Except for very high frequency clocks, this requirement is easily satisfied. This latter method is generally preferred because delay considerations are unimportant. In this case, the holding register produces an output that is asserted at the delayed state beginning (DSB) and deasserted at the delayed state ending

(DSE). Any glitches at the decoding gates would occur at the state beginning and would disappear prior to setting data into the holding register.

For conditional outputs, the data changes at the midpoint of the state time; thus, we need to allow the data line time to settle before loading the holding register. In this case, the inverter shown between clock signal and holding register in Figure 6.53 must serve as a delaying inverter. The output register is then loaded long enough after the midpoint of the state time to allow the data line to become stable and allow for propagation delay of the output gates.

When a holding register is used, the output glitches appearing at the gate outputs as state changes occur are ignored. Consequently, state assignments need not be made to avoid output glitches. Instead, the IFL or the OFL or both can be minimized by state choice. Reduction in this circuitry may offset a portion of the cost of the output register.

Creating Delays

There are several methods of creating the needed delay time in logic circuits. In the TTL series, several different families can be mixed. The propagation delay time for an inverter might be 4 ns for Schottky TTL, 10 ns for high-speed TTL, 15 ns for low-power Schottky, and 22 ns for standard TTL. Thus, some difference in delay time can be effected by using a different TTL family. A second method uses a series of cascaded inverters to increase delay time. As many as six or eight inverters are sometimes used in logic circuits to achieve longer delays. A third method to control delay time consists of adding a capacitor from gate input to ground to delay a signal. The output impedance of the driving stage forms an RC time constant with the capacitor to control the total delay time of the circuit. While additional capacitors can be costly, this method allows delay time control over a very wide range of values.

6.5.4 OUTPUT-FORMING LOGIC DESIGN EXAMPLES

We will now return to Examples 6.1 and 6.2 to discuss the design of the OFL for these systems. The state diagram of Figure 6.12 indicates that outputs Y and Z are unconditional outputs with Y asserted in state b and Z asserted in state c. If we consider the state assignment chosen to minimize IFL as shown in Figure 6.14(a), we note that output glitches due to transient crossing of output states can result. In switching from state a to c or from c to a, state b may be traversed. If we insist on using this assignment, we must either accept the output glitches (if glitches can be tolerated at the output) or we must use holding registers or clock suppression to solve this problem. The arrangement of Figure 6.54(a) will lead to possible glitches on output Y. This may be acceptable in some designs, but often these glitches must be removed. Figure 6.54(b) uses a holding register that sets the output data into the flip-flops at midstate time. The outputs in this case lag the outputs of the gates by one-half the period of the clock. In some rare cases, this time lag may be

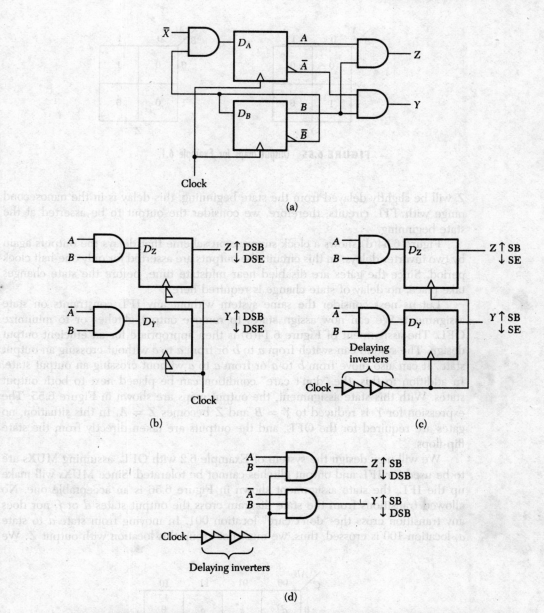

FIGURE 6.54 Unconditional outputs. (a) State machine with gates for OFL. (b) Holding register arrangement. (c) Alternate holding register arrangement. (d) Clock suppression.

unacceptable. If so, the arrangement of Figure 6.54(c) can be used. In this circuit, the clock is delayed long enough by the inverters to allow the glitches to die out before the information is set into the holding register. Although the outputs Y and

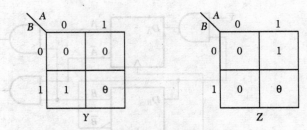

FIGURE 6.55 Output maps for Example 6.1.

Z will be slightly delayed from the state beginning, this delay is in the nanosecond range with TTL circuits; therefore, we consider the output to be asserted at the state beginning.

Figure 6.54(d) shows a clock suppression scheme that delays the outputs again by two inverter delays. In this circuit, the outputs are asserted for only one-half clock period. Since the gates are disabled near midstate time, before the state changes take place, no delay of state change is required here.

Let us next consider the same system without any IFL constraints on state assignment. We can now assign states to remove output glitches or to minimize OFL. The assignment of Figure 6.14(b) is then appropriate for an efficient output design. The system can switch from a to b or from a to c without crossing an output state. It can also move from b to a or from c to a without crossing an output state. In addition to this, the "don't care" condition can be placed next to both output states. With this state assignment, the output maps are shown in Figure 6.55. The expression for Y is reduced to $Y = B$ and Z becomes $Z = A$. In this situation, no gates are required for the OFL, and the outputs are taken directly from the state flip-flops.

We will now design the system of Example 6.2 with OFL, assuming MUXs are to be used for IFL and output glitches cannot be tolerated. Since MUXs will make up the IFL, the state assignment shown in Figure 6.56 is an acceptable one. No allowed transitions from the state diagram cross the output states d or f, nor does any transition cross the "don't care" location 001. In moving from state d to state a, location 100 is crossed; thus, we must not group this location with output Z. We

C \ AB	00	01	11	10
0	d^Y	e	a	θ
1	θ	b	c	f^Z

FIGURE 6.56 A present state map for Example 6.2.

C \ AB	00	01	11	10
0	1	1	$\bar{X}$	θ
1	θ	0	X	1

D_A

C \ AB	00	01	11	10
0	1	1	1	θ
1	θ	X	$\bar{X}$	1

D_B

C \ AB	00	01	11	10
0	0	0	1	θ
1	θ	0	X	0

D_C

C \ AB	00	01	11	10
0	1	0	0	0
1	θ	0	0	0

Y

C \ AB	00	01	11	10
0	0	0	0	0
1	0	0	0	1

Z

FIGURE 6.57 Pertinent maps for system design of Example 6.2.

need not use clock suppression or holding registers, and we can minimize the OFL (except for output Z). The next state maps or D input maps and the output maps are given in Figure 6.57. Application of these maps leads to the system of Figure 6.58.

Before leaving the design of Example 6.2, we will consider the OFL design for the state assignment given in Figure 6.24. This assignment was made to minimize the IFL of the circuit and is repeated in Figure 6.59. We note that the allowed transition from state c to state e may cause a glitch on output Y as could the transition from c to f. The transition from state d to a may cause a glitch on output Z. If this assignment is to be used and output glitches cannot be tolerated, either clock suppression or a holding register must be used. Since the state assignment is chosen in Figure 6.59 to minimize IFL, the clock suppression method of Figure 6.54(d) might be chosen to minimize OFL. Figure 6.60 shows the complete design based on JK flip-flops with a clock to output delay time of 8 ns. The two inverters each introduce a minimum delay of 6 ns. The expression for output Y is

$$Y = AC \cdot \text{(Delayed clock)}$$

and Z is given by

$$Z = A\bar{B} \cdot \text{(Delayed clock)}$$

The final example of this section will demonstrate an overall design with a conditional output.

EXAMPLE 6.4

Implement the state diagram of Figure 6.37 with the modification that output R is to be asserted for one complete clock time. Use D flip-flops and MUXs for the IFL.

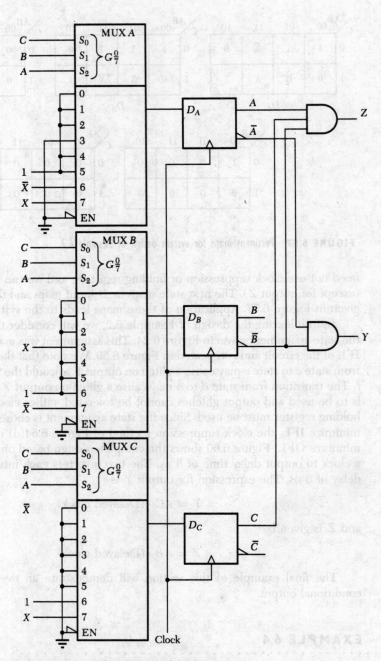

FIGURE 6.58 Complete system for Example 6.2.

FIGURE 6.59   State assignment to minimize IFL.

C \ AB	00	01	11	10
0	a	b	e	f^Z
1	θ	c	d^Y	θ

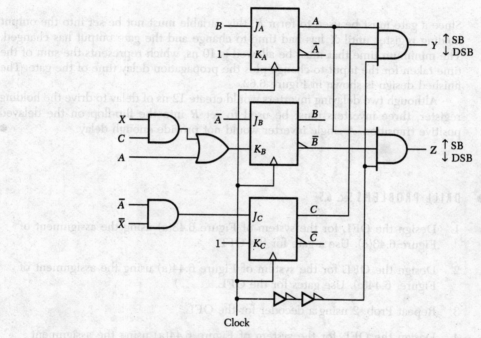

FIGURE 6.60   State machine design for Example 6.2 using clock suppression.

Assume that input X will change within 4 ns of the positive clock transition. The maximum propagation delay of any gate or flip-flop (clock to output) is 6 ns. The minimum delay of an inverter is 6 ns.

SOLUTION

Since a conditional output is required in this system, we need not be concerned with the assignment of proper states to avoid glitches. Delays will be used to eliminate possible glitch problems. Furthermore, since MUXs will be used for the IFL, no consideration need be given to IFL minimization. We are then free to assign states to minimize the OFL.

The specification that the output must last one clock time eliminates the possibility of using clock suppression, therefore we will choose to use a holding register. For an AST state machine, the output cannot be asserted until the midstate time to allow the input to change. In this design, the output must then be asserted at DSB and deasserted at DSE. One possible state assignment is shown in Figure 6.61.

The state change takes place on the negative clock transition and the input is updated on the positive transition. The output expression for R is

$$R = AX$$

Since a gate must be used to form R, this variable must not be set into the output holding register until X has had time to change and the gate output has changed. The minimum time that must be allowed is 10 ns, which represents the sum of the time taken for the input to change plus the propagation delay time of the gate. The finished design is shown in Figure 6.62.

Although two delaying inverters would create 12 ns of delay to drive the holding register, three inverters must be used to set R into the flip-flop on the delayed positive transition. A single inverter would not provide enough delay.   ●

● ● ●   DRILL PROBLEMS Sec. 6.5

1. Design the OFL for the system of Figure 6.43(a) using the assignment of Figure 6.43(c). Use a gate for the OFL.

2. Design the OFL for the system of Figure 6.44(a) using the assignment of Figure 6.44(c). Use gates for the OFL.

3. Repeat Prob. 2 using a decoder for the OFL.

4. Design the OFL for the system of Figure 6.44(a) using the assignment of Figure 6.44(b). Use gates for the OFL along with a holding register to eliminate possible glitches.

C $\overset{AB}{\diagdown}$	00	01	11	10
0	a	c	d^R	θ
1	b	e	θ	θ

FIGURE 6.61 State assignment for diagram of Figure 6.37.

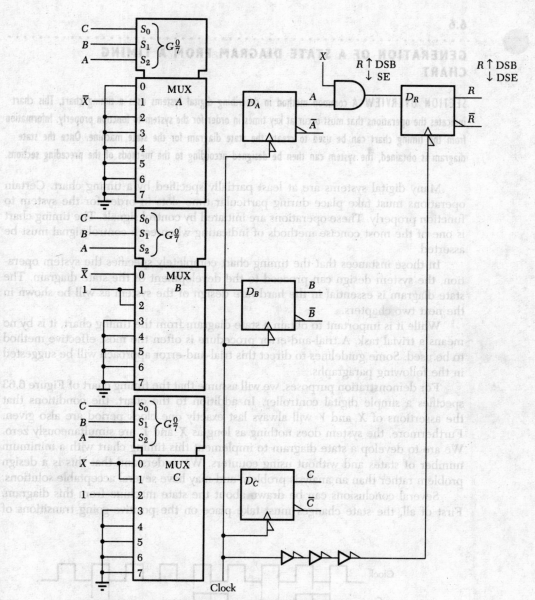

FIGURE 6.62 State machine with output-holding register.

5. Design the OFL for Figure 6.37 if state d is 011. Use clock suppression to eliminate possible glitches, and assume data changes on the positive clock transition.

6.6

GENERATION OF A STATE DIAGRAM FROM A TIMING CHART

SECTION OVERVIEW A common method in describing digital systems uses a timing chart. This chart indicates the operations that must occur at key times in order for the system to function properly. Information from the timing chart can be used to create the state diagram for the state machine. Once the state diagram is obtained, the system can then be designed according to the methods of the preceding sections.

Many digital systems are at least partially specified by a timing chart. Certain operations must take place during particular time slots in order for the system to function properly. These operations are initiated by control signals. The timing chart is one of the most concise methods of indicating when each control signal must be asserted.

In those instances that the timing chart completely specifies the system operation, the system design can proceed to the development of the state diagram. The state diagram is essential in the hardware design of the system as will be shown in the next two chapters.

While it is important to obtain a state diagram from the timing chart, it is by no means a trivial task. A trial-and-error procedure is often the most effective method to be used. Some guidelines to direct this trial-and-error approach will be suggested in the following paragraphs.

For demonstration purposes, we will assume that the timing chart of Figure 6.63 specifies a simple digital controller. In addition to the chart, the conditions that the assertions of X and Y will always last exactly one clock period are also given. Furthermore, the system does nothing as long as X and Y are simultaneously zero. We are to develop a state diagram to implement this timing chart with a minimum number of states and without using counters. We understand that this is a design problem rather than an analysis problem and may have several acceptable solutions.

Several conclusions can be drawn about the state machine from this diagram. First of all, the state changes must take place on the positive-going transitions of

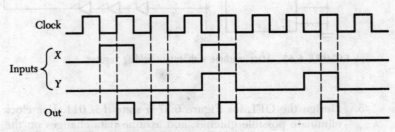

FIGURE 6.63 A digital system timing chart.

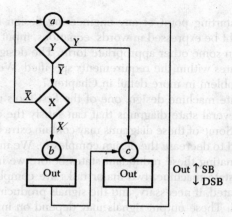

FIGURE 6.64 A state diagram.

the clock since the inputs change on the negative transition. We also notice that the output assertions begin at the positive transition of the clock and thus begin when certain states have been entered. This eliminates the possibility of conditional outputs that cannot be asserted at the state beginning.

We can also note from the timing chart that the output is asserted during the state times that input Y is found to be asserted without regard to the assertion of input X. With this basic knowledge and the fact that the state machine does nothing when neither X nor Y is asserted, the state diagram of Figure 6.64 is proposed.

The state machine remains in state a if neither input is asserted. If Y is asserted when the positive clock transition takes place, the system moves to state c, generates an output, and moves back to state a. When X is asserted, but Y is not asserted, the system moves to state b, generates an output, moves to state c, and generates a second output before returning to state a.

Several other possible state diagrams can be chosen to satisfy the specifications given by the timing chart. Time for development of the state diagram can be minimized if valid observations are made to guide the trial-and-error approach. The next chapter will demonstrate the application of these methods to actual digital systems.

6.7

REDUNDANT STATES

SECTION OVERVIEW In generating a state diagram, it is possible that more than one state will perform identical functions. If so, all but one of these states is redundant and can be removed from the diagram. The state machine designed from this reduced diagram will use minimal hardware. Methods of detecting and removing redundant states are discussed in this section.

The starting point of any engineering design is a set of specifications. These specs might be expressed in words, equations, input and output waveforms, a timing chart, or in some other appropriate form. The designer must then develop a system that operates within the requirements specified. We will consider the digital system design problem in more detail in Chapter 7.

In state machine design, one of the first tools used is the state diagram. There may be several state diagrams that can satisfy the original specifications of a given problem. Some of these diagrams may contain extra or redundant states that could be eliminated to decrease the design complexity. We must consider methods of locating and eliminating these redundant states before we implement the state diagram.

In a state machine, two major tasks are completed during each state. Outputs are generated, if necessary, and the signals producing the correct next state must be generated. These output signals may depend on input signals (conditional outputs) as also may the next state signals.

These considerations lead to the concepts of equivalent states and redundant states. The definitions of these types of states can be expressed as follows [2]:

● **Equivalent States:** If a state machine is started from either of two states and identical output sequences are generated from every possible set of input sequences, the two states are said to be equivalent.

● **Redundant State:** A state that is equivalent to another state is called a redundant state.

Obviously, if the machine reacts to all possible inputs when started from one state in the same way that it reacts when started in a second state, one of these equivalent states is redundant and can be removed. A reduced state diagram contains no equivalent states.

A theorem relating to equivalent states attempts to simplify this concept. It states that two or more states that generate identical output signals and identical next state signals for all possible input conditions are said to be equivalent states.

While this theorem is true, it does not encompass all possible equivalent states. For example, it is possible that states a and b are equivalent, yet state a moves to state c while state b moves to state d for identical inputs. This is possible only for the case that states c and d are also equivalent. Thus, we must apply this theorem with care in reducing the state diagram.

As an example of redundant states, we will again consider a system that is to sample a synchronous data line X. This system should take four samples before returning to its initial state to begin another 4-bit check. If the 4-bit sequence (word) 0111 is detected, an output is to be generated and the system should then begin the next check without missing a bit. One possible state diagram is shown in Figure 6.65.

If the sequence 0111 occurs, the system will proceed through states a, b, c, and d and generate an output prior to returning to state a. Any other sequence will

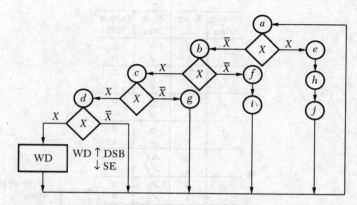

FIGURE 6.65 A state diagram for a word detector.

not generate an output, but the state diagram provides the proper delay of 4 bits before returning to state a. We note that a conditional output is required in this instance. An alert designer might note that there are several redundant states in the diagram of Figure 6.65. For example, states f and h are provided as timing states to ensure that if an incorrect bit has been detected in bit 1 or bit 2 of the word, two more samples will be required before returning to state a. States f and h could be combined into a single state. The same argument applies to states g, i, and j; that is, only one of these states is required.

There is an orderly method available to identify and eliminate redundant states. This method is a simplified version of a more complex method [4] that is generally applied to asynchronous systems. For practical synchronous systems, the following method is often sufficient to create the reduced state diagram. First, the next state table for the state diagram is constructed. This table consists of the present states, the various possible inputs, the next states, and the outputs of the system. Obviously, the state table for a given system may differ from one designer to another. Equivalent states are then identified. As equivalent states are identified, all but one are considered redundant and can thus be consolidated into a single state. This method will be demonstrated in Figure 6.66. This is the state table for the word detector. We note from the table that states g, i, and j are equivalent. While state d is similar to states g, i, and j, it has a different output specification and is therefore not equivalent. We next choose names for these equivalent states. We will designate states g, i, and j, by state g, replacing i and j by g in the table. This results in the modifications shown in Figure 6.66. Before we construct the reduced state table for the system, we check the chart to see if the modified states lead to any further equivalent states. States f and h are now seen to be equivalent. We will select state f as the name for the states f and h and replace the letter h with an f each time it appears in the table. We continue this process until no equivalent states exist and then develop the reduced state table. For this example, Figure 6.67 shows the reduced state table, and Figure 6.68 indicates the reduced state diagram.

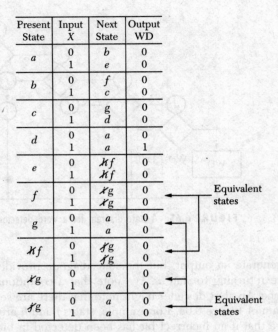

Present State	Input X	Next State	Output WD
a	0	b	0
	1	e	0
b	0	f	0
	1	c	0
c	0	g	0
	1	d	0
d	0	a	0
	1	a	1
e	0	~~e~~f	0
	1	~~e~~f	0
f	0	~~f~~g	0
	1	~~f~~g	0
g	0	a	0
	1	a	0
~~e~~f	0	~~f~~g	0
	1	~~f~~g	0
~~e~~g	0	a	0
	1	a	0
~~f~~g	0	a	0
	1	a	0

Equivalent states

Equivalent states

FIGURE 6.66   State table for a word detector.

Present State	Input X	Next State	Output WD
a	0	b	0
	1	e	0
b	0	f	0
	1	c	0
c	0	g	0
	1	d	0
d	0	a	0
	1	a	1
e	0	f	0
	1	f	0
f	0	g	0
	1	g	0
g	0	a	0
	1	a	0

FIGURE 6.67   Reduced state table.

Actually, the simplified method of reducing states does not guarantee a minimal number of states. As an example of a situation in which this method fails, consider the partial table of Figure 6.69.

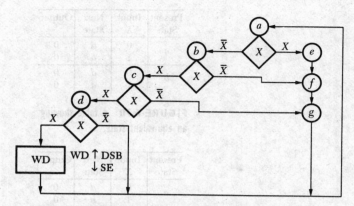

FIGURE 6.68 Reduced state diagram for word detector.

Present State	Input X	Next State	Output
a	0	d	0
	1	b	0
b	0	d	0
	1	a	0
⋮	⋮	⋮	⋮

FIGURE 6.69 A partial state table.

Applying the method previously discussed would not identify states a and b as being equivalent. If the assumption that a is equivalent to b results in equivalent next states for both present states, the two states are indeed equivalent. This, of course, applies only if both states have the same outputs. In this example, the next state for $X = 0$ is the same for both present states. The assumption that the two states are equivalent then results in the table of Figure 6.70. Obviously, the two states a and b are equivalent and one of them can be eliminated. Each state must be assumed to equal every other state that is a possible equivalent state. If the outputs are different, no comparison is necessary.

A more complex situation is shown in Figure 6.71. In this example, the assumption that states a and b are equivalent does not reveal any equivalent states. Nor does the assumption that states d and e are equivalent reveal any equivalent states. However, the assumption that states a and b are equivalent and the simultaneous assumption that states d and e are equivalent results in the table of Figure 6.72. This table is formed by replacing state b with a and state e with d in the next state column. It now becomes apparent that b is equivalent to a and e is equivalent to d.

Present State	Input X	Next State	Output
a	0	d	0
	1	a	0
b	0	d	0
	1	a	0

FIGURE 6.70   A table showing an equivalent state.

Present State	Input X	Next State	Output
a	0	d	0
	1	b	0
b	0	e	0
	1	a	0
·	·	·	·
·	·	·	·
d	0	e	0
	1	a	1
e	0	d	0
	1	b	1

FIGURE 6.71   A more complex state table.

Present State	Input X	Next State	Output
a	0	d	0
	1	a	0
b	0	d	0
	1	a	0
·	·	·	·
·	·	·	·
d	0	d	0
	1	a	1
e	0	d	0
	1	a	1

FIGURE 6.72   A table showing two pairs of equivalent states.

While the state diagram represented by the partial table of Figure 6.72 is somewhat impractical, a reduction method should be developed to handle this situation. This more complex method can be carried out by executing the following steps.

1. Reduce the state table as far as possible using the simple method outlined earlier.
2. Group states having equivalent outputs together.
3. Within each group, eliminate those states that are not potentially equivalent. For example, if state *a* is in group 1 and has a next state in group 3 and no other state in group 1 has a next state in group 3, state *a* can be eliminated as a potentially equivalent state.
4. Within each group, assume the equivalence of every possible pair of states. For each assumed pair, if the next states are in another group, and those next states are potentially equivalent, the assumption that these next states are equivalent should be made.

To demonstrate this example, we will consider the state diagram of Figure 6.73. The state table is shown in Figure 6.74. We first apply the simple method to the

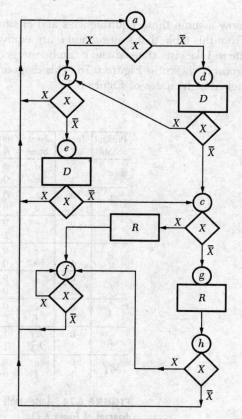

FIGURE 6.73 A state diagram.

state table, which reveals that states f and h are equivalent. After eliminating state h and replacing this letter by f in the table of Figure 6.74, it is not apparent that other equivalent states exist. We now group states having equivalent outputs together resulting in four groups:

- Group 1 a, b, f
- Group 2 c
- Group 3 d, e
- Group 4 g

Within Group 1, it is obvious that f cannot be equivalent to a or b because the next states of f are both in Group 1 while the next states of a and b include a state from Group 3. Thus, the only potential equivalencies are states a and b and states d and e.

We now assume these equivalencies and create the table of Figure 6.75. It is obvious from this table that states a and b are equivalent and can be merged into a single state a. Likewise, states d and e can be merged into state d. The final reduced state diagram is shown in Figure 6.76. This diagram has only five states compared to the original eight states of Figure 6.73.

Present State	Input X	Next State	Outputs R D
a	0	d	0 0
	1	b	0 0
b	0	e	0 0
	1	a	0 0
c	0	g	0 0
	1	f	1 0
d	0	c	0 1
	1	b	0 1
e	0	c	0 1
	1	a	0 1
f	0	a	0 0
	1	f	0 0
g	0	$\cancel{h}f$	1 0
	1	$\cancel{h}f$	1 0
$\cancel{h}f$	0	a	0 0
	1	f	0 0

FIGURE 6.74 State table for diagram of Figure 6.73.

Present State	Input X	Next State	Outputs R D
a	0	d	0 0
	1	a	0 0
b	0	d	0 0
	1	a	0 0
c	0	g	0 0
	1	f	1 0
d	0	c	0 1
	1	a	0 1
e	0	c	0 1
	1	a	0 1
f	0	a	0 0
	1	f	0 0
g	0	f	1 0
	1	f	1 0

FIGURE 6.75 State table used to find equivalent states.

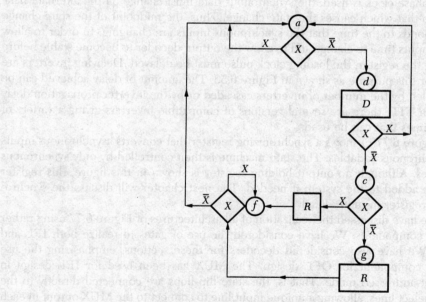

FIGURE 6.76 Reduced state diagram.

6.8

GENERAL STATE MACHINE ARCHITECTURE

SECTION OVERVIEW This section reviews several possible arrangements for state machines.

6.8.1   TRADITIONAL STATE MACHINES

Figure 6.77 shows the general block diagram of the state machine along with two possible variations. The output holding register of Figure 6.77(b) is added to latch the outputs into this register at the midpoint of the state time or after the state flip-flops settle. As explained earlier, this arrangement is used to filter out the glitches from the output decoder. The information shifted into the register is unaffected by any glitches since these transients would occur only when state changes take place. One point that is significant here is that if any outputs are conditional, that is, dependent on input variables, the transition that fills the holding register must be delayed slightly from the midpoint of the state time. We recall that when a single phase clock is used, the synchronous data must change on the alternate half-cycle to that which causes the state change. Thus, the midpoint of the state change corresponds to the time that the synchronous inputs are changing. In order to allow these inputs time to change and to allow the output decoder to become stable before loading the register, the loading clock pulse must be delayed. Delaying inverters are used for this purpose as shown in Figure 6.53. The amount of delay achieved can be controlled by the number of inverters cascaded or by the inverter propagation delay time. In TTL, there are several versions of compatible inverters giving a variety of delay times that can be used.

Figure 6.77(c) shows a synchronizing register that converts asynchronous inputs to synchronous variables. The state machine is then controlled by only synchronous variables. Although no output holding register is shown in this figure, this register may be added to the system if needed. The next chapter will discuss the synchronizing register in more detail.

We have discussed the realization of all architectures of Figure 6.77 using rather simple components. We have considered the use of gates to realize both IFL and OFL. We have also considered decoders for these sections, emphasizing the use of this component for OFL design. The MUX has been used for IFL design in a direct-addressed mode. That is, the state flip-flops are connected directly to the MUX select lines, allowing a unique input line to connect to the MUX output in each state. Generally, this method is used to minimize design time rather than component cost.

It is possible to implement the entire state machine using a programmable logic device. In this approach, a computer program directs the programming of a chip to create the state machine without interconnecting several IC chips. Chapter 8 will consider this method of state machine implementation.

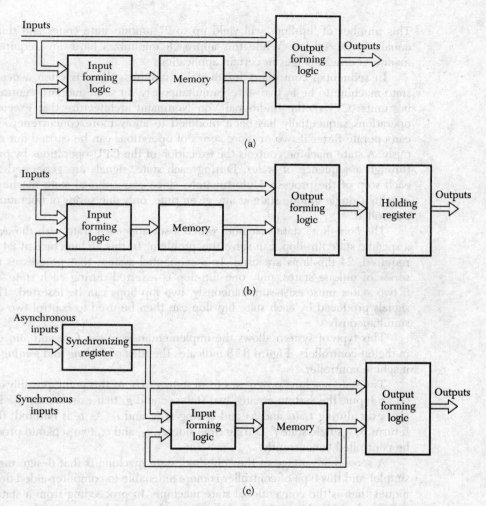

FIGURE 6.77 (a) General state machine. (b) State machine with output-holding register. (c) State machine with input-synchronizing register.

6.8.2   THE "ONE-HOT" STATE MACHINE

The methods of state machine design considered to this point are based on the use of the minimum number of state flip-flops. If m states are required to implement a sequential controller, the number of state flip-flops required, n, is found by selecting the minimum value of n to satisfy

$$2^n \geq m \tag{6.4}$$

This number of flip-flops will yield up to 2^n unique state codes and this number equals or exceeds m. While this approach minimizes hardware requirements, it manifests disadvantages in certain applications.

In sequential controllers for digital computers, there is often a need for the state machine to be in two states simultaneously. In some modern central processing units (CPUs), the traditional Von Neumann architecture that executes single operations sequentially has been modified to allow more concurrency. The CPU can operate faster if two or more series of operations can be carried out simultaneously. A state machine controls the execution of the CPU operations by progressing through a sequence of states. During each state, signals are generated to control each step of the process. Unfortunately, since conventional state machines always exist in a single unique state at any given time, only one series of operations can be controlled.

The "one-hot" state machine, which associates each state with the assertion of a specific state flip-flop, can solve this problem. In this architecture, if 24 states are required, 24 flip-flops are used. In a sequential system that progresses through a series of unique states, only one flip-flop is asserted during each state. However, if two states must exist simultaneously, two flip-flops can be asserted. The output signals produced by each state flip-flop can then be used to control two operations simultaneously.

This type of system allows the implementation of the *fork* and *join* operations in digital controllers. Figure 6.78 indicates the idea of *forking* and *joining* in a state machine controller.

The state machine progresses from state a to b. At this point, possibly depending on an input, the system creates two states, c and g, that exist during t_3. Two states also exist during t_4 (d and h) and during t_5 (e and i). As t_6 is entered, the system returns to a single state f. During time slots t_3, t_4, and t_5, two separate processes can be controlled concurrently.

A second advantage of the "one-hot" state machine is that design methods are simpler and this type of controller is more amenable to computer-aided design techniques than is the conventional state machine. In proceeding from a state diagram to a finished IC chip, automated design methods can be more easily applied to the "one-hot" state machine, leading to a shorter development time. This can be an obvious advantage in the highly competitive electronics field.

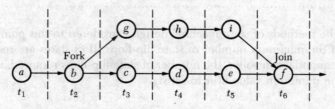

FIGURE 6.78   Schematic representation of forking and joining.

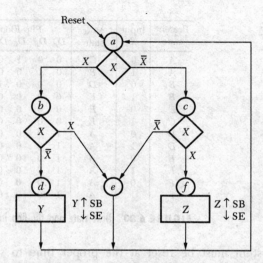

FIGURE 6.79 State diagram for a serial bit checker.

While more flip-flops are required in this system, an output state decoder is unnecessary and the IFL is often fairly simple. We will demonstrate a "one-hot" design based on the state diagram of Figure 6.22, which is reproduced in Figure 6.79. This system checks the signal X on a serial input line and generates output Z if an input of 01 occurs on X and generates output Y if 10 occurs. The circuit continues to check the data line after a delay of one state time after the previous check has been made.

In Example 6.2, this system was realized by a conventional state machine requiring three flip-flops. For a "one-hot" design, six flip-flops are required. We will label the flip-flop outputs A through F to correspond to states a through f. Assuming that D flip-flops are used, the next state and flip-flop input table is shown in Figure 6.80.

The equations for the D inputs are easily found from Figure 6.80 to be

$$D_A = D + E + F$$

$$D_B = AX$$

$$D_C = A\bar{X}$$

$$D_D = B\bar{X}$$

$$D_E = BX + C\bar{X}$$

$$D_F = CX$$

Present state	Input X	Next state	Flip-Flop Inputs D_A D_B D_C D_D D_E D_F					
A	0	C	0	0	1	0	0	0
A	1	B	0	1	0	0	0	0
B	0	D	0	0	0	1	0	0
B	1	E	0	0	0	0	1	0
C	0	E	0	0	0	0	1	0
C	1	F	0	0	0	0	0	1
D	0	A	1	0	0	0	0	0
D	1	A	1	0	0	0	0	0
E	0	A	1	0	0	0	0	0
E	1	A	1	0	0	0	0	0
F	0	A	1	0	0	0	0	0
F	1	A	1	0	0	0	0	0

FIGURE 6.80 Next state and flip-flop input tables.

The system must be reset at the proper time to state a. This can be accomplished by driving the direct flip-flop inputs as shown in the finished circuit of Figure 6.81.

An output decoder is unnecessary in this system and the IFL design is not overly complex. There is more potential for serious errors if such a system is used in controllers that must not allow the simultaneous existence of two states. For example, a traffic controller normally assigns states that drive the red, yellow, and green lights for both directions. If a conventional state machine moves to an erroneous state, the possibility of two green lights being asserted does not exist. In a "one-hot" system, both flip-flops that drive a green light could be erroneously asserted, leading to serious consequences. In such cases, the state flip-flop outputs can be connected to a detector that senses when more than one flip-flop is asserted. This can be accomplished with an n-input NEXOR gate that drives the system to a predetermined state when the erroneous condition is detected.

SUMMARY
• • • • • • • • • • • • • • •

1. Sequential counters can be designed as simple state machines.
2. Design techniques for state machines can minimize the number of flip-flops required for controllers and can minimize the number of gates used for input- and output-forming logic.
3. The elimination of unwanted glitches in controllers can be accomplished by following the outlined procedures for OFL design.
4. A state diagram can be produced from a consideration of the timing chart of a sequential system.
5. The basic concept of state machines leads to several possible methods of implementation.

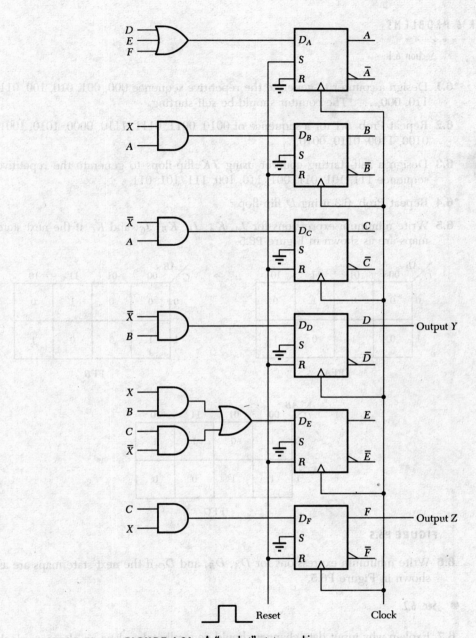

FIGURE 6.81   A "one-hot" state machine.

CHAPTER 6 PROBLEMS

● Section 6.1

°**6.1** Design a counter to generate the repetitive sequence 000, 001, 010, 100, 011, 110, 000, ... The counter should be self-starting.

6.2 Repeat Prob. 6.1 for a sequence of 0010, 0011, 0111, 1110, 0000, 1010, 1001, 0100, 1100, 0110, 0010, ...

6.3 Design a self-starting counter using JK flip-flops to generate the repetitive sequence 111, 101, 011, 001, 110, 100, 111, 101, 011, ...

°**6.4** Repeat Prob. 6.3 using D flip-flops.

6.5 Write minimum expressions for J_A, K_A, J_B, K_B, J_C, and K_C if the next state maps are as shown in Figure P6.5.

C \ AB	00	01	11	10
0	0	0	1	0
1	0	1	0	1

FFA

C \ AB	00	01	11	10
0	0	0	1	0
1	1	0	0	1

FFB

C \ AB	00	01	11	10
0	1	0	1	1
1	1	1	0	0

FFC

FIGURE P6.5

6.6 Write minimum expressions for D_A, D_B, and D_C if the next state maps are as shown in Figure P6.5.

● Sec. 6.2

6.7 Explain why input data changes should generally take place on alternate clock transitions to those causing the state changes.

6.8 If the delayed state transition method is used and the maximum delay time between clock transitions and input data changes is 6 ns, the maximum propagation delay time of the IFL is 12 ns, and the clock setup time is 3 ns, what is the minimum necessary delay between the clock signal and the state flip-flops?

● Sec. 6.3

6.9 How many state flip-flops would be required to implement the state diagram of Figure 6.68?

°**6.10** How many state flip-flops would be required to implement the state diagram of Figure 6.65 without eliminating equivalent states?

● Sec. 6.4

6.11 Design the IFL for a state machine to implement the waveform shown in Figure P6.11. Minimize the gates for the IFL. When the input is asserted, it will always have a duration of four clock periods. *Hint:* Use trial-and-error methods to generate a state diagram with four total states and two output states.

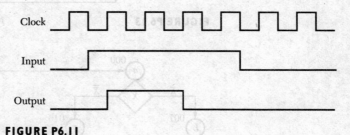

FIGURE P6.11

6.12 Repeat Prob. 6.11 using MUXs for the IFL.

● Sec. 6.5

°**6.13** Select states to eliminate output glitches for the diagram shown in Figure P6.13.

6.14 Implement the system shown in Figure P6.14. Show timing chart. Use JK flip-flops and gates for IFL.

6.15 Repeat Prob. 6.14 using D flip-flops and MUXs for IFL.

6.16 Implement the circuit described in Prob. 5.23. Show state diagram and timing chart. Use JK flip-flops and gates for IFL.

°**6.17** Repeat Prob. 6.16 using D flip-flops and MUXs for IFL.

6.18 Implement the circuit described in Prob. 5.24. Show state diagram and timing chart. Use D flip-flops and gates.

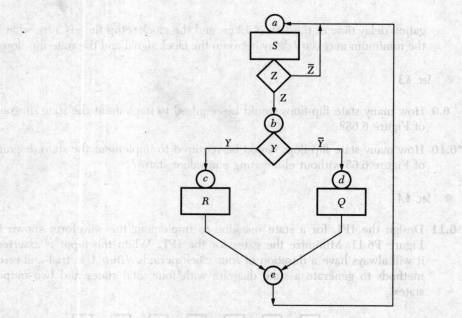

FIGURE P6.13

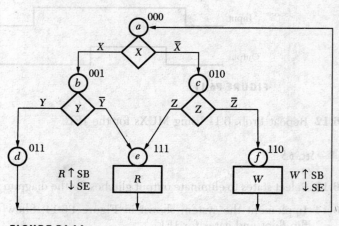

FIGURE P6.14

6.19 Assuming output glitches are unimportant, realize the state machine corresponding to the state diagram of Figure P6.19. Use MUXs for the IFL.

6.20 Realize the state machine represented by the diagram shown in Figure P6.20. Use MUXs for the IFL and use clock suppression to eliminate output glitches.

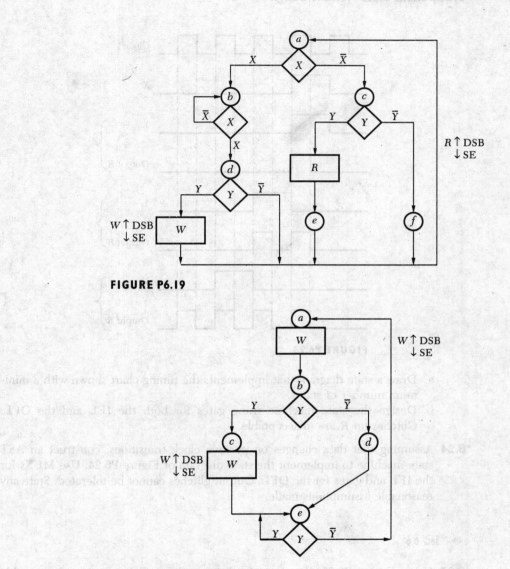

FIGURE P6.19

FIGURE P6.20

6.21 Repeat Prob. 6.19 using a holding register to eliminate output glitches. Assume R and W can remain asserted for a full clock period.

6.22 Construct a state machine to implement the state diagram of Prob. 5.26. State any reasonable assumptions made.

6.23 The input X initiates an output R. Another input Y is always asserted in one of three possible ways shown in Figure P6.23.

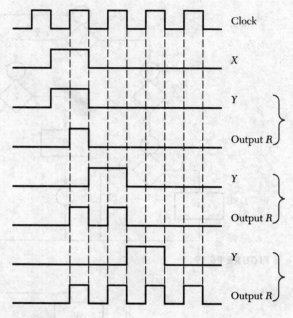

FIGURE P6.23

a. Draw a state diagram that implements the timing chart shown with a minimum number of states.

b. Design the state machine using gates for both the IFL and the OFL. Glitches on R are unacceptable.

°6.24 Assuming that data changes on positive clock transitions, construct an AST state machine to implement the state diagram of Figure P6.24. Use MUXs for the IFL and gates for the OFL. Output glitches cannot be tolerated. State any reasonable assumptions made.

● **Sec. 6.6**

6.25 Using Figure P6.25, draw a state diagram that will implement the possible outputs shown. Design the OFL using gates, but do no other design. Assume that X and Y will each be asserted only one cycle, but Y can be asserted in one of the three possible ways shown when X has been asserted.

● **Sec. 6.7**

6.26 This system has no output if RDY is not asserted, regardless of the values of X and Y. When RDY is asserted, the output signal will depend on the values of X

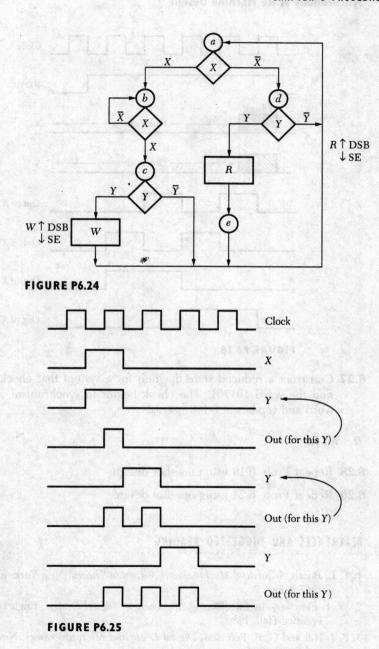

FIGURE P6.24

FIGURE P6.25

and Y that are present during the first clock period in which RDY is asserted. All other values of X and Y are unimportant. RDY always lasts for five clock periods. Draw a state diagram to produce the four possible output waveforms of Figure P6.26 with as few states as possible and using no counters.

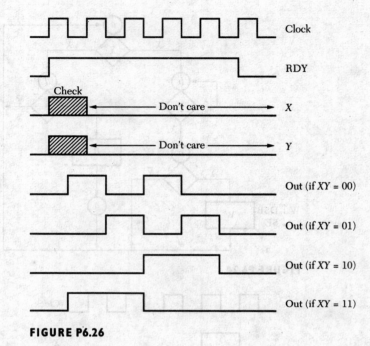

FIGURE P6.26

6.27 Construct a reduced state diagram for a system that checks an input line to find the word 101101. The check begins in synchronism with the start of a word and repeats at 6-bit intervals.

● **Sec. 6.8**

6.28 Repeat Prob. 6.19 using one-hot design.

6.29 Repeat Prob. 6.24 using one-hot design.

REFERENCES AND SUGGESTED READING
· ·

1. T. L. Booth, *Sequential Machines and Automata Theory*. New York: John Wiley and Sons, 1967, Chap. 3.

2. W. I. Fletcher, *An Engineering Approach to Digital Design*. Englewood Cliffs, NJ: Prentice-Hall, 1980.

3. F. J. Hill and G. R. Peterson, *Digital Logic and Microprocessors*. New York: John Wiley and Sons, 1984.

4. H. Lam and J. O'Malley, *Fundamentals of Computer Engineering*. New York: John Wiley and Sons, 1988.

5. M. M. Mano, *Digital Design*, 2nd ed. Englewood Cliffs, NJ: Prentice-Hall, 1991.

6. D. Winkel and F. Prosser, *The Art of Digital Design*, 2nd Ed. Englewood Cliffs, NJ: Prentice-Hall, 1987.

7

Interfacing and Design of Synchronous Systems

Chapters 5 and 6 introduced the sequential controller or state machine and developed principles that allow the design of these circuits. The purpose of a state machine is to control some digital system. To this point, the state machine has been considered as a separate entity without considering how to interface this circuit with the system it will control. This chapter introduces the techniques required to use the state machine in practical digital system design.

7.1

MAINLY SYNCHRONOUS SYSTEMS

SECTION OVERVIEW The mainly synchronous system is introduced in this section. This very important type of state machine requires special treatment of the asynchronous inputs. After establishing the problem of erroneous state transitions, three methods of solving this problem are discussed. The go-no go configuration is the first and best method to use whenever possible. Conversion of asynchronous to synchronous inputs is the second method considered. State assignment to avoid erroneous transitions is the third method suggested.

The state machines considered in the last two chapters were assumed to have input data that changed in synchronism with the system clock. These synchronous systems actually occur in practice, allowing the principles discussed to be applied to their design. Perhaps the state machine most often encountered in practice, however, is the mainly synchronous system. This system has clock-driven state changes and one or more asynchronous inputs. An asynchronous input has no fixed relationship to the clock signal and consequently may change values at any point of the clock period.

Two important categories of asynchronous systems are (1) systems requiring human interaction and (2) systems that must communicate with one another, but that are driven by different master clocks. A human operator must typically communicate with a digital system by switches, the assertion of which will have no relation to the clock signal. Operating a computer keyboard is an example of asynchronous human communication with a digital system (the computer). The computer's master clock transitions bear no relationship to the closure of the keyboard switches. The signals generated from the keyboard are then asynchronous inputs.

Two separate digital systems exchanging information over connecting lines will also transmit or receive asynchronous signals. Even if the two system clocks are adjusted to the same frequency, they cannot be controlled precisely enough to result in simultaneous clock transitions. Hence, data sent by one system in synchronism with that system's clock is asynchronous with respect to the other system's clock. Most terminals communicating serially with computers are examples of this type of system.

In this section, we will consider the problems that can arise in asynchronous systems along with some possible solutions.

7.1.1 ERRONEOUS STATE TRANSITIONS

When an asynchronous input changes just prior to the state-changing clock transition, the system may move to a state not indicated by the state diagram. To demonstrate this problem, we will consider the diagrams of Figure 7.1. If the present state of the system is a and $X = 0$, the IFL generates $D_A = 0$ and $D_B = 1$ to cause the next state to be 01. Now suppose that X changes to 1 just prior to the positive clock transition. The IFL will respond by generating $D_A = 1$ and $D_B = 0$ to set up a next state of 10 as indicated by the state diagram. Unfortunately, D_A and D_B will not change instantaneously due to propagation delays in the input logic. Perhaps D_A will move from 0 to 1 before D_B moves from 1 to 0. If the state change occurs at this instant, the next state will be 11 instead of 10. On the other hand, the propagation delays may be such that D_B responds to a change in X more rapidly than does D_A. If a state change occurs and D_B has changed from 1 to 0, but D_A has not completed the transition from 0 to 1, the next state will be 00 instead of 10.

If these states of 11 or 00 are reached instead of the proper state, an erroneous state transition has occurred. Although the probability of X producing an erroneous state transition may be very low, perhaps 1 in 1000, the one time that such a problem

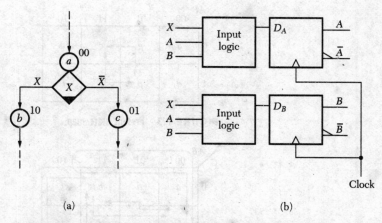

(a)   (b)

FIGURE 7.1 (a) Partial state diagram. (b) Block diagram of system.

occurs may lead to serious consequences. Therefore, we must consider methods to deal with this problem if we intend to produce a reliable system.

It is possible to predict the code of the erroneous states by using a K-map. The system of Figure 7.2 is a mainly synchronous system that can lead to erroneous states. The present state map is shown in Figure 7.3. The state machine switches from state a to b or c, depending on the asynchronous variable X. From the time

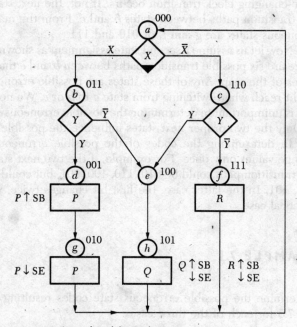

FIGURE 7.2 A mainly synchronous system.

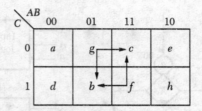

FIGURE 7.3   Present state map.

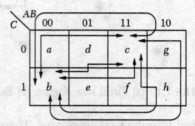

FIGURE 7.4   Another present state map.

state a is entered, the flip-flop inputs will drive the IFL to result in a next state of b or c. If the variable X changes, the gating conditions on the state flip-flop change toward the proper next state, but may not reach the proper levels before the state-changing clock transition occurs. If not, the next state will be one that lies on the transition paths between states b and c. From the map we see that the possible erroneous states are g and f, or 010 and 111.

Now let us assume a second state assignment as shown in Figure 7.4. In this map, there are six possible transition paths between b and c that traverse all six remaining states of the map. Any of these states are possible erroneous states that the system might reach when switching from state a to b or c. We note here that the state code of a is unimportant in determining the possible erroneous states when switching from a. Only the two proper next states influence the possible erroneous state codes.

In determining the codes of the possible erroneous states, a given bit may change values only once. For example, if the two next state codes are 010 and 101, the transition path could be 010, 110, 100, 101, but could not be 010, 110, 100, 000, 001, 101. In the latter case, the first bit changes twice, which cannot occur in the practical case.

EXAMPLE 7.1
· ·

Determine the possible erroneous state codes resulting from the diagram of Figure 7.5 for each of the three cases.

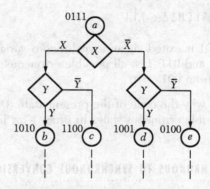

FIGURE 7.5 Partial state diagram for Example 7.1.

a. X asynchronous and Y synchronous,
b. X synchronous and Y asynchronous,
c. X and Y asynchronous.

SOLUTION

It is important in this example to recognize that the synchronous variable will change at midstate time in an AST system and all transients on the IFL outputs due to this variable will have settled when the state-changing transition occurs. Only the asynchronous variable can cause transient gating conditions at the time of the state change.

a. If $Y = 1$, X might change near the state change, causing the IFL to be in transition between states b and d. The possible erroneous state codes between 1010 and 1001 are 1000 and 1011. If $Y = 0$, the asynchronous switching of X will cause the IFL to create flip-flop input signals in transition between states c and e. There are no possible erroneous state codes between 1100 and 0100 since these codes are adjacent. Thus, the possible erroneous state codes are 1000 and 1011.

b. If $X = 1$ and Y changes asynchronously, the IFL outputs can be in transition between states b and c. The erroneous state codes are then 1000 and 1110. If $X = 0$ and Y changes asynchronously, the IFL can be in transition between states d and e. The erroneous codes are then 0000, 0001, 0101, 1000, 1100, and 1101.

c. When both X and Y are asynchronous, the IFL outputs can be in transition between any two states of the group b, c, d, and e. The possible erroneous state codes are then 0000, 0001, 0010, 0101, 0110, 1000, 1011, 1100, 1101, and 1110. Note that one of these codes corresponds to state c, but is nevertheless an erroneous state. ●

● ● ●   **DRILL PROBLEMS Sec. 7.1.1**

1. State 001 is exited via an asynchronous variable X. The two following states are 101 and 011. List all possible erroneous states to which the system might switch from 001.

2. Explain why the code of the present state (001) does not influence the code of the possible erroneous states in Prob. 1 (or in any mainly synchronous system).

●   **7.1.2   ASYNCHRONOUS TO SYNCHRONOUS CONVERSION**
. . . .

In theory, it is possible to convert all asynchronous input variables to synchronous variables by means of flip-flops. One latch can be used for each asynchronous input to create synchronized input signals for use by the state machine. This synchronizing register is driven by a signal derived from the state machine clock. The newly created synchronous signals can then be applied as inputs to the synchronous state machine. From a design standpoint, conversion to synchronous signals is very simple, allowing synchronous state machine design principles to be utilized.

The D latch shown in Figure 7.6 demonstrates the conversion of an asynchronous variable G to a synchronous variable called T driven by the negative clock transition.

There are two problems related to the use of a flip-flop in converting asynchronous inputs to synchronous. The first is the obvious increase in cost and complexity. One flip-flop is required for each asynchronous input variable. This is not often a serious problem as the cost of additional IC flip-flops is quite low.

In addition to this problem, however, there is a possibility of unreliable operation in the D flip-flop. This problem has been well-documented in the literature [1, 4] and occurs when the asynchronous input changes value very near the clock transition that loads the flip-flop. If the input data change occurs prior to the clock transition by an amount that exceeds the specified set-up time for the flip-flop, the clock pulse will reliably set the data into the flip-flop. If the input data changes after the clock transition, no change in the flip-flop will result. There is a very small window in time, beginning just after it is too late to satisfy set up time and ending near the clock transition, wherein an input data change can be troublesome. This window

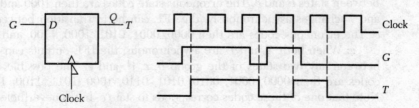

FIGURE 7.6   Asynchronous to synchronous signal conversion.

may be only 0.1 ns long for a flip-flop that can operate at a clock frequency of 100 MHz. A possible result of an input change occurring in this window is a strange output for the flip-flop.

Normally, the output of the flip-flop sets at either the high or low voltage level, but never at the midpoint of the two levels. If the input changes very near the clock transition, it is possible for the flip-flop output to move toward the opposite level, then stop moving somewhere between the levels. This midpoint level is called a metastable state and leads to an ambiguous output. The prefix "meta" means something that is in the middle of a transition region between two areas. We will refer to this voltage as a metastable level. In this case, the metastable level may exist near the midpoint of the 0 and 1 voltage levels. Figure 7.7 indicates the possible metastable level for the output of a flip-flop.

Typically, the metastable level will exist for a short period of time and then disappear as the output moves to one of the stable states. If the flip-flop drives succeeding circuitry such as a gate or another flip-flop input, the metastable state can lead to a fault in the system.

This peculiar behavior of clocked flip-flops with asynchronous inputs is a probabilistic event that depends on the clock frequency, the input data frequency, and the design of the flip-flop [1]. There is no known method of constructing a single flip-flop to avoid metastable levels when asynchronous inputs are present. The appropriate approach to the metastability problem is to recognize its existence and design systems that are unaffected by the occurrence of these unusual levels.

The effects of the occurrence of this metastable level can generally be minimized if the asynchronous input data is guaranteed to change at a lower frequency than the system clock. This means the input will always exist at a particular level for over a full clock period. If the input changes near the clock transition and causes a metastable output, the synchronizing latch could output a metastable level that may be interpreted as a 0 or a 1, followed by the correct output at the next clock transition. This situation is depicted in Figure 7.8. If timing is not a critical factor in the succeeding circuitry, the correct output is guaranteed to occur with the synchronizing latch when the data lasts longer than one clock period.

Although the synchronizing register is a reasonable solution in many applications, there is one situation that warrants closer attention. If the system state changes take place on the positive-going clock transitions, the input synchronizing register should be updated prior to this time. This register update should be far enough in advance of the state change to allow the input-forming logic outputs to settle before the state-changing clock transition. In some instances, the synchronizing register will

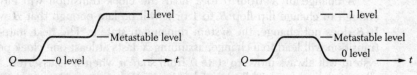

FIGURE 7.7   Possible metastable flip-flop outputs.

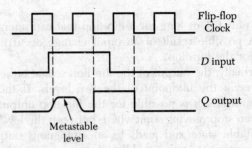

FIGURE 7.8 Input that lasts longer than one clock period.

be updated on the opposite clock transition. If the asynchronous variable changes just after the register is updated, this change does not affect the next state transition. It is not until the next update of the synchronizing register plus a half-period of the clock that this variable has an effect on the system.

If it is important to register the change of the asynchronous variables as soon as possible, the register update should take place just prior to the state change. This can be accomplished by changing the synchronizing register flip-flops and state flip-flops on the same clock transition. In this case, the state change must be slightly delayed to allow the IFL outputs to settle prior to this state change. Delaying inverters are often used to achieve this delay. This minimizes the delay between input variable assertion and effect on the system. Unfortunately, a maximum delay of almost one state time is still possible.

Figure 7.9(a) shows the use of an input synchronizing register, which is updated on the transition that is alternate to the state-changing transition. Figure 7.9(b) indicates the method of registering input changes nearer the state-changing transition. Either additional inverters or slower inverters may be required to achieve the proper delay.

7.1.3   THE GO-NO GO CONFIGURATION

In several instances, branching from a given state to only a single next state is controlled by an asynchronous variable. When this is true, the go-no go arrangement of Figure 7.10 is appropriate. In this configuration, states a and b must be logically adjacent to avoid erroneous state transitions. We will assume state a corresponds to $AB = 10$ and state b is 11.

A change in X from 0 to 1 near the clock transition will either occur early enough to change flip-flop B to 1 or it will be late enough that B will remain at 0. If B does not change, the system remains in state a. The next state-changing clock transition will lead to a change, assuming X lasts at least one clock period. Thus, the system will always move to state b from state a when X is asserted. In this method, all asynchronous variables must last more than one clock period to ensure proper switching. A shorter input pulse must be stretched to guarantee reliable operation.

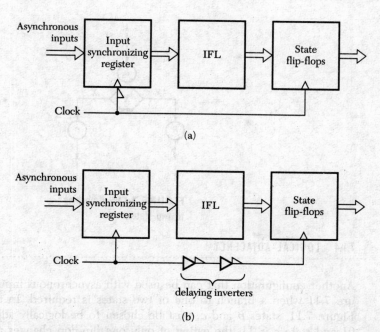

FIGURE 7.9 Asynchronous to synchronous conversion: (a) on alternate clock transition, (b) on same transition.

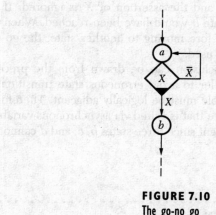

FIGURE 7.10
The go-no go
configuration.

The go-no go configuration is the most reliable scheme available in mainly synchronous system design. Although modification of a state diagram to achieve this configuration may result in additional states, the increased reliability is well worth the investment for most applications.

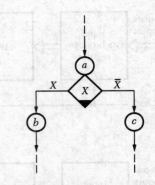

FIGURE 7.11   Branching
to two states.

7.1.4   LOGICAL ADJACENCY

Another configuration that can be used with asynchronous inputs is shown in Figure 7.11 when a branch to one of two states is required. In the arrangement of Figure 7.11, states b and c must be chosen to be logically adjacent. If state b is 01 and state c is 11, the gating of only one flip-flop changes as X changes. If X arrives early enough, the change is reflected by switching to the proper state. The disadvantage with this configuration is that if X changes from 0 to 1 near the state change time, the system may move to state c. If X remains equal to 1, the system never reaches state b and the assertion of X is ignored. If X had been asserted just slightly sooner, state b would have been reached. When it is important to register a change in X before moving to another state, the go-no go arrangement of Figure 7.11 should be used.

An important conclusion can be drawn from the preceding paragraphs. We have found that in order to avoid erroneous state transitions, the states following an asynchronous variable must be logically adjacent. Therefore, a maximum of two states may follow a state that is exited via asynchronous variables. Figure 7.12 shows an improper arrangement since three states b, c, and d cannot be mutually adjacent.

7.2

TOP-DOWN DESIGN

SECTION OVERVIEW This section defines top-down design and proposes a method to apply this type of design to digital systems.

The last two chapters discussed principles that allow the design of state machines. The remainder of this chapter will consider formal procedures for (1) the

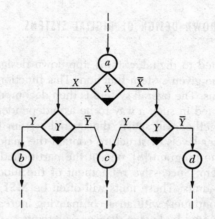

FIGURE 7.12 State diagram leading to erroneous state transitions.

design of state machines and (2) the design of digital systems that contain a state machine.

7.2.1 DEFINITION OF TOP-DOWN DESIGN

The term "top-down design" has grown out of the programming area and relates closely to the concept of structured programming. Niklaus Wirth, the developer of the PASCAL programming language, provides the following definition of structured programming [8]:

● Structured programming is the formulation of programs as hierarchical, nested structures of statements and objects of computation.

Implied in this definition is a decomposition or breaking down of a large problem into component parts. These parts are then decomposed into smaller problems with this successive refinement continuing until each remaining task can be implemented by simple program statements or a series of statements. As each statement or structure is executed, a part of the overall objective is accomplished. The program is decomposed into units called modules. These modules are decomposed into control structures or series of statements. The statement is the basic unit of the program.

A significant point in applying top-down design is that the modules should be selected to result in minimum interaction between these units. In general, complexity can be reduced with the weakest possible coupling between modules. Minimizing connections between modules also minimizes the paths along which changes and errors can propagate into other parts of the system [3, 8]. While complete independence is impossible since all modules must be harnessed to perform the overall program objective, interaction is minimal in a well-designed program.

7.2.2 TOP-DOWN DESIGN OF DIGITAL SYSTEMS

When applied to digital systems, top-down design proposes a broad, abstract unit to satisfy the given system function. This function will be characterized by a set of specifications. The overall system is then decomposed into modules. These modules are partitioned in such a way to be as independent of one another as possible, but working together will satisfy the overall system function. Just as in programming where higher levels of structure contain the major decision-making or control elements of the program [8], one of the partitioned modules will be the control unit of the system. Successive refinement of the modules leads to decomposition into operational units. These units will often be MSI circuits or groups of circuits. Refinement continues, with an accompanying increase in detail, until all sections can be implemented by known devices or circuits.

There are certain characteristics of the digital field that make top-down design an easy method to apply. For example, the large number of SSI and MSI circuits available allows operational units and even modules to be realized directly with chips. These chips are generally designed to require a minimal number of control inputs, and thus minimal interaction between modules is automatically achieved, at least to an extent.

Another advantage of the top-down method in digital design is that the control module can be realized as a state machine. This allows well-known design procedures to be used in designing a reliable control unit.

7.3
DESIGN PROCEDURES

SECTION OVERVIEW This section proposes two design procedures. The first procedure is for the design of an overall digital system. Among other things this procedure will lead to a control unit module. The second procedure proposes a method of designing this control unit using a state machine.

7.3.1 DIGITAL SYSTEM DESIGN

An important component in most digital systems is the control unit that oversees the operation of all other units in the system. This component accepts input information from an external source along with information from other internal units and generates appropriate output signals for controlling other units of the system. As mentioned earlier, this control unit can consist of a state machine. If a state machine is used, all design principles developed over the years for such devices can be applied to the controller design.

The overall system design can be broken down into two procedures. The first applies top-down design to decompose the system into a controlling state machine module and other sections easily implemented with digital circuits. The second procedure consists of designing the state machine.

We will now consider the accomplishment of the first procedure. Using the idea of top-down design, this procedure might consist of the following:

DD1. Determine, from a thorough study of the system specifications, the major goals of the system.

DD2. Decompose the system into functional modules required to implement the major goals. One module will be the controlling state machine.

DD3. Determine the next lower level of tasks that must be accomplished by each module.

DD4. Decompose all modules except the controlling state machine into operational units designed to accomplish these tasks. Often these subsections will correspond to MSI circuits.

DD5. Decompose the operational units until the entire system is reduced to components that can be realized with basic circuits.

DD6. Determine all control signals necessary to make the circuits operate.

DD7. Generate a timing chart to show the time relationship of all signals referenced to the clock signal.

DD8. Design the controlling state machine to implement the system.

DD9. Document your work.

We will emphasize that trial-and-error methods may play a part in several steps of this procedure. Furthermore, two or more different designs may be equally good; there is generally no single "best" design. Experience improves the efficiency of applying this or any design method. Some examples in the next section will demonstrate the application of the proposed procedure.

One of the major parts of the proposed design procedure is the development of the controlling state machine that will now be considered.

7.3.2 OVERVIEW OF STATE MACHINE DESIGN

As indicated previously, a state machine is a sequential circuit that controls the behavior of a digital system. Although there may be several different sequences of states that can be executed, the operation of the system is cyclic. The particular sequence executed depends on external conditions or signals applied along with internal states of the system. At a given state, the next state of the sequence always depends on the present state and often depends on the input signals. Outputs can be generated at appropriate states to drive or control other circuits or systems in order to perform useful functions. As a result of this control capability, the system is often called a controlling state machine.

The design of a state machine consists of the following:

1. Developing an appropriate state diagram,
2. Selecting the proper number of states,
3. Designing the input-forming logic,
4. Designing the output-forming logic.

While these four steps appear to be reasonably straightforward, there is a tremendous latitude of choice involved in each step that requires creativity on the part of the designer. There are, for example, many different state diagrams that generally can satisfy a given set of specifications. As we shall later see, states can often be exchanged for counter circuits to reduce the total number of states. In IFL and OFL design, the highest priority goes to developing a reliable system. Elimination of erroneous state problems and troublesome glitches must be accomplished before other matters such as minimization can be considered.

There are various quantities that a designer might attempt to minimize, depending on the circumstances related to the design. A state machine may be needed for a research or development project with no intention of marketing the resulting system. In such a case, minimizing the required design and construction time would be appropriate. Off-the-shelf SSI and MSI circuits would probably be used for this project with little consideration given to component cost. On the other hand, if a controller with a high-volume sales potential is to be marketed, production cost is the quantity that must be minimized. If SSI and MSI circuits are to be used, chip count and cost will determine overall system cost. Thus, attention to these factors must be given. If the system is to be integrated, the design should follow guidelines imposed by the IC fabricator to minimize cost.

While the design process of a state machine is not difficult, such factors as component cost, fabrication cost, company policy, market potential, design flexibility, labor cost, and other considerations tend to drastically influence the process.

7.3.3 STATE MACHINE DESIGN

The design of the controlling state machine is the second task that must be accomplished to complete the major goal of the system. This task also can be broken down into a series of steps. Because of the many degrees of freedom in designing a controlling state machine, no single procedure can be considered to be the best available. To provide a framework for solving this problem, the following steps are suggested.

SM1. Construct a state diagram for the state machine.

SM2. Eliminate all redundant states.

SM3. Construct the reduced state diagram.

SM4. Determine how to handle asynchronous inputs.

SM5. Determine whether output glitches are significant.

SM6. Assign proper states.

SM7. Construct next state table or maps along with input information for flip-flops.

SM8. Design input-forming logic.

SM9. Design output-forming logic.

One of the more difficult steps in this procedure is that of constructing a state diagram. The timing chart is very useful at this point and may be developed prior to or in conjunction with the construction of the state diagram. Section 6.6 discussed the generation of a state diagram from the timing chart. The following section will consider several examples of design demonstrating the application of these procedures.

7.4

DESIGN EXAMPLES

SECTION OVERVIEW This section demonstrates the two design procedures of the previous section. The first example applies the state machine design procedure. The next example is a simple digital system that applies both procedures. The remaining examples involve slightly more complex digital systems that again require the application of both procedures.

7.4.1 A TESTER CONTROL

A device receives input data from a set of sensors as shown in Figure 7.13. The tester generates a synchronous signal lasting exactly three clock periods. When this signal is applied to device 1, lines X and Y may become asserted depending on the sensor inputs. These lines will contain synchronous NRZ signals. If either of the combinations of X and Y shown in Figure 7.13 occur, the corresponding output should be generated on the Out line. Any combination of X and Y assertions other than that shown should result in no output. The tester cannot tolerate glitches on the T_{in} line.

This digital system is not completely specified, but the requirements on the controlling state machine are given. We have only to apply the state machine design procedure of Section 7.3.3 to complete the problem.

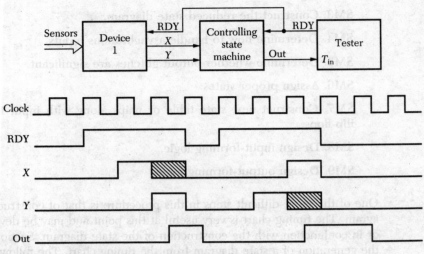

FIGURE 7.13 A test system controlled by a state machine.

STEP SM I

Construct a state diagram for the state machine.

Using a trial-and-error approach, we ultimately arrive at the diagram shown in Figure 7.14. The system will remain in state *a* until RDY is asserted. When RDY is asserted, the system checks inputs *X* and *Y*. If either of these signals is asserted during the first clock period following RDY assertion, the state machine should produce no output regardless of future values of *X* and *Y*. Since RDY lasts exactly three clock periods, delay state *d* is required to force the deassertion of RDY before the system returns to state *a*. If neither *X* nor *Y* is asserted during the initial clock period, the system moves to state *b*. This state checks the values of *X* and *Y* during the second clock period. If *X* is 1 and *Y* is 0, the system moves to state *c*. If *X* is 0 and *Y* is 1, the system moves to state *e*. Any other combination of *X* and *Y* moves the system to state *d*. In state *c*, the system checks *Y* during the third clock period. If *Y* = 1, the system moves to output state *f*. If *Y* = 0, the system moves to state *d*. If the system had reached state *e* instead of *c*, *X* would also be checked during the third clock period. If *X* = 1, output state *g* is reached. For this condition the output should be asserted for two clock periods. An output state could follow *g* to produce a longer output, but this state would be redundant. It would be equivalent to output state *f*. Therefore, it is eliminated and the system proceeds from state *g* to state *f* to produce the required length of output assertion.

STEP SM 2

Eliminate all redundant states.
This was accomplished as part of step SM1.

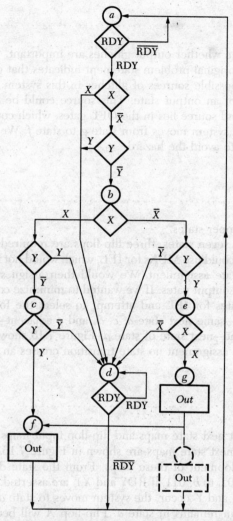

FIGURE 7.14 State diagram for the system of
Figure 7.13.

STEP SM3

Construct the reduced state diagram.
Removing the dashed lines in Figure 7.14 results in the reduced diagram.

STEP SM4

Determine how to handle asynchronous inputs.
There are no asynchronous inputs in this system.

STEP SM 5
. . . .

Determine whether output glitches are important.

The original problem statement indicates that glitches should not occur. There are two possible sources of glitches in this system. One source is that of transient crossing of an output state. This source could be eliminated by state assignment. The second source lies in the OFL gates, which could produce a static zero hazard when the system moves from state g to state f. We will choose states f and g to be adjacent to avoid the hazard.

STEP SM 6
. . . .

Assign proper states.

With seven states, three flip-flops are required. If we want to minimize design time, we could use MUXs for IFL, which would not impose any particular constraints on the state assignment. We would then assign states g and f to be adjacent to minimize output gates. If we wanted to minimize component cost, we would choose to use gates for IFL and attempt to select the following adjacencies: a, f, and d adjacent—same next state a; c, d, and e adjacent—next state of state b; a, b, and d adjacent—next state of state a. Figure 7.15 shows one possible state assignment. With this assignment no state transition crosses an output state.

STEP SM 7
. . . .

Construct next state maps and flip-flop input maps.

The next state maps are shown in Figure 7.16. We will consider only a part of the development of these maps. From the state diagram, the system moves from state a, 001, to b, 011, if RDY and $\bar{X}\bar{Y}$ are asserted. If RDY and any other combination of X and Y occur, the system moves to state d, 010. If RDY is never asserted, the system remains in state a. Flip-flop A will become 0 for any input condition.

C \ AB	00	01	11	10
0	c	d	e	$f°$
1	a	b	θ	$g°$

FIGURE 7.15 State assignment to eliminate output glitches.

C \ AB	00	01	11	10
0	Y	0	X	0
1	0	$\bar{X}Y$	θ	1

$$D_A = \bar{A}\bar{B}CY + ABX + BC\bar{X}Y + AC$$

C \ AB	00	01	11	10
0	$\bar{Y}$	RDY	$\bar{X}$	0
1	RDY	$\bar{X}+Y$	θ	0

$$D_B = \bar{A}\bar{B}\bar{C}\bar{Y} + \bar{A}\bar{B}CRDY \\ + BC\bar{X} + BCY + AB\bar{X} + \bar{A}\bar{B}\bar{C}RDY$$

C \ AB	00	01	11	10
0	0	$\overline{RDY}$	X	1
1	$\overline{RDY} + \bar{X}\bar{Y}$	0	θ	0

$$D_C = \bar{A}B\bar{C}\overline{RDY} + \bar{A}\bar{B}C\bar{X}\bar{Y} \\ + \bar{A}\bar{B}C\overline{RDY} + A\bar{C}X + A\bar{B}\bar{C}$$

FIGURE 7.16 Next state and D input maps.

Flip-flop B will become 1 if state b or d is to be entered, and flip-flop C becomes 1 if the next state is a or b. The entries in the 001 location of the next state maps reflect these conditions. Each location of the map is completed using similar considerations.

STEP SM 8
....

Design the IFL.

We recall that the next state maps also correspond to the input maps for D flip-flops. The reduced expressions for D_A, D_B, and D_C are shown in Figure 7.16. When using gates for IFL, it is sometimes more economical to use JK flip-flops. Accordingly, the input maps for this type of device are given in Figure 7.17. These apply the excitation table for a JK flip-flop in the map development. The resulting expressions are no simpler than those for the D flip-flops, and so D flip-flops will be used.

STEP SM 9
....

Design the OFL.

The output map is shown in Figure 7.18 along with the expression for Out. The final circuit is shown in Figure 7.19.

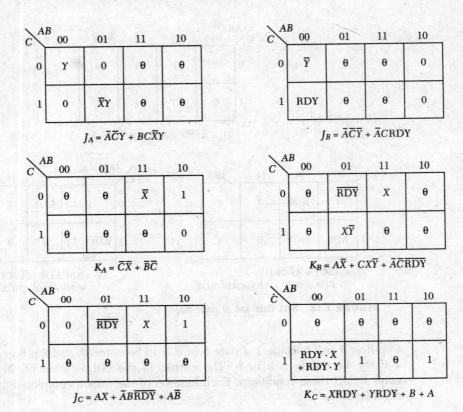

FIGURE 7.17 *JK* input maps.

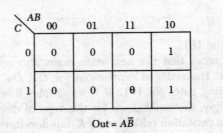

Out = $A\bar{B}$

FIGURE 7.18 Output map.

7.4.2 A FRAME COUNTER

We will now demonstrate the design of a digital system using the procedures given in the preceding section. We will consider a rather simple frame counter design. This system monitors a serial line for a particular word. Each time this word is detected, a counter is incremented to record the total number of occurrences of the word.

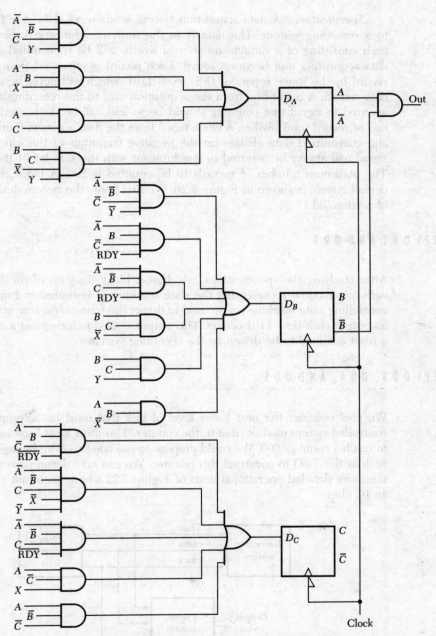

FIGURE 7.19 Controlling state machine for the system of Figure 7.13.

Specifications: A data acquisition system sends serial data at a 1 kHz bit rate to a recording system. The data is in the form of 4-bit words. Several records, each consisting of a variable number of words, will be transmitted each time the data acquisition unit becomes active. Each record is separated from the following record by the frame separator (FS) word 1101, which will never be contained in a data record. A count line from the acquisition unit to the recording system will be asserted to signal that counting should begin and will be deasserted to signal the end of record transmission. A clock signal from the data acquisition unit is available, and transmitted data change on the negative transition of this clock. The count signal will always be asserted in synchronism with the first bit of the first record. The maximum number of records to be counted is 200. A block diagram of the overall system is shown in Figure 7.20. We first apply the system design procedure of section 7.3.1.

STEPS DD1 AND DD2

After studying the specifications, we draw a block diagram of the frame counter with one module representing the state machine as indicated in Figure 7.21. The controlling state machine is also used to detect the frame separator word, producing an output each time 1101 occurs. This output is used to increment a counter having a reset assumed to be driven by the recording system.

STEPS DD3, DD4, AND DD5

We now consider the next lower level of task that must be accomplished by the controlled system module, that is, the counter. This must be at least an 8-bit counter to reach a count of 200. We could propose to cascade two 4-bit binary counter chips such as the 7493 to construct this counter. We can now decompose this block into the more detailed operational units of Figure 7.22 where each unit corresponds to an IC chip.

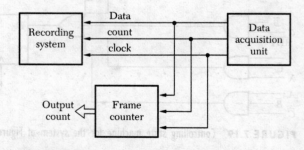

FIGURE 7.20   A data acquisition/recording system.

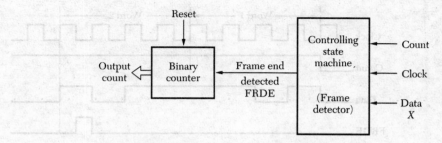

FIGURE 7.21 Block diagram of frame counter modules.

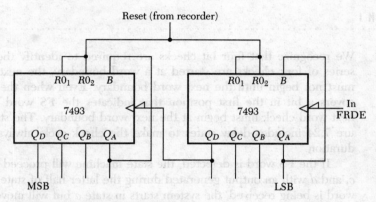

FIGURE 7.22 An 8-bit binary counter.

STEP DD6

The only control signal produced by the state machine is the frame end detected signal FRDE. Each time the frame separator word is detected, a pulse should appear at the input of the counter.

STEP DD7

A timing chart may not be necessary for this simple system, but one is shown in Figure 7.23.

STEP DD8

Here we must design the controlling state machine by applying the procedure of Section 7.3.3.

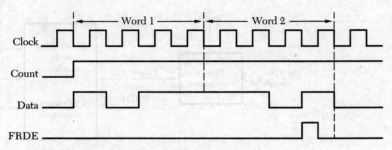

FIGURE 7.23 Timing chart for frame counter.

STEP SM1

We recognize that four bit checks are required to identify the FS word. Once a series of four checks are started at a word boundary, the next set of four checks must not begin until the next word boundary. Even when the state machine discovers a bit in the first position that indicates the FS word is not present, the next word check must begin at the next word boundary. The state diagram of Figure 7.24 includes delay states to make the check cycle always equal four bits in duration.

If the FS word is detected, the state machine will proceed through states a, b, c, and d with an output generated during the latter half of state d. When any other word is being received, the system starts in state a but will move into the sequence of states e, f, or g depending on when the first bit occurs that differs from the FS word. Assertion of the count line deasserts the reset, allowing the system to leave state a.

STEPS SM2 AND SM3

There are no redundant states in this diagram.

STEP SM5

Output glitches on FRDE will increment the counter causing a false count to result. We will eliminate output glitches by clock suppression.

STEP SM6

We will choose MUXs for the IFL and a gate for the OFL. Since clock suppression will be used to eliminate glitches, state assignment is arbitrary. We could decrease

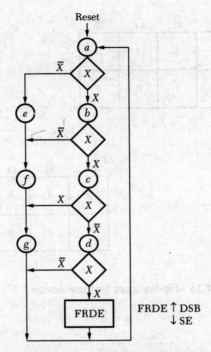

FIGURE 7.24 State diagram for frame detector.

C $\overset{AB}{\diagdown}$	00	01	11	10
0	a	b	c	d°
1	e	f	g	θ

FIGURE 7.25 State assignment for frame detector.

the number of inputs to the OFL gate by arranging state d next to the "don't care" state. One possible assignment is shown in Figure 7.25.

STEP SM7

Since MUXs will be used, D flip-flops are appropriate. The flip-flop input (and next state) maps are shown in Figure 7.26.

C \ AB	00	01	11	10
0	0	X	1	0
1	0	1	0	θ

$$D_A$$

C \ AB	00	01	11	10
0	X	1	X	0
1	1	1	0	θ

$$D_B$$

C \ AB	00	01	11	10
0	$\bar{X}$	$\bar{X}$	X	0
1	1	1	0	θ

$$D_C$$

FIGURE 7.26   Flip-flop inputs for frame detector.

STEP SM 8

We will use three 8:1 MUXs for the IFL connecting the state flip-flop outputs to the select lines. The maps of Figure 7.26 will be used to determine input connections.

STEP SM 9

The output map is shown in Figure 7.27 along with the expression for FRDE. Figure 7.28 indicates the final state machine.

C \ AB	00	01	11	10
0	0	0	0	X
1	0	0	0	θ

$$FRDE = A\bar{B}X$$

FIGURE 7.27   Output map for frame detector.

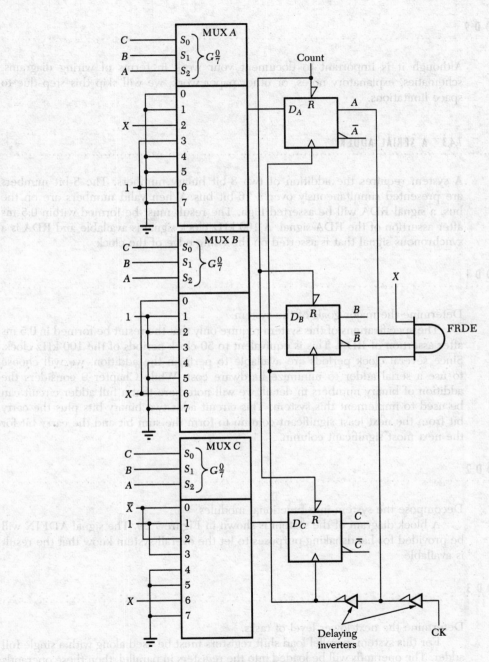

FIGURE 7.28 State machine for frame counter.

STEP DD9

Although it is important to document your work in terms of wiring diagrams, schematics, explanatory notes, or other paper work, we will skip this step due to space limitations.

7.4.3   A SERIAL ADDER

A system requires the addition of two 8-bit binary numbers. The 8-bit numbers are presented simultaneously over a 16-bit bus. When valid numbers are on the bus, a signal RDA will be asserted high. The result must be formed within 0.5 ms after assertion of the RDA signal. A 100 kHz clock signal is available and RDA is a synchronous signal that is asserted on the rising edge of the clock.

STEP DD1

Determine the major goals of the system.

The specifications of this system require only that the result be formed in 0.5 ms after assertion of RDA. This is equivalent to 50 clock periods of the 100 kHz clock. Since several clock periods are available to perform the addition, we will choose to use a serial adder to minimize hardware cost. While Chapter 9 considers the addition of binary numbers in detail, we will note here that a full adder circuit can be used to implement this system. This circuit adds two binary bits plus the carry bit from the next least significant column to form the sum bit and the carry bit for the next most significant column.

STEP DD2

Decompose the system into functional modules.

A block diagram of the system is shown in Figure 7.29. The signal ADFIN will be provided for handshaking purposes to let the overall system know that the result is available.

STEP DD3

Determine the next lower level of tasks.

For this system, parallel load shift registers must be used along with a single full adder. The operands will be loaded into the registers in parallel, then these operands will be presented serially to the full adder. The results will be stored serially into one of the shift registers as the operand shifts out of the register. In order to have a carry

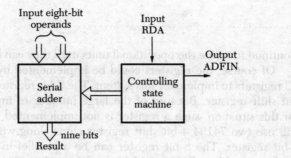

FIGURE 7.29 An 8-bit serial adder.

bit and also to provide a buffer between the outgoing operand and the incoming result, a 9-bit register will be used to store the result.

STEP DD4

....

Decompose the modules into operational units.

Figure 7.30 shows the operational units required to implement the serial adder.

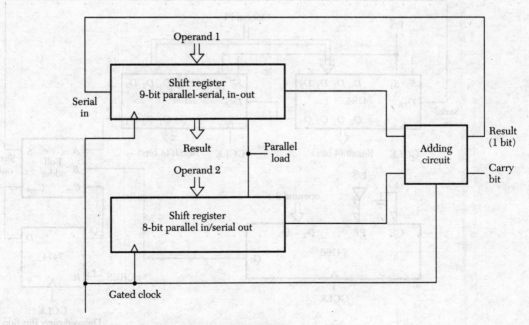

FIGURE 7.30 Operational units for serial adder.

STEP DD5

Continue to refine the operational units until they can be realized with basic circuits.

Of course, the registers could be implemented by D flip-flops. We will choose IC registers to implement this system. The 9-bit register must be a parallel in/parallel out shift register. Because of the large number of input and output pins required for this situation, such a register is not implemented in standard TTL circuits. We will use two 74194 4-bit shift register chips along with a single D flip-flop for the 9-bit register. The 8-bit register can be a parallel in/serial out device and will be implemented by a 74166, 8-bit shift register chip. The full adder can be constructed from gates. The major components of the system in terms of basic circuits are indicated in Figure 7.31. We note that the ninth bit of the upper register can be contained in the delayed carry flip-flop of the adding circuit.

STEP DD6

Determine all necessary control signals.

For this system, the following control signals are necessary.

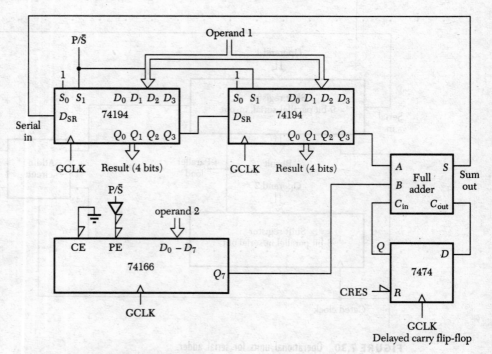

FIGURE 7.31 Basic circuit configuration.

1. A signal, P/$\bar{S}$, is required to switch the registers between the parallel and serial modes of operation. This signal must be high until the RDA signal is asserted and a positive transition is applied to the gated clock line. This procedure is necessary to load the 74166 and 74194 chips with the incoming operands.

2. A reset signal, CRES, must be applied to the delayed carry flip-flop. This must be a low-asserted signal for the 7474 chip.

3. The gated clock signal, GCLK, must first complete the parallel loading process and then provide 8 shift pulses to the system. This will cause the 8 bits to be sequentially presented to the adder with the results being stored in the 9-bit register (the two 74194 chips plus the D flip-flop).

4. A signal to indicate that the addition has been completed must be asserted at the proper time. This signal, designated ADFIN, must be deasserted when the two new operands are presented, that is, when RDA is asserted.

STEP D D 7

. . . .

Generate a timing chart.

After some careful consideration relative to timing problems, the timing chart of Figure 7.32 is proposed. The assertion of the RDA signal initiates the transfer and addition process. The ADFIN signal must be deasserted to indicate that new operands are being received to form a new sum. The timing of $\overline{\text{ADFIN}}$ is such that it

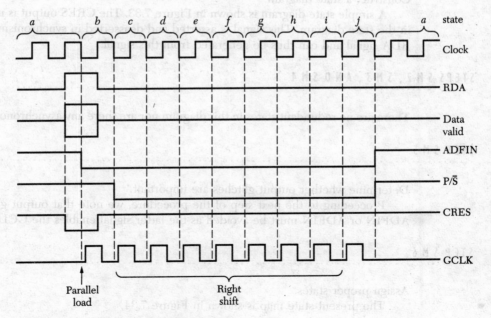

FIGURE 7.32 Timing diagram for serial adder.

can be used to enable an AND gate to create GCLK from the 100 kHz clock signal. The other input to the AND gate will be the inverted clock signal. The inverted clock signal should be delayed slightly to avoid glitches in the GCLK signal.

The P/$\overline{\text{S}}$ signal is high when the first transition of GCLK occurs. This positive transition of GCLK loads the two operands into the two registers in parallel. This loading takes place during the RDA assertion, which guarantees that the operand data lines contain valid information. The P/$\overline{\text{S}}$ line then goes low prior to the next transition and remains low until eight more positive transitions occur on GCLK. The last signal to be considered is the reset for the delayed carry flip-flop, CRES. This signal must be asserted low to clear the flip-flop prior to the occurrence of the first shift pulse and then returned high. Shifting begins with the second positive transition of GCLK, thus the CRES signal shown satisfies this requirement.

STEP DD8

Implement the state machine.

We will use the alternate state transition (AST) method, which allows state changes on the alternate clock transition to that of the input changes.

STEP SM1

Construct a state diagram.

A simple state diagram is shown in Figure 7.33. The CRES output is not shown in the state diagram. This signal is asserted and deasserted in synchronism with the RDA signal and can thus be generated from this signal.

STEPS SM2, SM3, AND SM4

There are no redundant states in this diagram nor are there any asynchronous inputs.

STEP SM5

Determine whether output glitches are important.

Proceeding to the next step of the procedure, we note that output glitches on ADFIN or $\overline{\text{ADFIN}}$ must be avoided as the latter signal enables the GCLK signal.

STEP SM6

Assign proper states.

The present state map is shown in Figure 7.34.

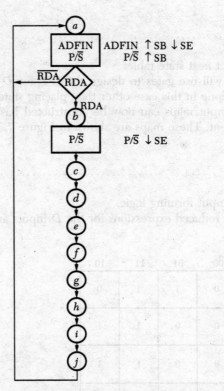

FIGURE 7.33 State diagram for serial adder.

	AB			
CD	00	01	11	10
00	$a°$	d	e	j
01	$b°$	c	f	i
11	θ	θ	g	h
10	θ	θ	θ	θ

Output states

FIGURE 7.34 Present state map for serial adder.

STEP SM 7
. . . .

Construct next state tables.
We will use gates to design the IFL for D flip-flops. Very little minimization can be done in this case other than placing states a and b adjacent. The next state and D input maps can now be constructed based on the state diagram and state assignment. These maps are shown in Figure 7.35.

STEP SM 8
. . . .

Design input forming logic.
The reduced expressions for the D inputs are given in Figure 7.35.

CD\AB	00	01	11	10
00	0	1	1	0
01	0	0	1	1
11	θ	θ	1	1
10	θ	θ	θ	θ

$$D_A = B\bar{D} + AD$$

CD\AB	00	01	11	10
00	0	1	1	0
01	1	1	1	0
11	θ	θ	0	0
10	θ	θ	θ	θ

$$D_B = B\bar{C} + \bar{A}D$$

CD\AB	00	01	11	10
00	0	0	0	0
01	0	0	1	0
11	θ	θ	1	0
10	θ	θ	θ	θ

$$D_C = ABD$$

CD\AB	00	01	11	10
00	RDA	0	1	0
01	1	0	1	0
11	θ	θ	1	1
10	θ	θ	θ	θ

$$D_D = C + AB + \bar{A}\bar{B}(\text{RDA}) + \bar{A}\bar{B}D$$

FIGURE 7.35   Next state and D input maps.

CD \ AB	00	01	11	10
00	1	0	0	0
01	0	0	0	0
11	θ	θ	0	0
10	θ	θ	θ	θ

$\text{ADFIN} = \overline{A}\,\overline{B}\,\overline{D}$

CD \ AB	00	01	11	10
00	1	0	0	0
01	1	0	0	0
11	θ	θ	0	0
10	θ	θ	θ	θ

$P/\overline{S} = \overline{A}\,\overline{B}$

FIGURE 7.36 Output maps for serial adder.

STEP SM 9

Design the output forming logic.

Output maps for ADFIN and $P/\overline{S}$ are shown in Figure 7.36.

The final control circuit is shown in Figure 7.37. The delaying inverters must delay the inverted clock signal more than the sum of the clock-to-output delay of the flip-flop plus the propagation delay of the gate that generates ADFIN plus the inverter that produces $\overline{\text{ADFIN}}$. The 100 kHz clock is used to drive the state machine.

7.4.4 A REACTION TIMER

In this example, we will assume that our company is approached by a prominent agent for athletes. This individual makes a living by negotiating contracts between professional sports organizations and his clients. Over the years he has been retained by some highly successful athletes and has recently become very selective in the clients he agrees to represent. He now requires each prospective client to complete a series of physical tests. From these tests the agent hopes to evaluate the probability of success in professional sports for each athlete. Of course he will represent only those athletes with the greatest potential.

Among other devices, the agent would like our company to build a reaction timer to measure human reflexes. After a few meetings between the agent and one of our engineers, the following set of specifications are agreed upon:

1. The reaction timer will have a RESET-READY toggle switch. In one position the system is reset. When the switch is set to the READY position, the system will initiate the measurement process.

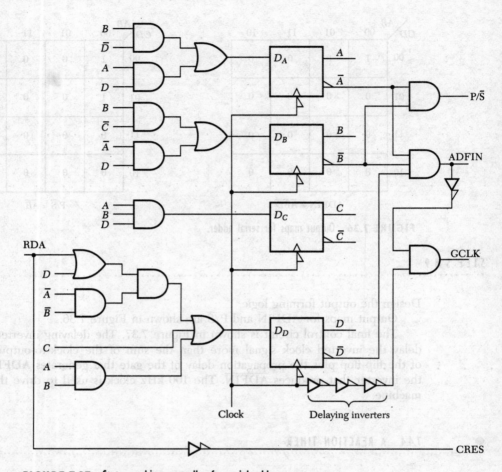

FIGURE 7.37 State machine controller for serial adder.

2. Some random time between 1 and 9 seconds after READY is asserted, the GO LED (light-emitting diode) is asserted. The assertion of the GO LED signals the start of the reaction timing process. The random time period before lighting the GO LED removes the possibility of anticipating the start of the timed period.
3. A pushbutton STOP switch is depressed to end the reaction measurement by stopping the timing process.
4. The reaction time in milliseconds must be displayed by three 7-segment displays. The error should be less than or equal to ±2 ms.
5. If an individual depresses the STOP switch prior to assertion of the GO LED, a CHEAT LED should be asserted.
6. The 7-segment displays are to be blanked until a valid measurement is completed.

Reaction time (ms)	LED	Classification
000–149	◯	Tennis player
150–179	◯	Basketball player
180–199	◯	Baseball player
200–299	◯	Quarterback
300–399	◯	Offensive lineman
400–499	◯	Referee
500–599	◯	Athlete's spouse
600–699	◯	Golfer
700–799	◯	Head coach
800–899	◯	Assistant coach
900–998	◯	Team owner
999 or over	◯	Dummy

FIGURE 7.38   Range LEDs.

7. An LED display is to be available that classifies various ranges of reaction time. When reaction time falls within a given range, the corresponding LED should be asserted. These ranges are given in Figure 7.38 along with the display layout.

In this situation, we again do not have the system block diagram. We must first do the system design after which the controlling state machine will be designed. We will apply the procedure of Section 7.3.1 to produce the overall system, but we will not carry out the design of the state machine.

Digital System Design

STEP DD1

Determine the major goals of the system.

The major goals are to measure the reaction time, display the results, check for cheating, and activate the proper LED to identify the range of results.

STEP DD2

Decompose the system into functional modules.

A block diagram that implements the broad functions of the system is shown in Figure 7.39. The controlling state machine will exchange signals with the timing section and will transmit to the display section to indicate when the display should be active. The classifier is a combinational circuit that is driven by the timing section and drives a portion of the display.

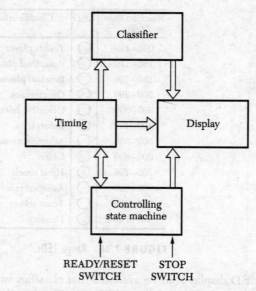

READY/RESET   STOP
SWITCH   SWITCH

FIGURE 7.39   Block diagram of reaction timer.

STEP DD3

Determine the next lower level of tasks.

This step considers the functions that each block should accomplish. The timing section must generate a random time period between the time that the READY switch is depressed and the GO LED assertion. At this point, it must begin timing the period taken for the person to press the STOP switch. The display section must contain means for indicating the number of milliseconds taken to respond to the GO LED assertion. The range here is 0 to 999 ms. This section must also contain the GO LED along with 12 range LEDs. The classifier section must determine in which range the reaction time falls.

STEP DD4

Decompose the modules into operational units.

This is a significant step that requires creativity and experience to perform effectively. A knowledge of available SSI and MSI circuits is necessary if these devices are to be used. The timing section can use three decade counters driven by a 1-kHz clock to perform the timing function. The random time can be generated with a fourth decade counter. The display section can be made up of three 7-segment displays and drivers with blanking. Fourteen LEDs and twelve 2-input AND gates are also required. The classifier section uses two 4-bit comparators, gates, and a decoder to identify the range of results. Each module broken down into operational units is shown in Figures 7.40 through 7.42.

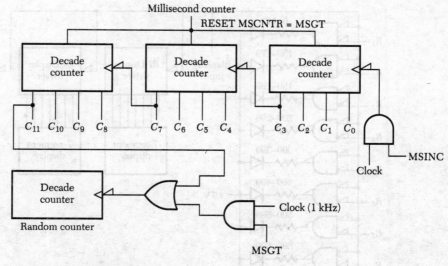

FIGURE 7.40 Timing module.

STEP DD5
. . . .

Continue to refine the operational units until they can be realized with basic circuits.

This step has essentially been completed as step 4 was carried out. The decade counters can be realized with 7490 chips. The 7447A chip is appropriate for the display drivers including the blanking function. A 7442 chip can be used for the BCD-to-decimal decoder, while two 7485 comparator chips will complete the classifier module.

This module uses a decimal decoder to determine the number of hundreds of milliseconds contained by the millisecond counter. The most significant 4 bits of the counter drives this decoder. Comparator 1 generates a 1 on its $A > B$ line if the number of hundredths of a second (tens of milliseconds) is greater than 4. Comparator 2 generates a 1 on its $A > B$ line if this number is greater than 7. If the millisecond counter stops below 150 ms, either output line 0 or 1 of the decoder will be asserted. When one of these lines is asserted and comparator 1 is not asserted, the result must be less than 150 ms. These conditions assert the line R_1 to activate the correct LED. The assertion of R_2 requires that output line 1 of the decoder is asserted, the output of comparator 1 is asserted, and the output of comparator 2 is not asserted. This condition is true only when the counter contains over 149 and less than 180 ms. The conditions for assertion of R_3 through R_{11} can be found in a similar manner. Since the counter stops if 999 is reached, R_{12} will be asserted for all times equal to or greater than this value.

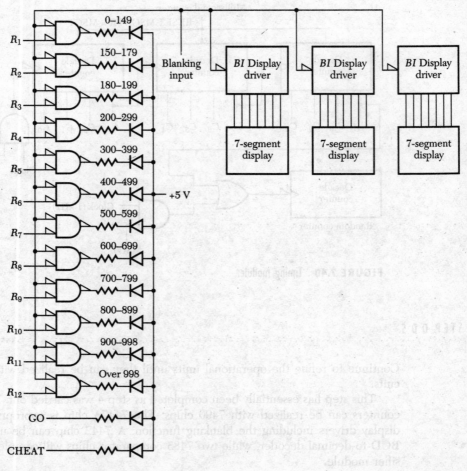

FIGURE 7.41 Display module.

STEP DD 6

Determine all necessary control signals.

This step requires that we not only consider the control signals needed, but also consider the proper assertion levels related to the actual circuits selected in the previous step.

The first signal needed is the RESET, which can be derived from the READY/ RESET switch. Although the RESET signal need not be debounced since it will connect to the direct sets of the state flip-flops, the READY signal should be. We will debounce this switch and also convert READY to a synchronous variable with a JK flip-flop. Prior to the assertion of READY, the signal MSGT must be as-

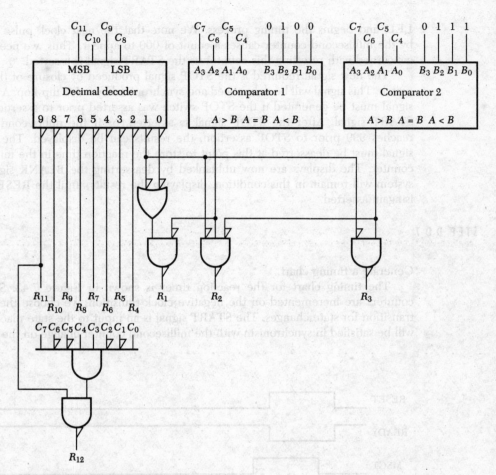

FIGURE 7.42 Classifier module.

serted to increment the random counter at 1 ms intervals. When the READY switch is asserted, MSGT is deasserted. This results in a random number being contained in the random counter when the READY switch is asserted. At this time, the signal incrementing this counter changes from a period of 1 ms to 1 s. The signal MSGT is also applied to the direct reset of the millisecond counter and will be removed from this counter when READY is asserted. MSINC is now asserted to apply the input clock signal. This 1-kHz input signal to the millisecond counter produces an output signal that increments the random counter once per second.

When the random counter reaches a count of 8 or 9 and the millisecond counter reaches a count of 999, the START signal is asserted. This signal will be created by an AND gate asserted by the correct counter conditions. This activates the GO

LED and begins the timing process. We note that the next clock pulse received by the millisecond counter causes a count of 000 to appear. Thus, we need not be concerned with resetting this counter as the START signal is asserted.

The next signal required is the STOP signal produced by closure of the STOP switch. This signal will be debounced and synchronized by a JK flip-flop. A CHEAT signal must be generated if the STOP switch was asserted prior to assertion of the START signal. After the STOP signal is asserted, or if the millisecond counter reaches 999 prior to STOP assertion, the results can be displayed. The MSINC signal must be deasserted at this point to store the reaction time in the millisecond counter. The displays are now unblanked by deasserting the BLANK signal. The system will remain in this condition, displaying the results, until the RESET signal is again asserted.

STEP DD7

Generate a timing chart.

The timing chart for the reaction timer is shown in Figure 7.43. Since the counters are incremented on the negative clock transition, we choose the positive transition for state changes. The START signal is an input to the state machine and will be satisfied in synchronism with the millisecond counter change on the negative

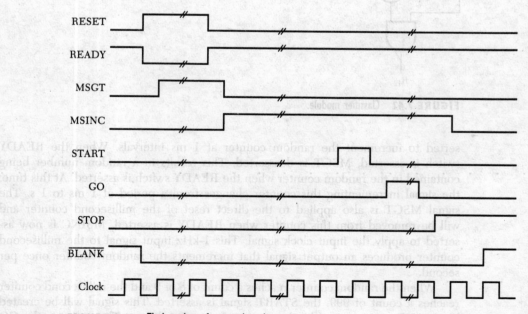

FIGURE 7.43 Timing chart for reaction timer.

clock transition. The conversion of READY and STOP to synchronous signals is also done on the negative clock to avoid conflicts with state changes.

The timing chart should help resolve any timing problems that might exist as a result of our choice of times of signal generation. As an example of this, we note that as MSINC is deasserted to deactivate the millisecond counter input, the clock signal is simultaneously going positive. Depending on circuit delays, this could allow a sliver to pass on to the counter input. To avoid this, we might consider delaying the clock signal to the MSINC gate to allow MSINC to drop prior to the positive transition of the clock.

STEP DD8

Implement the state machine.

This step can be broken down into the nine steps required to design the state machine. We will not carry out these steps, but will propose a possible state diagram as shown in Figure 7.44.

The asynchronous input for READY need not be converted to a synchronous variable in this system. Branching from all states controlled by this variable appears in the go-no go configuration, and logically adjacent assignments could be made. The switch signals for both READY and STOP must be debounced in either instance; thus, JK flip-flops can be used to both debounce and synchronize the signals as shown in Figure 7.45. Since manual assertion of these switches will last for several clock periods, no problem from ambiguous metastable states will result.

7.4.5 AN ASYNCHRONOUS RECEIVER

Many digital systems communicate with other systems in a serial mode [3]. For example, the cathode ray tube or CRT terminal has become prominent in computer input/output applications. This device contains a keyboard that allows an operator to transmit data to the computer and a CRT on which to display both received and transmitted information. The communication takes place serially over two lines, using considerably fewer lines than does a parallel system. Not only are wiring costs reduced, but also communication can take place over telephone lines when the terminal and computer are separated by long distances.

When transmission over phone lines is necessary, a device called a *modem* (*modulator-dem*odulator) is used at each end of the line. The modem accepts a serial bit stream and converts the two voltage levels into two corresponding frequencies. These sinusoidal signals are transmitted over the phone lines, received by a second modem, and converted back to the voltage levels of the original signal. This frequency-keyed transmission method is much less susceptible to phone line noise and signal degradation problems than are voltage-level methods.

The computer is designed to receive and transmit parallel data rather than serial data. In order to take advantage of serial transmission methods, a device called

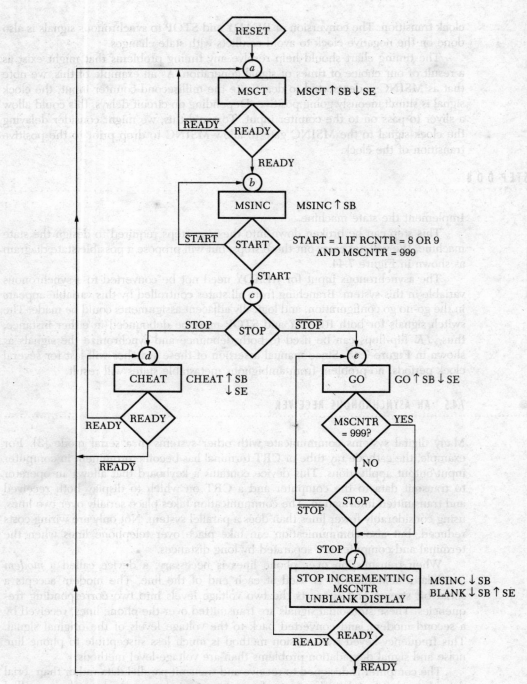

FIGURE 7.44 State diagram for reaction timer.

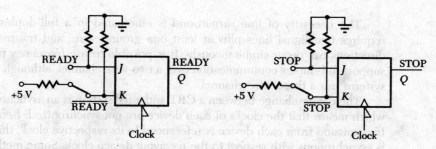

FIGURE 7.45 Debouncing and synchronizing with a flip-flop.

the universal asynchronous receiver/transmitter or UART is used. This device can receive parallel data from one system, convert to serial, and transmit over the serial pair of lines. It can also receive serial information over the pair of lines and convert this data to parallel form for use by the digital system. One UART is associated with each digital system, and these UARTs then handle the serial communication over the serial lines.

Serial communication is accomplished in one of three configurations: simplex, half duplex, and full duplex. These three channel types are demonstrated in Figure 7.46. A simplex system requires one signal line plus a ground wire, and information flows in one direction only. Point B receives information from point A but never transmits to point A. A data acquisition system at point A transmitting to a computer at point B might use a simplex channel.

A half-duplex system uses the same number of wires (2) as does the simplex channel, but the channel is bidirectional. In this system, transmission cannot take place simultaneously in both directions because the line must be "turned around" to reverse the direction of the information flow. This means that the transmitter at point A must be replaced by a receiver, and the receiver at point B must be replaced by a transmitter to move information from B to A. Of course, the bidirectional capabilities of integrated circuits using three-state outputs make the turnaround problem easy to solve.

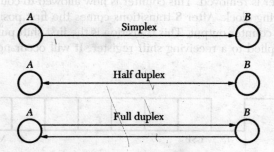

FIGURE 7.46 Communication channels.

The necessity of line turnaround is eliminated in a full-duplex system, as it requires two signal lines plus at least one ground wire, and transmission in both directions can occur simultaneously. It is possible to use frequency multiplexing to support full-duplex communication over a two-wire channel, although most CRT-μP systems use a three-wire channel.

The data exchange between a CRT and a μP system is an asynchronous process, which means that the clocks of each device are not synchronized. Because the serial transmission from each device is referenced to its respective clock, this transmission is asynchronous with respect to the receiving device clock. Some method must then be used to receive each transmitted bit at the proper time.

A popular code used in digital systems is the ASCII code (American Standard Code for Information Interchange) mentioned in Chapter 1 [3]. This code uses seven bits to represent characters and one bit for parity. In order to identify the beginning of a character transmission, a start bit is used. As a holdover from the days of the teletypewriter, two bits are reserved after the character transmission for stop bits. A typical signal format is shown in Figure 7.47, which assumes low-asserted bits.

The receiving register is a shift register. Incoming data are applied to the gate of the first flip-flop. A shift pulse must be applied to the register's shift input at the proper time to receive each incoming bit. Ideally, the shift pulses would occur at the midpoint of the incoming bit time, which would be simple if the transmitting clock were driving the shift input of the receiving register. Because the transmitting clock has shifted the data onto the line, it is in synchronism with the data bits. However, the transmitting clock signal is unavailable at the receiver, thus a receive clock must be generated to produce the shift pulses. A UART contains the shift register and control circuitry necessary to generate this required pulse train.

The main problem of creating an accurate pulse train is solved by referencing the start of this series of pulses to the beginning of the word's start bit. Every transmitted word must be preceded by a start bit, which then allows the receiving clock to reference the pulse train to this event. The receiving clock train is derived from a much higher frequency clock than is the transmitting clock, typically 16 to 64 times higher. Figure 7.48 indicates how this pulse train is created. In this case, the receiver clock has a frequency equal to 16 times that of the transmitting clock. When the leading edge of the start bit is detected by the UART, the reset of a divide-by-16 counter is removed. This counter is now allowed to count negative transitions of the receiving clock. After 8 transitions comes the first positive transition of the divide-by-16 counter output. This transition is the first shift pulse of the pulse train and can be applied to a receiving shift register. It will occur near the midpoint of the start

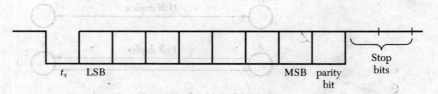

t_s LSB   MSB   parity bit   Stop bits

FIGURE 7.47 Serial signal format.

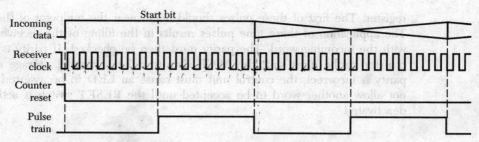

Incoming
data

Receiver
clock

Counter
reset

Pulse
train

FIGURE 7.48 Generation of shift pulse train.

bit with a maximum error of one receiving clock period, or 1/16 of the transmitting clock period. Succeeding positive transitions of the pulse train will occur at 1-bit time intervals. The difference in transmitting clock frequency and pulse train frequency leads to an error as the remaining bits of the word are received. The accumulated error from this source must not exceed 1/2 bit minus 1/16 bit time, or 7/16 of a bit time.

In addition to receiving the incoming serial data, the UART performs many other functions. The received word is retained until the computer reads it in parallel over the data bus. It also receives parallel information from the computer to transmit serially to the terminal. All handshaking signals required to exchange these data are also provided by the UART.

This example of design will not consider the entire UART, but will address the design of an important component of the UART, the receiving register. This register should accept serial information at a 1600-Hz bit rate. Each word to be received will consist of nine bits: a start bit, a 7-bit ASCII code word, and a parity bit. The receiver should be capable of checking for even or odd parity, depending on the position of the parity switch. If incorrect parity is detected, an error signal should assert an LED indicator. No further data should be received after a parity error is detected until a RESET switch is activated and deactivated.

If the parity of a word is correct, a strobe is generated to transfer the ASCII data in parallel from the shift register to a storage register. This strobe should occur within two bit times of the filling of the receiving register. The incoming word will consist of high-asserted, NRZ code. There will be a minimum of two bit times between the last bit of one word and the start bit of the next word. It is assumed that the word transferred to the parallel storage register will be used before the succeeding word overwrites the contents.

Digital System Design

STEP DD1

Determine the major goals of the system.

The control circuitry of the receiving register must be activated by the start bit of the incoming data word and provide nine equally spaced shift pulses to the receiving register. The first of these pulses should begin near the midpoint of the start bit. The application of these nine pulses results in the filling of the receiving register with the incoming word. The parity must then be checked. If parity is correct, a transfer strobe must shift the ASCII data to a 7-bit, parallel, storage register. If parity is incorrect, the control unit must cause an LED to be asserted and must not allow another word to be accepted until the RESET switch is activated and deactivated.

STEP DD 2
....

Decompose the system into functional modules.

Figure 7.49 shows the functional modules of the system. The incoming line drives the receiving register and the controlling state machine. The controller then determines when the shift pulses are applied to fill the receiving register. The parity check is then performed. The controller will consider this result before either driving the parity LED or generating a strobe to transfer the word to the storage register.

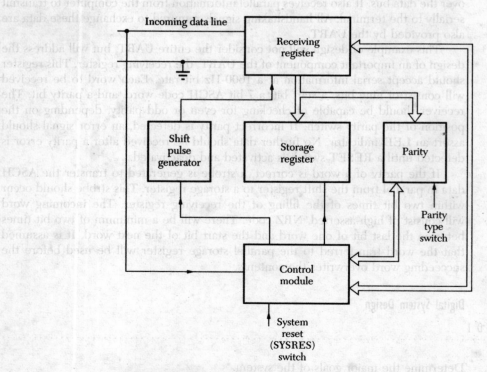

FIGURE 7.49 Functional modules of a receiving register.

STEP DD 3

Determine the next lower level of tasks.

The receiving register must be a 9-bit, serial in/parallel out register that has a master reset. Upon application of the ninth shift pulse, the start bit will enter the ninth flip-flop of the receiving register. The $Q9$ pin is an input to the state machine indicating when the ninth shift pulse has occurred. The shift pulse generator must generate the nine equally spaced shift pulses under the direction of the control module. The storage register is a parallel in/parallel out register that stores seven bits. The parity module must check parity and indicate parity errors.

STEP DD 4

Decompose the modules into operational units.

From the description of the previous step, the modules are decomposed into the operational units of Figure 7.50. The receiving register can be realized by the 8-bit 74164 shift register chip plus a single clocked flip-flop for the ninth bit. It could also be implemented with the 7495 4-bit register chip plus the 7496 5-bit register chip. Two quad D flip-flop chips such as the 7475 can be used for the storage register. The 74180 chip can be used for the parity checker. The 7493 binary counter chip can be used for the divide by 16 counter.

STEP DD 5

Continue to refine operational units.

The operational units have been defined in step 4. The schematics of the operational units are shown in Figure 7.51.

STEP DD 6

Determine all necessary control signals.

The assertion of the system reset or SYSRES line must keep the system inactive. When this line is deasserted, the controller is prepared to receive a serial word and the receiving register is reset. The controller must detect the start bit signal on the incoming line indicating that a serial word is being transmitted. In response to this signal, the counter reset line must be deasserted to allow the counter to begin dividing the clock signal by a factor of 16. The signal $\overline{\text{COUNT}}$ must be asserted low to remove the reset signal. The counter output will generate the shift pulses to be applied to the serial receiving register. When nine bits have been shifted into the register, the output of FF9 will be asserted. This signals that the register is full and no more shift pulses should be applied. The controller is driven by the RFULL line

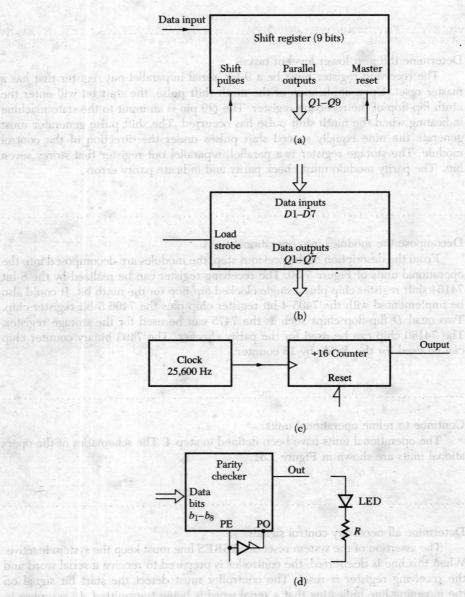

FIGURE 7.50 Operational units. (a) Receiving register. (b) Storage register. (c) Shift pulse generator. (d) Parity module.

and must deassert the $\overline{\text{COUNT}}$ line in response to this input. The counter is now driven to the reset condition and no longer generates shift pulses.

The controller must also check the parity after RFULL is asserted. If the PCHK line is low, an error is indicated and the PERROR line must be asserted low to

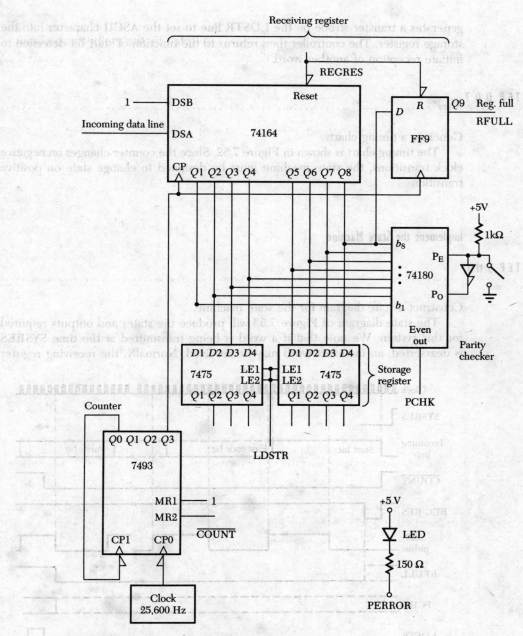

FIGURE 7.51 Schematic of operational units.

drive the error LED. The system must remain at this point until the SYSRES line is asserted. The system will then resume normal operation after the SYSRES line is deasserted. If no parity error is indicated by the parity checker, the controller generates a transfer strobe on the LDSTR line to set the ASCII character into the storage register. The controller then returns to the function of start bit detection to initiate reception of another word.

STEP DD7

Generate a timing chart.

The timing chart is shown in Figure 7.52. Since the counter changes on negative clock transitions, the state machine must be designed to change state on positive transitions.

Implement the State Machine

STEP SM1

Construct a state diagram for the state machine.

The state diagram of Figure 7.53 will produce the states and outputs required for this system. We note that if a word is being transmitted at the time SYSRES is deasserted, an incorrect word may be received. Normally, the receiving register

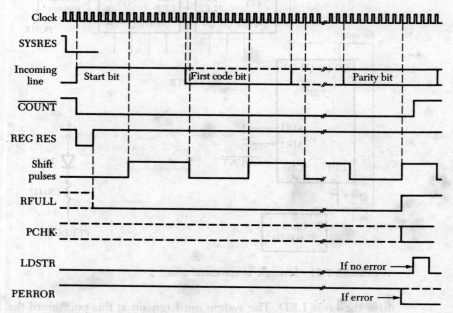

FIGURE 7.52 Timing chart for receiving register.

would be in state *b* before incoming data is allowed to be applied to the receiving register. We will not consider the design of the circuitry that would eliminate this problem.

STEPS SM2 AND SM3
....

Eliminate all redundant states.

There are no equivalent states in the diagram of Figure 7.53.

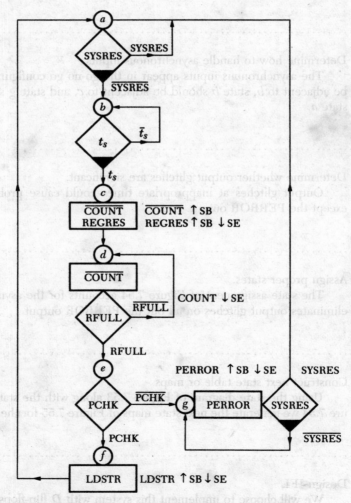

FIGURE 7.53 State diagram for receiving register.

C \ AB	00	01	11	10
0	a	d°	e	g°
1	b	c°	f°	θ

°indicates output states

FIGURE 7.54 **State assignment for receiving register.**

STEP SM4

Determine how to handle asynchronous inputs.

The asynchronous inputs appear in the go-no go configuration. State a should be adjacent to b, state b should be adjacent to c, and state g should be adjacent to state a.

STEP SM5

Determine whether output glitches are significant.

Output glitches at inappropriate times could cause problems on all outputs except the PERROR output.

STEP SM6

Assign proper states.

The state assignment of Figure 7.54 accounts for the asynchronous inputs and eliminates output glitches on all but the PERROR output.

STEP SM7

Construct next state table or maps.

Using the state diagram of Figure 7.53 along with the state assignment of Figure 7.54 we generate the next state maps of Figure 7.55 for the three state flip-flops.

STEP SM8

Design IFL

We will choose to implement this system with D flip-flops and 8:1 MUXs. The overall circuit is shown in Figure 7.56.

C \ AB	00	01	11	10
0	0	RFULL	1	SYSRES
1	0	0	0	θ

FFA

C \ AB	00	01	11	10
0	0	1	PCHK	0
1	t_s	1	0	θ

FFB

C \ AB	00	01	11	10
0	SYSRES	0	PCHK	0
1	1	0	0	θ

FFC

FIGURE 7.55 Next state maps for receiving register.

STEP SM 9

Design OFL.

Gates are used to generate the outputs as shown in Figure 7.56. The output maps appear in Figure 7.57.

7.5

MISCELLANEOUS ASPECTS OF STATE MACHINE DESIGN

SECTION OVERVIEW This section discusses the reliability of state machines and then proceeds to a consideration of timing problems in these systems.

7.5.1 STATE MACHINE RELIABILITY

There are two major sources of error in any digital system: extraneous noise and timing inaccuracies. The theory of noise is complex and will not be developed here. We will briefly mention some practical noise sources that might be encountered in actual systems.

One source of noise comes from within the circuit itself and is a result of ground loops. Figure 7.58(a) indicates schematically the general problem of allowing too many ground points in a circuit. As a result of the different currents through

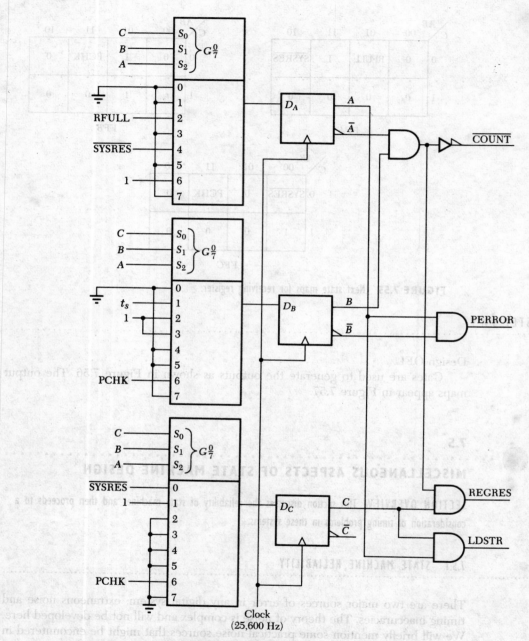

FIGURE 7.56   State machine for receiving register.

There are two major sources of error during program: extraneous noise and timing inaccuracies. The theory of noise is complex and will not be developed here. We will briefly mention some practical noise sources that might be encountered in actual systems.

The source of noise comes from within the circuit itself and is a result of ground loops. Figure 7.66 a) indicates schematically the general problem of allowing too many ground points in a circuit. As a result of the different currents through

C \ AB	00	01	11	10
0	0	1	0	0
1	0	1	0	θ

$$\text{COUNT} = \bar{A}B$$
$$\overline{\text{COUNT}} = \bar{A}B$$

C \ AB	00	01	11	10
0	0	0	0	0
1	0	1	0	θ

$$\text{REGRES} = \bar{A}BC$$

C \ AB	00	01	11	10
0	0	0	0	0
1	0	0	1	θ

$$\text{LDSTR} = AC$$

C \ AB	00	01	11	10
0	0	0	0	1
1	0	0	0	θ

$$\text{PERROR} = A\bar{B}$$

FIGURE 7.57 Output maps for state machine.

each section of the circuit, the voltages V_1, V_2, V_3, and V_4 are not equal. Although these points are connected to ground via wires or printed circuit conductors, these conductors each possess a small amount of ohmic resistance. Small voltages exist at the points indicated by V_1, V_2, V_3, and V_4 in Figure 7.58(a). As gates switch on and off, the currents and voltages at these points change. These small changes can be amplified and may ultimately affect the overall circuit operation adversely. This problem can be minimized by the common point ground shown in Figure 7.58(b). Although the voltage of this point may vary slightly as current into the node varies, the reference voltages for all sections of the circuit are identical. There are now no fluctuations of reference voltage between sections, and ground loop problems are minimized. While this method need not be applied to the positive voltage feed, it is good practice to use a common point feed also. It is also significant not to allow loops to be completed in the grounding circuit. Line a in Figure 7.58(a) is removed as the common point ground is established in part (b).

Another source of noise within the circuit is introduced by resistors and transistors. This is one source that can be treated on a theoretical basis and related fairly accurately to practice. We will only say here that the larger the circuit bandwidth or the higher the temperature, the greater is the resulting noise.

External radiation sources can also lead to noise in a given circuit. There are many radio and TV stations transmitting continuous energy into the environment. Closing or opening solenoids radiate electromagnetic waves as do spark plug wires in an automobile when the motor is running. Even though the Federal Communications Commission imposed the high-resistance carbon spark plug wire on the auto industry to minimize radiation, a significant amount of energy continues to radiate from a running engine. Long wires in a circuit may act as antennas to pick

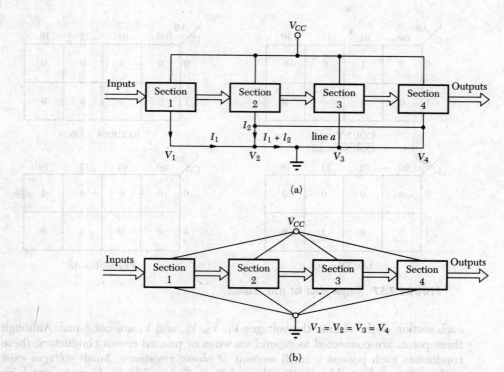

FIGURE 7.58 Grounding circuits: (a) with possible ground loops, (b) common-point grounding.

up radiated energy from these sources. In general, the longer the wires, the more extraneous noise is received and conducted to the circuit elements. Coaxial conductors with an external shield are often used to reduce this type of noise pickup when longer conductor lengths are required.

The preceding brief description of noise sources is given to emphasize the fact that a digital system is not immune to noise problems. On the other hand, certain methods of design lead to circuits that are less susceptible to noise than other circuit configurations. As mentioned in the Appendix, the one-shot multivibrator is quite susceptible to noise and therefore is not often used in circuits that must be highly reliable. This is particularly true if the trigger input must connect to a long transmission line. The SR flip-flop is likewise not a good element to receive signals over a potentially noisy line.

State machines are inherently more reliable than asynchronous devices such as one-shots, or SR flip-flops. Not only must a noise pulse with large amplitude occur, it must do so precisely at a time that just precedes the state-changing clock transition. Otherwise, the clocked flip-flop will not respond to the noise pulse. It should be obvious at this point why a "ones-catching" device (master-slave FF) is inappropriate for state machine design.

There is another reason why the master-slave flip-flop cannot be used in state machine design. When input data to the IFL change, hazards may be generated at the IFL output. Since we force the state-changing clock transition to be a half-period later than data changes, these short hazards do not affect the edge-triggered device, but they would cause erroneous state changes with the master-slave device. Since hazards are ignored by the edge-triggered flip-flop, hazard covers need not be used for the IFL.

We conclude that the state machine is more immune to extraneous noise than other types of system and also ignores hazards that are generated by the IFL.

7.5.2 TIMING CONSIDERATIONS

We found in Chapter 6 that output glitches may result from differences in flip-flop or gate switching times. A holding register or clock suppression methods can be used to eliminate these problems. The necessary delays can be introduced in various ways to avoid timing conflicts. At higher frequencies, timing problems become more critical. A 1 MHz clock frequency results in 500 ns half-periods, which allow reasonable delay times to be introduced with ease. A 20-MHz clock leads to 25 ns half-periods; thus, delay times now make up a significant part of the clock period. In such a case, higher frequency devices such as ECL or Schottky TTL must be used to minimize required delays.

In calculating maximum clock frequency for the state machine, we must consider the IFL and the state FFs. Since there is one-half clock period between data change and flip-flop change for the AST system, the IFL must settle in less than a half-period. The equation relating the clock frequency to IFL is

$$t_{cd} + t_{pIFL} + t_{su} < \frac{1}{2f} \tag{7.1}$$

where t_{cd} is clock to data stable time, t_{pIFL} is propagation delay time through the IFL, t_{su} is the set-up time for the flip-flop, and f is the clock frequency. Isolating f in this inequality results in

$$f < \frac{1}{2(t_{cd} + t_{pIFL} + t_{su})} \tag{7.2}$$

It would be possible to increase this upper limit by allowing less than one half-period between state change and data change. A period greater than one-half period is then allowed for data changes to set up the proper flip-flop inputs. In the general case, however, the upper limit of f is calculated from the above inequality. For ECL circuits, this figure can exceed 100 MHz; thus, the state machine can operate at very high frequencies.

SUMMARY

1. Design is a complex process requiring a knowledge of analysis.
2. Top-down design can be applied to digital system design. In this method, successive refinement is used to decompose the system, first into modules and then into operational units. These units are realized with familiar circuits.
3. One module of the digital system is always a controlling state machine.
4. A design procedure for digital systems is suggested.
5. A design procedure for state machines is suggested.
6. The reliability of state machines and maximum frequency of operation are considered.

CHAPTER 7 PROBLEMS

● Sec. 7.1.1

7.1 An erroneous state transition from state a to state d can occur in the system of Figure P7.1 if X is asserted near the time of the state change. What are the possible codes for state d?

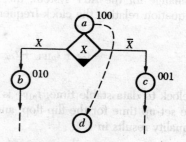

FIGURE P7.1

°7.2 What are the possible codes for the erroneous state e in Figure P7.2? Does the code for state a affect the erroneous state codes?

7.3 Repeat Prob. 7.2 if state b is changed to a code of 1010.

7.4 List all possible erroneous states as the system of Figure P7.4 moves from state a. X is an asynchronous variable while Y is synchronous.

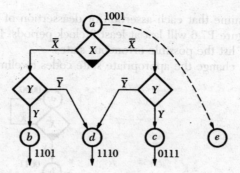

FIGURE P7.2

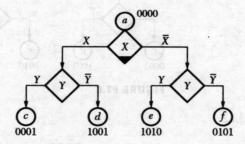

FIGURE P7.4

7.5 Find all possible erroneous state codes in Figure P7.5 if
    a. *X* is asynchronous and *Y* is synchronous,
    b. *X* is synchronous and *Y* is asynchronous,
    c. *X* and *Y* are both asynchronous.

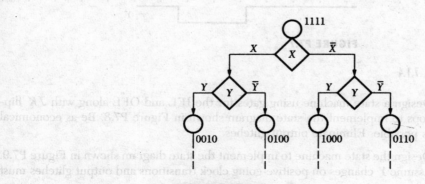

FIGURE P7.5

°**7.6** Assume that each assertion or deassertion of the asynchronous variable X in Figure P7.6 will last at least 3 clock periods. If Y is synchronous,
   a. list the possible erroneous states.
   b. change the appropriate state codes to eliminate possible erroneous states.

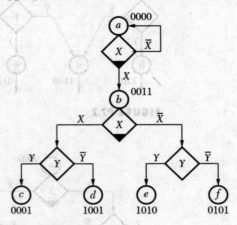

FIGURE P7.6

● Sec. 7.1.2

7.7 The signals X and Y of Figure P7.7 are asynchronous variables. Show how to convert these signals to X_S and Y_S that are synchronized to the negative clock transition. X_S is to be a high asserted signal while Y_S is to be a low asserted signal.

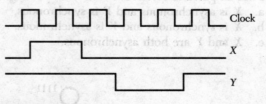

FIGURE P7.7

● Sec. 7.1.4

°**7.8** Design a state machine using gates for the IFL and OFL along with JK flip-flops to implement the state diagram shown in Figure P7.8. Be as economical as possible. Eliminate output glitches.

7.9 Design the state machine to implement the state diagram shown in Figure P7.9. Assume Y changes on positive-going clock transitions and output glitches must not occur.

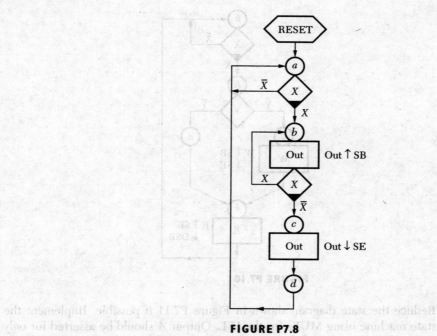

FIGURE P7.8

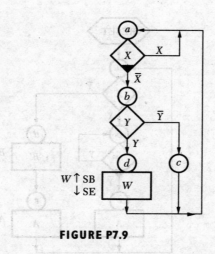

FIGURE P7.9

7.10 The state diagram for a mainly synchronous system is shown in Figure P7.10.
  a. Select proper states to eliminate erroneous state transitions and output glitches. Show your present state map.
  b. Design the state machine to implement the state diagram.

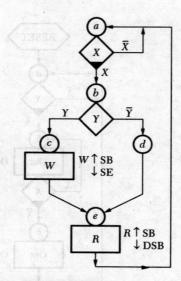

FIGURE P7.10

7.11 Reduce the state diagram shown in Figure P7.11 if possible. Implement the state machine using MUXs for the IFL. Output A should be asserted for only one-half clock period while B is asserted for a full period.

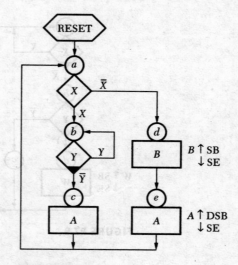

FIGURE P7.11

● **Secs. 7.2–7.3**

7.10 The state diagram for a ma̶i̶n̶l̶y̶ synchronous system is shown in Figure P7.10. a. Select proper states to eliminate erroneous s̶t̶a̶t̶e̶s̶ and output glitches. Show your present state map.

7.12 A system is to be designed to check two incoming data lines, X and Y. These lines will continuously present bits in NRZ code, that change on the rising edge of a clock signal. When the sequence $X = 100$ and $Y = 110$ have appeared coincidentally on the lines, a code of 101 is to be transmitted on output line M. The two sequences can occur anywhere in the data stream, without regard to word boundaries. When the sequences have been detected, the system should resume checking without missing a bit. Assume that a synchronous start bit lasting one clock period initiates the checking.

a. Develop the block diagram of the functional modules for this system.
b. Implement all functional modules except the control module.
c. Generate an appropriate timing chart.
d. Design the controlling state machine.

7.13 This problem consists of designing a synchronous serial transmission system. Seven bits of a word (ASCII code) are presented in parallel over seven input lines, DI1-DI7. Valid data is guaranteed to be present on these lines during the assertion of an input signal, PLOAD, which is asserted for one clock cycle. A 25,600-Hz clock signal is available along with a signal derived from dividing this signal by 16 (1600 Hz).

The 7-bit word is to be loaded in parallel into a shift register. A parity bit is to be generated and added to the shift register as the last bit of the now 8-bit word. An input signal, EPAR, should result in even parity generation when high and odd parity when low.

The 8-bit word should then be transmitted onto a serial line, preceded by a high-asserted start bit at a rate of 1600 bits/second. When transmission of the nine bits is completed, the serial line must go low and remain low until the start bit of the succeeding word is transmitted.

Successive assertions of the PLOAD signal, indicating a new word is available for transmission, will be separated by at least 12 cycles of the 1600-Hz clock.

a. Develop the block diagram of the functional modules for this system.
b. Implement all functional modules except the control module.
c. Generate an appropriate timing chart.
d. Design the controlling state machine.

7.14 An 8-bit data collection system is to be interfaced to a computer. The collection system often transmits the parallel words faster than the computer can receive them, thus a buffer or first in-first out memory (FIFO) must be connected between the two systems. The FIFO should be capable of storing 32 8-bit words.

Along with the data lines there are also some control lines designed to ensure that the data collection system does not attempt to write to the FIFO at the same time the computer is reading out. The data collection system must not be allowed to write to the FIFO when the buffer is full and the computer must not be allowed to read from the FIFO when the buffer is empty. In order to accomplish the interlocking just described, certain handshaking signals are required. The FIFO should provide an input ready signal ($\overline{IR}$) that remains high until the FIFO is ready to receive data, at which time $\overline{IR}$ goes low. The data collection system must respond by presenting valid data on the eight input lines and then transmitting a high-to-low transition on its write line ($\overline{WR}$). If no data is available when $\overline{IR}$ is asserted (low), the $\overline{WR}$ line will not be asserted (low). The $\overline{IR}$ line must remain low until data is written as signified by the assertion of $\overline{WR}$. After $\overline{WR}$ has been asserted, the $\overline{IR}$ line can be deasserted (high). After this deassertion, the $\overline{WR}$ line will be deasserted (high). Only after $\overline{WR}$ goes high can the $\overline{IR}$ line again go low.

The same type of handshaking must also be implemented on the output of the FIFO. The output ready signal ($\overline{OR}$) should remain high until the FIFO is ready to transmit data to the computer. Valid data should appear on the output lines when $\overline{OR}$ is asserted (low). The computer will generate a read signal ($\overline{RD}$) that will go low when the computer accepts the data. The data should be valid for a minimum of 30 ns after $\overline{RD}$ is asserted (low). After $\overline{RD}$ goes low, $\overline{OR}$ can be deasserted (high). The computer will only allow $\overline{RD}$ to be deasserted after $\overline{OR}$ is deasserted.

A master reset signal MR must be included in the FIFO. When asserted it should clear the memory and set an empty flag. The data out lines should interface to the computer through three-state buffers.

The FIFO should have a 500-kHz throughput, that is, when the buffer has data in it, a new word can be written from the data collection system and another word can be read out in less than 2 μs.

a. Develop the block diagram of the functional modules for this system.
b. Implement all functional modules except the control module.
c. Generate an appropriate timing chart.
d. Design the controlling state machine.

7.15 A large reservoir of water serves several users. In order to keep the level of water sufficiently high, four sensors are placed vertically at 4-inch intervals. When the water level is above the highest sensor, the input flow rate should be zero. When the level is below the lowest sensor, the flow rate should be maximum. The flow rate when the level is between the upper and lower sensors is determined by two factors: the water level and the water level previous to the last sensor change. Each water level has a nominal flow rate associated with it. If the previous sensor change indicates the previous level was lower than the present level, the nominal flow rate should take place. If the previous sensor change indicates the previous level was higher than the present level, the flow rate is increased by ΔFR above the nominal flow rate.

You are to design the state machine to control this system. Figure P7.15 indicates the sensor arrangement and input flow rate controllers. The sensors produce high-asserted, TTL-compatible outputs and the flow rate controllers require high-asserted TTL inputs. If all of the inputs to the flow rate controllers are deasserted, the flow rate is zero. If all inputs to the flow-rate controllers are asserted, the flow rate is maximum. All necessary values are indicated in the table.

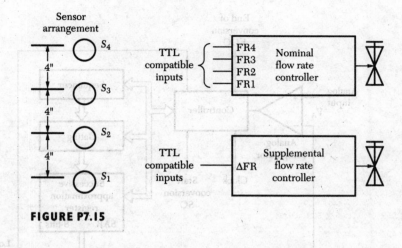

FIGURE P7.15

Water Level	Sensors Asserted	Nominal Flow Rate Inputs to Be Asserted
Above $S4$	S1, S2, S3, S4	None
Between $S3$ and $S4$	S1, S2, S3	FR1
Between $S2$ and $S3$	S1, S2	FR1, FR2
Between $S1$ and $S2$	S1	FR1, FR2, FR3
Under $S1$	None	FR1, FR2, FR3, FR4, ΔFR

Assume that the minimum time for a change of one level is 10 seconds. Show all important design steps and state any reasonable assumptions made.

7.16 The block diagram for an 8-bit successive approximation analog-to-digital converter is shown in Figure P7.16.

When the SC signal (start conversion) is asserted high, it indicates that the process of converting the analog signal to a corresponding digital value is ready to begin. The SC signal can only go high if the EOC signal (end of conversion) is high. After positive assertion of SC, the EOC line goes low and stays low until the converted value is transferred from the SAR to the output register. The SC assertion lasts for a short time, then goes low to begin the conversion process.

The process begins by transferring a 1 from the sequencer to the MSB of the SAR. This represents the first guess (10000000) of a code that represents the analog value. This estimated value is then converted to analog with a digital-to-analog converter (DAC) and compared to the input analog signal, V_A. If V_G is smaller than V_A, the analog comparator produces a zero volt output. If V_G is larger than V_A, the comparator produces a 4-V output. If the guess is too small ($V_G < V_A$), the sequencer allows the 1 to remain in the MSB of the SAR and places a 1 in the next most significant bit of the SAR. If the guess is too

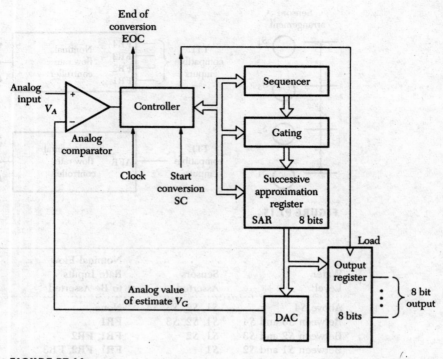

FIGURE P7.16

large ($V_G > V_A$), the sequencer resets the MSB of the SAR to 0 and places a 1 in the next most significant bit position. This process is repeated for each of the positions of the SAR until all 8 bits are determined. The contents of the SAR are then transferred to the output register. The EOC signal returns to the high level at this point.

The analog comparator requires 10 μs after application of the inputs to produce a valid output level. The clock frequency is 500 kHz.

a. Implement each block with operational units.
b. Create a timing chart.
c. Design the controller.

7.17 Design the controller for a three-number digital combination lock. The lock has a 4-bit register that accepts the entered numbers (0–9) from the keypad. Each number will be stable for a minimum of 200 ms from the time a number valid signal NV is asserted after the entered number has become stable. The system has a 1-kHz clock available. If the correct sequence of three numbers is entered, an output signal UNLK is to be asserted for 1 second. If three incorrect sequences of numbers are entered successively, the control circuit should become inactive for approximately 5 minutes. Show all important design steps and state any reasonable assumptions made.

REFERENCES AND SUGGESTED READING

1. P. Alfke and P. Wu, "Metastable Recovery," *The Programmable Gate Array Data Book.* San Jose, CA: XILINX, 1992, Chap. 6.

2. D. J. Comer, "Application of Top-Down Principles to Digital System Design," *IEEE Transactions on Education*, Vol. E-26, No. 4, November 1983.

3. D. J. Comer, *Microprocessor-Based System Design.* New York: Holt, Rinehart and Winston, 1986.

4. W. I. Fletcher, *An Engineering Approach to Digital Design.* Englewood Cliffs, NJ: Prentice-Hall, 1980.

5. H. Lam and J. O'Malley, *Fundamentals of Computer Engineering.* New York: John Wiley and Sons, 1988.

6. M. M. Mano, *Digital Design*, 2nd ed. Englewood Cliffs, NJ: Prentice-Hall, 1991.

7. C. H. Roth, Jr., *Fundamentals of Logic Design,* 4th ed. St. Paul, MN: West, 1993.

8. N. Wirth, "On the Composition of Well-Structured Programs," *Computing Surveys*, Vol. 6, No. 4, December 1974.

Chapter 8

Programmable Logic Devices

P rogrammable circuits have become extremely important to the digital systems area in the last decade. At the present time, more practical design is done using programmable circuits than using conventional circuits. This chapter explains what a programmable circuit is and how it is used to design digital systems.

8.1

INTRODUCTION TO PROGRAMMABLE LOGIC DEVICES

SECTION OVERVIEW The programmable semiconductor circuit is introduced and considered in this section. The history and evolution of these important components are reviewed. Methods of programming these chips are introduced.

8.1.1 DEFINITION OF PROGRAMMABLE CIRCUITS

Chapters 5 through 7 cover the basic operating principles of state machines or sequential control circuits. The operation of these circuits depends entirely on the IC chips used and the electrical connections between chips. The designer has no access to the internal interconnections of the IC chips. In designing a digital system, we must specify each IC to be used and indicate a wiring diagram to show how each circuit is to be connected. Once the design is completed, the system performs the function intended. If it is desired to modify the function of the circuit, the design must be modified. New circuits may be needed and some connections will certainly require changes. This type of system is often referred to as a *hardwired* system.

Another approach to circuit or system design uses programmable components. Such a device includes arrays of logic elements on a chip and allows the user to specify or program many internal connections between these components on the chip. The logic elements could be various gates, inverters, buffers, and even flip-flops. A system configuration can be created on the chip, simply by programming the chip or telling the chip where the interconnections are to be made.

While there are many specific names for programmable devices, most are classified within the broad area of programmable logic devices (PLDs). A PLD can be defined as follows.

● A PLD is an IC chip that includes arrays of logic elements and allows a user to specify the connections among many of these elements.

The user does not have access to every connection on the chip. Each type of PLD may specify a different set of programmable interconnections while several other interconnections on the chip are fixed.

PLDs have evolved over the years so that many different types are presently available. This chapter will discuss five popular types of PLDs in conjunction with the creation of a state machine: the read-only memory (ROM), the programmable logic array (PLA), the programmable array logic chip (PAL®), the programmable logic sequencer (PLS), and the field-programmable gate array (FPGA).

The ROM or combinational PLA or PAL® is used as one component of a state machine. Additional flip-flops are required to use these chips in sequential circuit design. Such devices implement combinational logic functions and can therefore be used to realize the input forming and output forming logic of the state machine. Use of the ROM or combinational PLA or PAL® can minimize chip count of a circuit and provide more flexibility to the system as discussed in a later section.

The PLS adds flip-flops to a PLA chip. So also does the registered PLA or registered PAL® chip. This allows all sections of the sequential circuit to be created on the chip. An entire state machine can be made from a single PLS, PLA, or PAL® chip. The same can be said of the FPGA. A state machine with an input synchronizing register and an output holding register can be created with the FPGA chip. This is becoming a very popular method of state machine design, especially when time to production is to be minimized.

Advantages of Programmable Circuits

The PLD belongs to a larger group of programmable chips called the application specific integrated circuit or ASIC. This category of IC can include analog as well as digital circuits and even mixed analog and digital circuits on a chip. Two advantages of the ASIC and the PLD are (1) chip count and physical size of a system can be minimized and (2) the time from conception of a system to marketing of the system can be minimized.

The keen competition among digital system manufacturers makes it desirable for a company to move from a prototype or a simulation to the marketplace as quickly as possible. Chip count of the system may also be important and, if so, must be minimized.

Generally, the most effective means of decreasing chip count is to integrate the system on a single chip or small number of chips. In this approach, a prototype system may be designed and debugged using conventional hardware, then the circuit is realized as a custom integrated circuit. Unfortunately, the process of integrating a prototype circuit may require four to six weeks. If the finished circuit does not operate up to specifications, the design is modified and several more weeks are required for a second fabrication. While custom integration minimizes chip count, it requires a large amount of time to produce a marketable system.

Rather than implement a logic circuit by IC fabrication, many companies use an ASIC or, more specifically, a PLD. The circuits on the chip can be connected together in one of several ways, but any of the methods used is easier and faster than developing a custom integrated circuit. Consequently, ASICs play an important role in minimizing the time between completion of a prototype system and the introduction of a finished product on the market. This ability to get a product on the market quickly has led to the popularity of the PLD in recent years.

Disadvantages of Programmable Circuits

While the PLD minimizes chip count of a digital system, a disadvantage is that the interconnections between elements on the chip must be specified or programmed. Unlike conventional circuits, even after PLDs are wired into a system, they will not function properly unless they have been programmed. This matter will be treated in succeeding paragraphs.

Before we discuss programmable chips in greater detail, we will review the historical development of programmable devices.

8.1.2   HISTORY

One of the most significant components in early digital computers was the magnetic core. This tiny doughnut-shaped ferrite material was used from the 1950s through the 1970s to construct the main memory of large computers. Each of these cores could store a binary bit of information by using the direction of magnetization of the core to indicate a 0 or a 1. The direction of the magnetic field inside the core could be changed by controlling the direction of current through wires that were wound around a portion of the core. For many years, magnetic core storage was the dominant type of main memory for the computer.

As the technology for core storage improved, the price continued to drop and some impressive computers became available. The IBM System 360 appeared in 1965 with one scientific model capable of storing about 64 million bits in its main memory. This system was sold for a price that varied between $1,000,000 and $2,000,000, depending on several options. The magnetic core cost about 1 to 2 cents per bit, wired into the memory. The 64-megabit storage system added approximately $700,000 to the cost of the system. It was obvious that a reduction of main memory costs would greatly reduce the overall cost of a computer. Furthermore, the core storage system required a set of high-current driver circuits that led to high power dissipation and expensive circuit components.

Ironically, this same year (1965) saw the first proposal to use semiconductor memory [1]. The obvious size benefits of integration led some engineers to believe that perhaps the integrated circuit might be used to produce low cost storage components. The first IC memories were more expensive and had much less storage capacity than the core memory and did not immediately replace this workhorse of the computer industry. One of the first commercial uses of a small semiconductor main memory was in the IBM 360/85 in 1969.

As IC fabrication and design techniques improved over the years, the semiconductor memory became smaller and cheaper, leading to the demise of core storage. Without this development, it would be difficult to produce the highly capable personal computers and workstations that are now available.

Before the semiconductor memory was made large enough to replace the main memory of the computer, it became obvious that the small IC memory would be useful in circuit applications. Several companies implemented small memories such as 64-bit devices that were targeted for use in digital circuits rather than in computer memories. One of the first such devices was the read-only memory or ROM. Small IC read-write memories, called semiconductor RAMs, also appeared at about the same time. As price dropped and size increased, semiconductor memories began replacing core memories. In the late 1970s, the semiconductor memory was used almost exclusively in the personal computer. By the early 1980s, even large mainframe computers were produced with exclusively semiconductor main memories.

It became obvious in the late 1970s that ROMs were also useful in logic function realization. As small ROMs were used for this purpose, the combinational PLA and PAL® chips were developed to reduce the number of devices needed on a chip. Fabrication methods improved to allow the inclusion of flip-flops on PLA and PAL® chips in the 1980s. As industry looked for faster methods of developing digital products, the registered PLA and PAL®, the PLS, and the FPGA were conceived. These devices became very popular in the late 1980s and continue to be significant in digital logic system design.

8.1.3 TYPES OF PROGRAMMABLE LOGIC DEVICES

There are two types of programming used for PLDs: user-specified programming or field programming. User-specified programming allows the user to tell the manufacturer how to interconnect the logic elements on the chip. The manufacturer then makes the specified connections using a process called mask programming. The process involves the creation of photographic masks that, after reduction, are used to determine the areas on which vaporized metal is deposited to form conductors. It is not uncommon for the manufacturer to require a few weeks to program the chips and deliver them to the user. The ROM is an example of a chip that can be a mask-programmable device. The ROM can be mask programmed much faster than a custom integrated circuit can be fabricated, but there is still a significant time required by the ROM manufacturer to complete the specified programming.

Field programming allows the user to program the device and leads to a much quicker finished product. Field programming requires that certain voltages or currents be applied to the appropriate pins of the chip in a sequence that is specified by the manufacturer. This can be accomplished by an electronic device called a programmer. Most modern programmers receive instructions from a computer and create the appropriate voltages and currents to program the chip. Field programming has become a rather simple design step to accomplish using a personal computer with this programming device.

There are two popular approaches used in creating the proper interconnections between logic elements of a field-programmable chip. PLDs have many circuits on a chip, but either have no connections between logic circuits or have many connections between circuits. In the devices with no connections between components, the proper connections must be added to the circuit. In devices with many connections, the unnecessary connections must be removed, leaving only the desired connections.

A method for adding connections that has become popular in the last decade is the placement of transistor switches between logic components. These switches can be turned on to create a connection or turned off to break the connection between components. An on-chip read/write memory or RAM can store the program that specifies the proper interconnections. The RAM outputs then drive the connecting transistors to the "on" state or the "off" state in accordance with the connections to be made. This method of programming can be repeated without limit by the designer. Furthermore, the programming can be done quickly by the user, minimizing production time. The Xilinx FPGA is an example of this reconfigurable user programmable device.

In a system using reconfigurable programming, a computer will load the on-chip RAM with the proper information when the system is powered up. Until the system is powered down or the contents of the RAM are changed, the FPGA will continue to be interconnected as specified.

The other method of obtaining the proper interconnections between logic components on the chip starts with many connections between these components. Programming now consists of removing all unnecessary connections or links between circuits to make the chip operable. The removal of the links can be accomplished by the designer and these devices are thus also field programmable. The metal conductors used to interconnect the components on this chip are called fusible links. They are designed to carry normal currents, but will evaporate if large amounts of current flow. By applying specified voltages along with appropriate addresses, the fusible links can be selectively evaporated to program the chip with the desired connections. The current through the link generates heat, causing a rise in temperature to the point that the link vaporizes. Programming equipment is available for the user to program or specify these interconnections in an efficient way. The evaporation or removal of links is an irreversible process and any connection that is broken can never be restored.

Fusible-link devices can be programmed quite rapidly and production time can be minimized with these devices. They are a little more costly than the mask-programmed devices when programming cost is included, but once the system is into production, the mask-programmed device can then be used if appropriate.

There are other types of programmable chips such as the ultraviolet erasable ROM (EPROM) and the electrically erasable ROM (EEROM) that are important in computer applications. These devices are not significant in logic function realization and will not be discussed further here.

8.2

READ-ONLY MEMORY

SECTION OVERVIEW The structure of the ROM is discussed along with the application of this device to logic function realization.

The read-only memory or ROM finds its major use as a storage unit for fixed programs in a computer. Generally, these ROMs are fairly large, storing several thousands of binary bits. Smaller ROMs are available that can be used to realize logic functions. These devices lead to more expensive function implementations than hardwired implementations.

A significant advantage of ROM implementation occurs in systems requiring changeable functions. If function X is to be generated on some line during a certain time period for a given input combination, and if function Y is to be generated during some other time period for the same input combination, then a ROM system is very useful. When function X is required, a ROM producing this function can be plugged into the circuit. A second ROM for function Y can be plugged into the circuit when appropriate. No wiring changes are required for this change of logic function. It is not even necessary to change chips since both ROMs can be used in the system with inputs in parallel and multiplexed outputs. The correct set of outputs are then selected by the multiplexing circuit.

8.2.1 ORGANIZATION OF THE ROM

A block diagram of a ROM is shown in Figure 8.1. The address lines point to locations within the ROM that store words of m bits. Since 2^n unique binary codes can be set up on the n input lines, 2^n different locations can be addressed. The ROM size is defined by the number of locations and the word size. A 256×4 ROM indicates that the device has 256 storage locations each holding a 4-bit word. This ROM would require eight address lines to access 256 locations. When a binary code is applied to the address lines, the contents of the location specified appear on the output lines. A smaller ROM size might be 32×8, while a size of 8192×8 represents a large ROM.

8.2.2 COMBINATIONAL LOGIC USING THE ROM

The ROM can be visualized in terms of a minterm decoder followed by a set of OR gates. For n address lines, a total of 2^n AND gates, each having n inputs, would be required to implement the minterm decoder. For an m-bit output word, m output OR gates with up to 2^n inputs would be necessary. Figure 8.2 shows this configuration for an 8×4 ROM. Three input lines lead to 2^3 or 8 AND gates to decode the 8 minterms. When a given minterm or location is accessed by the address lines, one AND gate output will be asserted. The connections between inputs and AND gates are fixed connections and represented in the figure by solid dots.

The output word generated when an AND gate is asserted depends on the connections from this gate output to the OR gate inputs. These connections are programmable and each is represented in the figure by the symbol $\times$.

Programmed as shown, this ROM would generate the following outputs for the given inputs.

Input		Output			
		X_0	X_1	X_2	X_3
$\bar{A}\bar{B}\bar{C}$	(000)	0	0	1	0
$\bar{A}\bar{B}C$	(001)	1	0	1	0
$\bar{A}B\bar{C}$	(010)	0	1	0	1
$\bar{A}BC$	(011)	1	0	0	1
$A\bar{B}\bar{C}$	(100)	1	1	0	1
$A\bar{B}C$	(101)	0	1	1	0
$AB\bar{C}$	(110)	1	0	0	0
ABC	(111)	0	1	1	0

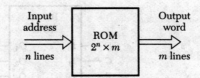

FIGURE 8.1 Block diagram of ROM.

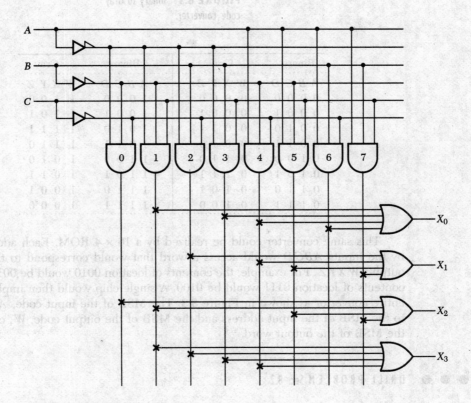

- fixed connection
- × programmed connection

FIGURE 8.2 An 8×4 ROM.

The ROM is particularly useful in applications requiring multifunction generation. Use of the ROM in multifunction logic synthesis can be demonstrated by reconsidering the code converter of Example 3.2. In this example, a decoder and four gates were used to convert binary code to Gray code.

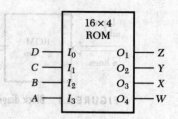

FIGURE 8.3   Binary to Gray code converter.

Binary	Gray	Binary	Gray
$A\,B\,C\,D$	$W\,X\,Y\,Z$	$A\,B\,C\,D$	$W\,X\,Y\,Z$
0 0 0 0	0 0 0 0	1 0 0 0	1 1 0 0
0 0 0 1	0 0 0 1	1 0 0 1	1 1 0 1
0 0 1 0	0 0 1 1	1 0 1 0	1 1 1 1
0 0 1 1	0 0 1 0	1 0 1 1	1 1 1 0
0 1 0 0	0 1 1 0	1 1 0 0	1 0 1 0
0 1 0 1	0 1 1 1	1 1 0 1	1 0 1 1
0 1 1 0	0 1 0 1	1 1 1 0	1 0 0 1
0 1 1 1	0 1 0 0	1 1 1 1	1 0 0 0

This same converter could be realized by a 16×4 ROM. Each address given by the inputs $ABCD$ would access a word that would correspond to the desired outputs $WXYZ$. For example, the contents of location 0010 would be 0011, and the contents of location 0111 would be 0100. A single chip would then implement the entire converter as shown in Figure 8.3. The MSB of the input code, A, connects to the MSB of the input address and the MSB of the output code, W, connects to the MSB of the output word.

● ● ●   DRILL PROBLEM Sec. 8.2

Realize the expressions $F1 = A\bar{B}C + AB\bar{C} + \bar{A}BC$, $F2 = \bar{A}BC + \bar{A}B\bar{C} + A\bar{B}\bar{C}$, and $F3 = \bar{A}\bar{B}\bar{C} + ABC$ with an 8×4 ROM. Show the contents of all locations.

8.3

PROGRAMMABLE LOGIC ARRAYS

SECTION OVERVIEW This section describes the structure of the PLA and its application in logic function realization. The related PAL[①] is also considered.

8.3.1 ORGANIZATION OF THE PLA

The programmable logic array or PLA can be thought of as a ROM that has had a large percentage of its locations or minterm gates deleted. A ROM with 16 input address lines and 8 output lines is a $2^{16} \times 8$ ROM. This $2^{16} \times 8$ ROM would contain 2^{16}, or 65,536, storage locations of 8-bit width. There is a PLA manufactured with 16 address lines and 8-bit word size, but instead of storing 65,536 words, it stores only 96. From a hardware standpoint, the ROM must decode every possible minterm of the input variables, while the PLA decodes a very small percentage of the minterms. In the actual PLA just mentioned, less than 0.2% of the input variable minterms are decoded.

Since the number of storage locations or minterms decoded in the PLA is important, the size of this device includes this number in the specifications. A PLA is specified by the number of input lines, the number of minterms that can be realized, and the number of output lines in that order. A $16 \times 96 \times 8$ PLA describes the device of the preceding paragraph.

A major difference between the ROM and the PLA is in the programming flexibility of the PLA. In the ROM, since each possible minterm AND gate was present, the connections to each of these many gates are fixed. Only the inputs to the OR gates are programmable in the ROM. In the PLA, since a limited number of AND gates are available, the connections to each gate are programmable. Not only can the inputs to the OR gates be programmed, so also can the inputs to the AND gates.

The primary application of the ROM is that of word storage in programmable systems. Logic function realization is an area in which the ROM can be applied, but it may be inefficient in terms of unused storage locations. The PLA is used almost exclusively for logic function realization and therefore does not require a large storage capacity. While the ROM can perform any function performed by the PLA in logic function realization, the ROM is more expensive and may require much more programming. PLAs can be mask-programmed or field-programmed (FPLA) in the same ways that ROMs are programmed.

The PLA contains an AND gate to decode chip inputs followed by an OR gate to produce each bit of the output word. The inputs of the AND gates are programmed to decode the desired minterms while the outputs of these gates become inputs to the OR gates to form a sum of products.

Because of the large number of interconnections on the PLA chip, the simplified schematic representation of Figure 8.4 is used. Only a single line to each AND gate is shown, but each connection dot along that line represents a separate line from that variable to the AND gate. The same scheme is used to represent connections from the AND gate outputs to the OR gate inputs. Each single line to an OR gate represents some number of separate connections equal to the number of connecting dots along the single line.

The FPLA of Figure 8.5 is an $N \times M \times K$ device with the capability of true or complement output. The EXOR gate passes the true value of the function generated by the OR gate if one input is connected to ground, as indicated by the connec-

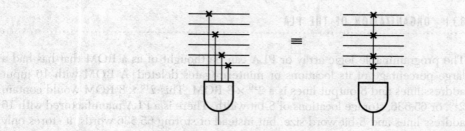

FIGURE 8.4 Simplified schematic representation of programmable connections on a PLA.

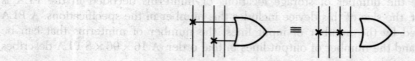

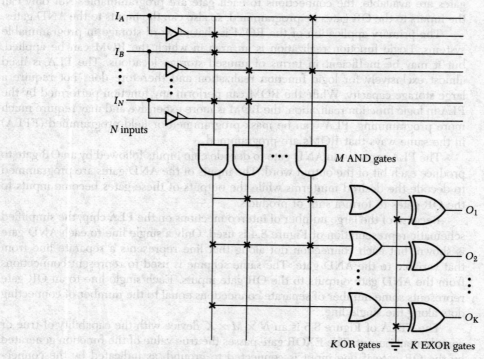

I_A

I_B

I_N

N inputs

M AND gates

O_1

O_2

O_K

K OR gates K EXOR gates

FIGURE 8.5 A typical FPLA configuration.

tion ×. If the fusible link between ground and the EXOR input is blown, as indicated by the O_2 EXOR gate, this input floats high and the output is the complement of the OR gate output.

8.3.2 COMBINATIONAL LOGIC USING THE PLA

When we are attempting to realize several logic functions with gates, we are careful to minimize the total number of gates or chips required. If we are to realize some set of logic functions with a given PLA, and the PLA is sufficiently large to synthesize the functions, minimization is not an important issue. On the other hand, if we are not constrained to use a given PLA, we should minimize the logic expressions to allow the smallest or lowest cost PLA to be used. For this reason, we will briefly consider the minimization of multifunction logic expressions. As we do so, we must note that the ability to invert the outputs of any OR gate allows the possibility of realizing the complement of a desired function and then inverting it.

In dealing with the PLA, we are not as interested in minimizing total gates and inputs as we are in minimizing the number of AND gates required. The number of inputs to each AND gate of a PLA equals twice the number of input variables. This number, in general, cannot be minimized. The number of OR gates is determined by the number of outputs and this number cannot be minimized. The only remaining variable that can be minimized is the number of AND gates. For a reasonably large number of variables, the minimization problem becomes so difficult that a computer is necessary to complete the problem in a reasonable time. Programs that minimize these functions are available from several PLD manufacturers. The following example will demonstrate some of the details involved in the minimization of the number of AND gates for a four-variable system.

EXAMPLE 8.1

A system transmits four variables, A (MSB), B, C, and D. A circuit is to be designed to produce three different outputs, F_1, F_2, and F_3. The function F_1 should be asserted when the input code equals 1, 3, 12, 13, 14, 15; F_2 should be asserted for an input code representing 5, 10, 11, 12, 13, 15; and F_3 should be asserted for input codes of 0, 3, 4, 7, 8, 10, 11. This information can be expressed in a shorthand method as sums of minterms, that is

$$F_1 = \sum (1, 3, 12, 13, 14, 15)$$

$$F_2 = \sum (5, 10, 11, 12, 13, 15)$$

$$F_3 = \sum (0, 3, 4, 7, 8, 10, 11)$$

Minimize the number of AND gates required.

SOLUTION

The three expressions can be written as

$$F_1 = \bar{A}\bar{B}\bar{C}D + \bar{A}\bar{B}CD + AB\bar{C}\bar{D} + AB\bar{C}D + ABC\bar{D} + ABCD$$

$$F_2 = \bar{A}B\bar{C}D + A\bar{B}C\bar{D} + A\bar{B}CD + AB\bar{C}\bar{D} + AB\bar{C}D + ABCD$$

$$F_3 = \bar{A}\bar{B}\bar{C}\bar{D} + \bar{A}\bar{B}CD + \bar{A}B\bar{C}\bar{D} + \bar{A}BCD + A\bar{B}C\bar{D} + A\bar{B}CD$$

and are plotted in the K-maps of Figure 8.6.

CD \ AB	00	01	11	10
00	0	0	1	0
01	1	0	1	0
11	1	0	1	1
10	0	0	1	0

F_1

CD \ AB	00	01	11	10
00	0	0	1	0
01	0	1	1	0
11	0	0	1	1
10	0	0	0	1

F_2

CD \ AB	00	01	11	10
00	1	1	0	1
01	0	0	0	0
11	1	1	0	1
10	0	0	0	1

F_3

FIGURE 8.6   K-maps for Example 8.1.

All possible minimal expressions for F_1, $\bar{F}_1$, F_2, $\bar{F}_2$, F_3, and $\bar{F}_3$ are then written. This results in

$$F_1 = AB + \bar{A}\bar{B}D \qquad\qquad \bar{F}_1 = \bar{A}B + A\bar{B} + \bar{B}\bar{D}$$
$$\bar{F}_1 = \bar{A}B + A\bar{B} + \bar{A}\bar{D}$$

$$F_2 = AB\bar{C} + B\bar{C}D + ACD + A\bar{B}C \qquad \bar{F}_2 = \bar{A}\bar{D} + \bar{A}C + B\bar{C} + BC\bar{D}$$
$$F_2 = AB\bar{C} + B\bar{C}D + ABD + A\bar{B}C$$
$$F_3 = \bar{A}\bar{C}\bar{D} + A\bar{B}C + \bar{A}CD + \bar{B}\bar{C}\bar{D} \qquad \bar{F}_3 = AB + \bar{C}D + \bar{A}C\bar{D}$$
$$F_3 = \bar{A}\bar{C}\bar{D} + A\bar{B}C + \bar{A}CD + A\bar{B}\bar{D}$$
$$F_3 = \bar{A}\bar{C}\bar{D} + A\bar{B}C + \bar{A}CD + \bar{B}\bar{C}\bar{D}$$
$$F_3 = \bar{A}\bar{C}\bar{D} + A\bar{B}\bar{D} + \bar{A}CD + \bar{B}CD$$

In determining whether F or $\bar{F}$ should be used, we consider not only the functions that have the fewest sum terms, but also product terms that can be used in more than one function. For example, F_1 requires only two AND gates while $\bar{F}_1$ requires three AND gates. Furthermore, the AB term of F_1 is common to $\bar{F}_3$, thus $\bar{F}_3$ can be implemented with only two additional gates. For this example, either F_2 or $\bar{F}_2$ could be used. If $\bar{F}_2$ is selected, the three functions to be implemented are F_1, $\bar{F}_2$, and $\bar{F}_3$. A total of 8 AND gates would be required to synthesize this function. The finished PLA is shown in Figure 8.7.

8.4

PROGRAMMABLE ARRAY LOGIC OR PAL®

A device that is related to the PLA is the PAL®, which is a mnemonic for programmable array logic. The trademark is owned by American Micro Devices (AMD). This component inserts a few OR gates between the AND gate outputs and the chip outputs. The OR gate inputs make fixed connections to certain AND gate outputs. These OR gate inputs are not programmable as in the PLA. For example, a PAL® may have 14 inputs, eight AND gates, and four output OR gates. Each OR gate may be driven by four AND gates with fixed connections. Another version may drive two OR gates with two AND gate outputs and the other two OR gates with four AND gate outputs. The PAL® is less flexible than the PLA, but is less expensive and easier to program. Generally, enough AND gates and OR gates are provided to allow the realization of any practical logic function.

8.4.1 CONFIGURATION OF THE PAL®

The simplified schematic of Figure 8.8 indicates a small PAL® configuration. In this PAL® there are four inputs (plus complemented values), four AND gates, and four OR gates. The input connections to the AND gates can be programmed while the connections between the AND gate outputs and the OR gate inputs are fixed.

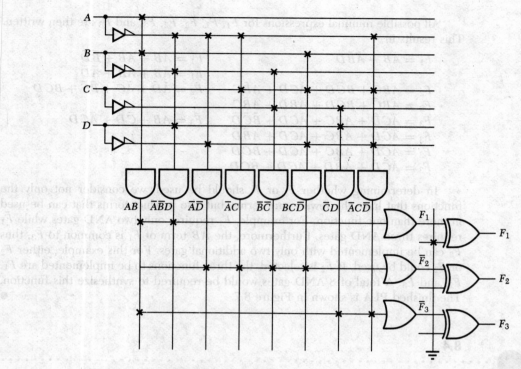

Practical PAL® chips are considerably larger than the simplified version of Figure 8.8. A typical AMD PAL® is the 16L8, which specifies 16 inputs to the AND gates and 8 outputs, each driven by an OR gate. The L specifies that this is a combinational logic chip. We will later see that some PLA or PAL® chips add flip-flops to the circuit. These are called *registered* devices. The AMD designation for such a chip would use R rather than L in the chip number.

Figure 8.9 shows the schematic of the AMD 16L8 [2]. This schematic is again abbreviated to conserve space. Each small AND gate is a 32-input gate connected by fusible links to all 32 vertical lines. The upper AND gate of each group does not drive an OR gate input, but drives the enable of a three-state inverting buffer. This AND gate can be programmed to enable and invert the OR gate output based on an AND relation of the inputs. The inverting output buffer can also be enabled if all 32 fusible links connected to the AND gate input are blown. In this latter instance, each input that is not connected floats high and is interpreted by the AND gate as a logic 1. The output of the AND gate is then asserted and the three-state buffer is continuously enabled.

There are ten dedicated input pins, I_0 through I_9. The normal and inverted value of each of these inputs connect to a vertical line. There are six other lines, I/O_2 through I/O_7, that can be used for input or output lines. When used for input

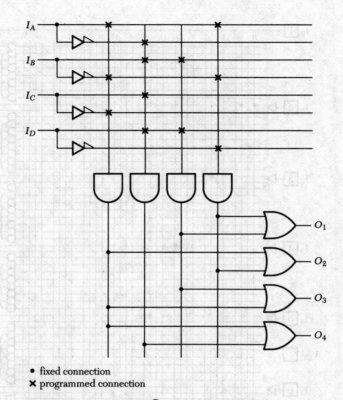

• fixed connection
✗ programmed connection

FIGURE 8.8 A small PAL[®].

lines, the three-state buffer should not be enabled. The 32 vertical lines can be driven by up to 16 inputs plus the inverted values of these inputs.

The other seven AND gates of a group are ORed to form a logic expression. Often only a few of these gates will be involved in the implementation of an expression and the remaining AND gates of the group must be deactivated. In order to keep the output of an AND gate deasserted, all input connections to the vertical lines are left intact. Since one input to the AND gate will be driven by an input variable and one gate input will be driven by the complement of this variable, the AND condition will not be satisfied. It is recommended that unused input lines be tied to ground or to the power supply voltage. This also guarantees that the AND condition cannot be satisfied.

As an example of a simple function realization using the 16L8 PAL[®], we will consider a situation that has four input variables A, B, C, and D connected respectively to inputs I_1, I_0, I_3, and I/O_5 of Figure 8.9. We want to implement the two output expressions

$$\bar{Y} = A\bar{B}\bar{C}D + BC\bar{D}$$

and

$$\bar{Z} = \bar{A}B\bar{C}D$$

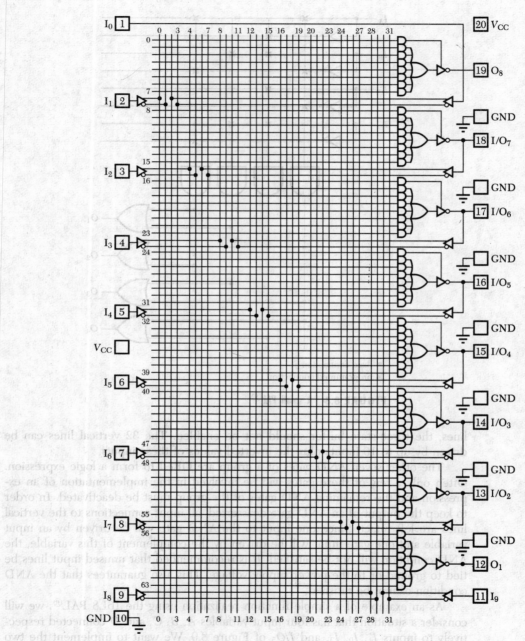

FIGURE 8.9 The 16L8 AMD PAL.

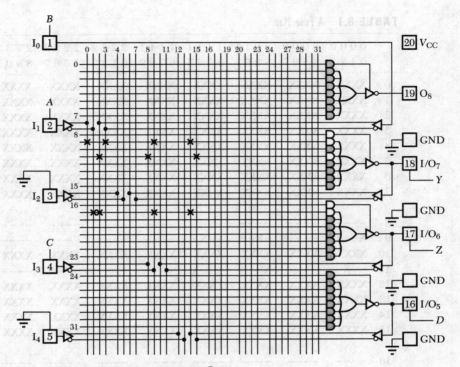

FIGURE 8.10 A portion of a 16L8 PAL®.

Figure 8.10 shows the portion of the 16L8 chip that realizes these two expressions. The unshaded gates represent circuits that have been programmed. The presence of a fuse or connecting link is represented by an × while no mark on a junction of two lines indicates that the corresponding fuse has been blown. The shaded AND gates represent circuits that have had no fuses blown. Even though no ×s appear on these inputs, we understand that all inputs are connected to the shaded AND gates.

The two output, three-state buffers for $\bar{Y}$ and $\bar{Z}$ are enabled by blowing all fuses to the AND gates that drive these two buffers. The variable D is input on I/O_5 and the three-state buffer driving this pin must be disabled. This is done by leaving all fuses that drive the corresponding AND gate intact. Since inactive input pins I_2 and I_4 are tied to ground, the enabling AND gate will have some inputs connected to ground and will never enable the three-state buffer. This is also true of all shaded AND gates, that is, the outputs of all of these gates will remain at the unasserted level and will not influence the outputs $\bar{Y}$ or $\bar{Z}$. The proper inputs to realize the two expressions remain connected to the gate inputs while all other inputs are disconnected by blowing the fuses.

Rather than redrawing a complete schematic for each problem to be solved with the PAL®, a fuse map is generally used. The fuse map for the circuit of Figure 8.10

TABLE 8.1   A Fuse Map.

	0000 0123	0000 4567	0011 8901	1111 2345	1111 6789	2222 0123	2222 4567	2233 8901	Expla- nation
0	XXXX	XXXX	XXXX	XXXX	XXXX	XXXX	XXXX	XXXX	Unused
1	XXXX	XXXX	XXXX	XXXX	XXXX	XXXX	XXXX	XXXX	group
2	XXXX	XXXX	XXXX	XXXX	XXXX	XXXX	XXXX	XXXX	of
3	XXXX	XXXX	XXXX	XXXX	XXXX	XXXX	XXXX	XXXX	gates
4	XXXX	XXXX	XXXX	XXXX	XXXX	XXXX	XXXX	XXXX	
5	XXXX	XXXX	XXXX	XXXX	XXXX	XXXX	XXXX	XXXX	
6	XXXX	XXXX	XXXX	XXXX	XXXX	XXXX	XXXX	XXXX	
7	XXXX	XXXX	XXXX	XXXX	XXXX	XXXX	XXXX	XXXX	
8	----	----	----	----	----	----	----	----	Enable $\bar{Y}$
9	X--X	----	-X--	--X-	----	----	----	----	$A\bar{B}\bar{C}D$
10	XXXX	XXXX	XXXX	XXXX	XXXX	XXXX	XXXX	XXXX	Unused
11	--X-	----	X---	---X	----	----	----	----	$BC\bar{D}$
12	XXXX	XXXX	XXXX	XXXX	XXXX	XXXX	XXXX	XXXX	Unused
13	XXXX	XXXX	XXXX	XXXX	XXXX	XXXX	XXXX	XXXX	Unused
14	XXXX	XXXX	XXXX	XXXX	XXXX	XXXX	XXXX	XXXX	Unused
15	XXXX	XXXX	XXXX	XXXX	XXXX	XXXX	XXXX	XXXX	Unused
16	----	----	----	----	----	----	----	----	Enable $\bar{Z}$
17	-XX-	----	-X--	--X-	----	----	----	----	$\bar{A}B\bar{C}D$
18	XXXX	XXXX	XXXX	XXXX	XXXX	XXXX	XXXX	XXXX	Unused
19	XXXX	XXXX	XXXX	XXXX	XXXX	XXXX	XXXX	XXXX	Unused
20	XXXX	XXXX	XXXX	XXXX	XXXX	XXXX	XXXX	XXXX	Unused
21	XXXX	XXXX	XXXX	XXXX	XXXX	XXXX	XXXX	XXXX	Unused
22	XXXX	XXXX	XXXX	XXXX	XXXX	XXXX	XXXX	XXXX	Unused
23	XXXX	XXXX	XXXX	XXXX	XXXX	XXXX	XXXX	XXXX	Unused
24	XXXX	XXXX	XXXX	XXXX	XXXX	XXXX	XXXX	XXXX	Unused
25	XXXX	XXXX	XXXX	XXXX	XXXX	XXXX	XXXX	XXXX	Unused
26	XXXX	XXXX	XXXX	XXXX	XXXX	XXXX	XXXX	XXXX	Unused
27	XXXX	XXXX	XXXX	XXXX	XXXX	XXXX	XXXX	XXXX	Unused
28	XXXX	XXXX	XXXX	XXXX	XXXX	XXXX	XXXX	XXXX	Unused
29	XXXX	XXXX	XXXX	XXXX	XXXX	XXXX	XXXX	XXXX	Unused
30	XXXX	XXXX	XXXX	XXXX	XXXX	XXXX	XXXX	XXXX	Unused
31	XXXX	XXXX	XXXX	XXXX	XXXX	XXXX	XXXX	XXXX	Unused

is shown in Table 8.1. The dash represents no connection or a blown fuse link while an X represents a connection. The 32 vertical lines are numbered from 0 to 31. The

AND gates are numbered with row numbers 0 to 31. The first group of AND gates are not used in this realization. All inputs to AND gate number 8 are removed to enable the inverting output buffer for $\bar{Y}$. Variable A is connected to vertical line 0 while $\bar{A}$ connects to vertical line 1. B connects to vertical line 2 and $\bar{B}$ connects to vertical line 3. C connects to vertical line 8 and $\bar{C}$ connects to vertical line 9. D connects to vertical line 14 and $\bar{D}$ connects to vertical line 15. AND gate number 9 realizes the first term of $\bar{Y}$ while AND gate number 11 realizes the second term. These two terms are ORed to form Y. The inverting output buffer inverts this value to form $\bar{Y}$. The decision to skip AND gate number 10 in favor of AND gate number 11 was arbitrary. AND gate number 17 implements $\bar{Z}$.

8.4.2 PROGRAMMING THE PAL℗

In recent years, the programmable device has become so popular that programming methods specified by various companies are becoming very similar. We will consider the PAL® device and discuss the programming of these devices using AMD methods. The PLA can also be programmed in a similar manner.

AMD has developed a software package designed to simplify the programming of PAL®s or PLAs. The program is called PALASM®, which is an acronym for programmable array logic-assembler. It can be executed on several types of personal computers or workstations. PALASM® is designed to accomplish the following tasks:

1. Accept specifications of a logic circuit and develop a fuse map from this information. The fuse map can then drive a chip programmer following a standard developed by the Joint Electron Device Engineering Council (JEDEC) to implement the logic circuit on a chip.
2. After accepting the logic circuit specifications and the type of chip that will be used for the hardware implementation, PALASM® can run a simple simulation to verify the behavior of the finished circuit. This allows debugging to be done prior to programming the actual chip.

The simplest method of specifying a logic function to be implemented is by Boolean equations. PALASM® accepts Boolean expressions written in terms of AND, OR, EXOR, or INVERT operations.

The simulation does not consider transient outputs. It allows input conditions to be changed and the output is calculated after all variables have become stable. While information to use PALASM® can be obtained from a handbook, it is instructive to consider a simple program that demonstrates the basic capabilities of this design tool.

EXAMPLE 8.2

· ·

Synthesize the function

$$F = \bar{A}\bar{B}C + \bar{A}B\bar{C} + A\bar{B}C + AB\bar{C} + ABC$$

using a 16L8 PAL®. Use PALASM® to generate a fuse map and simulate the resulting circuit for all possible values of A, B, and C.

SOLUTION

The source file to specify the logic functions and the simulation is shown in Table 8.2. The first five lines of the file are optional segments that are useful to the designer in identifying this project. The CHIP segment of the file is required and consists

TABLE 8.2 A PALASM® Source File.

TITLE	ex2.pds								
REVISION	A								
AUTHOR	DC								
COMPANY	BYU								
DATE	5-17-94								
CHIP	ex2 PAL16L8								
;PINS	1	2	3	4	5	6	7	8	9	10
	A	B	C	NC	NC	NC	NC	NC	NC	GND
;PINS	11	12	13	14	15	16	17	18	19	20
	NC	NC	NC	NC	NC	NC	NC	NC	F	VCC

EQUATIONS
F = /A°/B°C+A°/B°/C+A°/B°C+A°B°/C+A°B°C

SIMULATION
TRACE_ON A B C F
SETF /A /B /C
SETF /A /B C
SETF /A B /C
SETF /A B C
SETF A /B /C
SETF A /B C
SETF A B /C
SETF A B C
TRACE_OFF

· ·

of three pieces of necessary information. The first part is the name of the file that is being used for the design. The second part is the chip type to be used. Both of these specifications appear on the same line. For this example, the file name is ex2, and the chip name is PAL16L8.

The third piece of information in the CHIP segment is the specification of the pin connections. While there are other formats for pin information, this example uses a comment line for clarity. The comment line is identified by the beginning semicolon. Information following this symbol will be ignored when the program is run. Pin numbers 1 through 10 are listed on this line with connection information on the following line. The 16L8 chip uses pins 1 through 3 as input pins, thus variables A, B, and C are connected to these pins. Pin 10 is a ground line and the abbreviation GND specifies this pin. A space is required between each pin specification. Another comment line allows pins 11 through 20 to be labeled followed by the line defining the connections to these pins. Pin 19 is an output line and is connected to the desired output variable F. The power supply voltage is connected to pin 20 and all other pins are specified as having no connection by the abbreviation NC.

For combinational logic, the EQUATION segment is used to specify the function to be implemented. This information must be specified in terms of the following operations:

Operation	Designation
Invert	/
OR	+
AND	°
EXOR	:+:

After the equations are expressed, the SIMULATION segment can be included. This segment allows the input variables to be set to any possible set of values and the program calculates the resulting output value. The TRACE_ON statement defines the signals to be included in the simulation. In this program, the variables A, B, and C are specified along with the output variable F. The input variables are then set by the use of SETF statements. This statement uses the variable names or inverted variable names to express a set of values. The statement SETF A /B C sets the input variables to the values of $A = 1$, $B = 0$, and $C = 1$. The last statement is the TRACE_OFF statement that indicates no more values are to be calculated.

When this source file is compiled and simulated the results of Tables 8.3 and 8.4 are obtained. The remaining rows up to row 63 have no fuses blown.

The first row in Table 8.3 enables the output inverting buffer while rows two and three implement the minimized logic expression. From the pin specifications and the schematic of the 16L8, we see that B connects to column 00, $\bar{B}$ to 01, A to 02, $\bar{A}$ to 03, C to 04, and $\bar{C}$ to 05. Row 1 implements the expression

$$\bar{A}B$$

TABLE 8.3   Partial Fuse Map for Example 8.2.

	0000	0000	0011	1111	1111	2222	2222	2233
	0123	4567	8901	2345	6789	0123	4567	8901
0	----	----	----	----	----	----	----	----
1	X--X	----	----	----	----	----	----	----
2	---X	-X--	----	----	----	----	----	----
3	XXXX	XXXX	XXXX	XXXX	XXXX	XXXX	XXXX	XXXX
4	XXXX	XXXX	XXXX	XXXX	XXXX	XXXX	XXXX	XXXX
5	XXXX	XXXX	XXXX	XXXX	XXXX	XXXX	XXXX	XXXX
6	XXXX	XXXX	XXXX	XXXX	XXXX	XXXX	XXXX	XXXX
7	XXXX	XXXX	XXXX	XXXX	XXXX	XXXX	XXXX	XXXX

TOTAL FUSES BLOWN = 92

while row 2 implements

$$\bar{A}\bar{C}$$

These expressions are ORed and inverted to give an output of

$$F = \overline{\bar{A}B + \bar{A}\bar{C}}$$

Using DeMorgan's laws and other logic relations we can write this function as

$$F = \overline{\bar{A}(B + \bar{C})} = A + \overline{(B + \bar{C})}$$

$$= A + \bar{B}C$$

This is the reduced expression for F as indicated by the K-map of Figure 2.27.   ●

The simulation results of Table 8.4 indicate the values F takes on as the input variables assume all possible sets of values. The letter g above each set of values merely indicates the presence of a SETF command.

TABLE 8.4   Simulation of Circuit.

	g	g	g	g	g	g	g	g
A	L	L	L	L	H	H	H	H
B	L	L	H	H	L	L	H	H
C	L	H	L	H	L	H	L	H
F	L	H	L	L	H	H	H	H

There are several points that relate to the use of PALASM® that should be considered. The program automatically reduces the logic expression to the simplest form. It is unnecessary to minimize functions with K-maps since the program performs this task. Entry of the specifications need not be done with Boolean equations. Schematic entry can also be done in the newer versions of the program. Input and output pins need not be specified by the designer. As long as allowable numbers of inputs or outputs are not exceeded, the program can make these assignments.

PALASM® is a much more powerful program than indicated by this simple example. One feature allows state machines to be created using registered PAL®s in conjunction with this program. A handbook should be consulted to observe the many powerful features of this particular design aid for programmable logic devices.

8.4.3 REGISTERED PROGRAMMABLE DEVICES

Many manufacturers of programmable combinational circuits such as PLAs and PAL®s now offer versions of these chips with flip-flops added. The flip-flops can be used as storage registers, as state machine components, or both. These devices are still called PLAs or PAL®s even though they add considerable capability to the chips. Often these types of devices are referred to as registered PLAs or registered PAL®s. The confusion resulting from referring to all registered or unregistered devices as PLAs or PAL®s makes it necessary to examine specification sheets to distinguish between the two types of devices. In general, when flip-flops are included on the chip, the chip will be called a registered device.

8.5

COMBINATIONAL PLD-BASED STATE MACHINES

SECTION OVERVIEW The required combinational logic for a state machine can be realized with a ROM or with a combinational PLA or PAL®. Such systems have the capability of changing the state machine output sequence by changing the program of the PLD. These state machines may also result in a smaller chip count.

In some applications, the sequences executed by a state machine may need to be modified. This requires a change in the IFL in order to control a new sequence. In such a case, it may be appropriate to use programmable devices for the IFL. A PLD can be programmed to create the correct signals to drive the state flip-flops. A change in sequence of the state machine can be effected simply by replacing the PLD with a new chip containing a different program. Changing the machine with conventional IFL would require a rewiring job rather than a chip change.

The concept of changeable ROM cartridges has gained great acceptance in the last several years, both in state machine and microprocessor control. Several handheld calculators provide the capability of canned programs or subroutines that can be executed under ROM control. One of several programs can be used, simply by plugging the proper ROM into the program socket. Video games use a similar approach. The game to be played depends on the ROM plugged into the program socket. Control sequences are different for each ROM, providing a quick method of changing games.

A second advantage of ROM-controlled state machines occurs in system development work. A complex digital system will often require a large amount of debugging and modification during initial development. A hardwired system may lead to many necessary wiring changes during this phase. Wiring changes are costly and can be time consuming in a larger system. Some firms use the EPROM to avoid this problem.

An EPROM (erasable programmable read-only memory) is a ROM that can be programmed upon application of the correct electrical signals. Once programmed, the EPROM retains this information indefinitely or until the chip is exposed to intense ultraviolet radiation for approximately 15 minutes. The package contains a quartz window through which the ultraviolet rays can impinge on the chip. After erasure, the EPROM can be reprogrammed. There is no upper limit on the number of program-erase cycles for the EPROM.

Using this device to design a ROM-controlled state machine allows a system to be debugged and modified under software control. The EPROM can initially be written into by a programming device under control of a computer. The state machine system is then exercised to locate any problems in operation. As problems are encountered, the system is modified simply by rewriting the contents of the EPROM. This procedure continues until the system operates according to specs. A production system can then be created simply by replacing the more expensive EPROM with a mask-programmable device containing a duplicate program. Most EPROMs have a companion ROM that is pin compatible.

The configuration for a state machine using the direct-addressed ROM is shown in Figure 8.11. The next state and the outputs of a state machine are completely determined at any time by the present state of the system and the values of the input variables. The ROM uses the system input variables and present state (flip-

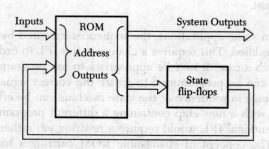

FIGURE 8.11 A direct-addressed ROM state machine.

flop outputs) as input addresses. The device is programmed to produce the correct next state conditions and correct system output variables as ROM outputs. The ROM then becomes both IFL and OFL for the system. The number of system inputs plus the number of state flip-flops determine the width of the input address, and hence the number of ROM locations needed. The number of outputs plus the number of state flip-flops determine the width of the output or the word size. A 5-input, 4-output system with three state flip-flops would require an 8-input ROM. This leads to a 256-word capacity ROM (2^8). Each word must be at least 7 bits wide. If a standard ROM is used, a 256×8 device would be selected.

The two state diagrams of Figure 8.12 represent systems with the same possible sequences of states, but with different output states. We will realize these systems with direct-addressed ROMs indicating the ROM programming for each system. One system can then be converted to another simply by exchanging ROMs. We emphasize that the modification of a nonprogrammable state machine would require wiring or PC board changes. Another choice existing for the ROM implementation is to wire both ROMs using multiplexed outputs to drive the flip-flops and output signals. A voltage level change that selects the proper ROM outputs can then select which sequence of operations will be carried out. This configuration results in a voltage-controlled state machine sequence.

With six states in the systems three flip-flops are required. Two system inputs, X and Y, bring the total number of ROM inputs needed to five. Since there are three system outputs, the number of ROM outputs is six. The minimum ROM size is then 32×6. A standard 32×8 ROM can be used. Figure 8.13 shows the realization of

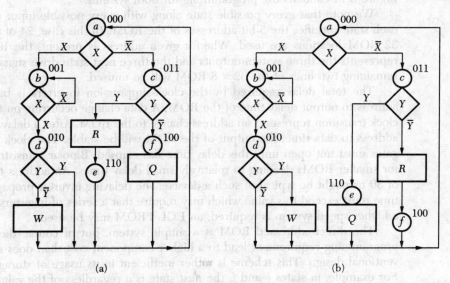

FIGURE 8.12 State diagrams of similar systems.

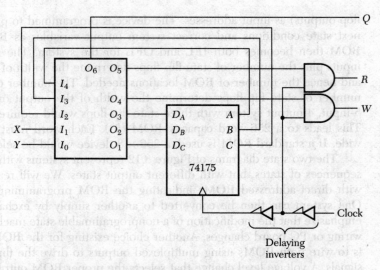

FIGURE 8.13   Direct-addressed ROM controlled state machine.

either of these systems using clock suppression to eliminate output glitches on the conditional lines R and W.

In order that the system perform properly, the ROM must be programmed correctly. While not every location must be programmed, those locations corresponding to each state and every possible input combination require programming. Figure 8.14 contains the programming for both systems.

We note that every possible state along with every possible input condition at each state specifies the 5-bit addresses of the ROM. In this case, 24 of the possible 32 ROM locations are used. When a given address is present, the ROM output represents the three system outputs and the three next state drive signals while the remaining two lines of the 32×8 ROM will be unused.

The total delay required by the clock-suppression inverters is based on the address to output delay time of the ROM. A data change occurring on the negative clock transition represents an address change to the ROM. After a delay specified as address to data time, the output of the ROM will be stable. The clock suppression gates must not open until this delay time has elapsed. Bipolar transistors are used for smaller ROMs, leading to relatively small delay times. An address to data time of 30 ns might be typical of such a device. The delaying inverter propagation delay time must exceed this value which may require that a series of inverters be used. If a higher speed system is required, an ECL PROM may be used.

The direct-addressed ROM is a simple system, but of course the ROM and programming requirement lead to a higher component cost than does a more conventional design. This scheme is rather inefficient in its usage of storage locations. For example, in states e and f, the next state is a regardless of the values of inputs X and Y. Since each of the four input combinations leads to a different address,

ROM Address					ROM Outputs						ROM Outputs					
Present State			System Inputs		System Outputs			Next State			System Outputs			Next State		
A	B	C	X	Y	Q	R	W	A	B	C	Q	R	W	A	B	C
I_4	I_3	I_2	I_1	I_0	O_6	O_5	O_4	O_3	O_2	O_1	O_6	O_5	O_4	O_3	O_2	O_1
0	0	0	0	0	0	0	0	0	1	1	0	0	0	0	1	1
0	0	0	0	1	0	0	0	0	1	1	0	0	0	0	1	1
0	0	0	1	0	0	0	0	0	0	1	0	0	0	0	0	1
0	0	0	1	1	0	0	0	0	0	1	0	0	0	0	0	1
0	0	1	0	0	0	0	0	0	0	1	0	0	0	0	0	1
0	0	1	0	1	0	0	0	0	0	1	0	0	0	0	0	1
0	0	1	1	0	0	0	0	0	1	0	0	0	0	0	1	0
0	0	1	1	1	0	0	0	0	1	0	0	0	0	0	1	0
0	1	0	0	0	0	0	1	0	0	0	0	0	0	0	0	0
0	1	0	0	1	0	0	0	0	0	0	0	0	1	0	0	0
0	1	0	1	0	0	0	1	0	0	0	0	0	0	0	0	0
0	1	0	1	1	0	0	0	0	0	0	0	0	1	0	0	0
0	1	1	0	0	0	0	0	1	0	0	0	1	0	1	0	0
0	1	1	0	1	0	1	0	1	1	0	0	0	0	1	1	0
0	1	1	1	0	0	0	0	1	0	0	0	1	0	1	0	0
0	1	1	1	1	0	1	0	1	1	0	0	0	0	1	1	0
1	0	0	0	0	1	0	0	0	0	0	0	0	0	0	0	0
1	0	0	0	1	1	0	0	0	0	0	0	0	0	0	0	0
1	0	0	1	0	1	0	0	0	0	0	0	0	0	0	0	0
1	0	0	1	1	1	0	0	0	0	0	0	0	0	0	0	0
1	1	0	0	0	0	0	0	0	0	0	1	0	0	0	0	0
1	1	0	0	1	0	0	0	0	0	0	1	0	0	0	0	0
1	1	0	1	0	0	0	0	0	0	0	1	0	0	0	0	0
1	1	0	1	1	0	0	0	0	0	0	1	0	0	0	0	0

FIGURE 8.14 ROM contents: (a) for system of Figure 8.12(a); (b) for system of Figure 8.12(b).

four locations corresponding to each of these states must be programmed. Furthermore, each of the four locations has the same contents. Another disadvantage of the direct-addressed ROM is that the size is determined by the number of inputs required rather than by the number of unique words that must be developed at the ROM output.

A PLA or PAL® can replace the ROM in the direct-addressed state machine with little effort. In systems that require a small ratio of programmed locations to total number of input combinations, the cost of these devices can be somewhat less than that of the ROM. In order to compare a PLA to a ROM, we will consider a system requiring a 10-input ROM needing only 48 programmed locations of 8-bit width. The ROM size would be 1024 × 8, whereas a PLA size of 10 × 48 × 8 could be used. The cost of a 48-location PLA may be 30% lower than the cost of a 1024-location ROM. While the savings may not always be this great, the PLA offers a reasonable alternative to the ROM. Unfortunately, the choice of PLA size is rather limited, and the ideal size needed for a given application may not be available.

Most PLAs and PAL®s are field-programmable using fusible-link technology. There are a few ultraviolet-erasable PLAs such as the Signetics PLC473 PLA, which is a $20 \times 24 \times 11$ unit and is pin-compatible with the PLHS473. The flexibility of creating two different state machines simply by exchanging one PLA chip for another chip having a different program makes this system very similar to the direct-addressed ROM scheme. Of course, these devices can also be simulated to assist in the debugging process.

● ● ●   **DRILL PROBLEMS** Sec. 8.5

1. Realize the system represented by the diagram of Figure 6.44(a), p. 272, using a direct-addressed ROM. Use the state assignment of Figure 6.44(b). Specify the contents of the ROM.

2. Repeat Prob. 1 using the state assignment of Figure 6.44(c).

8.6

STATE MACHINES ON A CHIP

SECTION OVERVIEW Registered PLDs and FPGAs allow the creation of a complete state machine on a single chip. These devices are discussed in this section.

8.6.1 NOMENCLATURE OF PLDS

Before we discuss the construction of state machines on a single chip, we should comment on some of the terms applied to programmable logic. Historically, there is little standardization of terms within the industry. The next few paragraphs will comment on the situation as it stands in the mid 1990s.

We have looked at the combinational PLA as a chip that contains an array of AND gates followed by OR gates. The combinational PAL® has a similar arrangement with AND gates followed by OR gates with fixed connections to the OR gates. Actually, the mnemonic PAL® was registered as a trademark of Monolithic Memories Incorporated and is now owned by American Micro Devices or AMD. As mentioned previously, the chip architecture that includes an AND gate array and an OR gate array (ROMs, PLAs, and PAL®s) can be referred to as a PLD.

When flip-flops are added to the chip, the capability to create state machines on the chip exists. A typical configuration allows the flip-flop inputs to be driven by the AND-OR gates and also provides a feedback path from flip-flop outputs to the inputs of the AND array. These devices are also referred to as registered PLDs, PLAs, or PAL®s. One manufacturer, Signetics, refers to their PLD chip that includes flip-flops as a programmable logic sequencer or PLS.

Adding to the general confusion in nomenclature is gate array logic. While the obvious mnemonic for this chip is the GAL®, this particular mnemonic is also a registered trademark of Lattice Semiconductor Corporation used to represent the words "generic array logic."

The term gate array has become associated with the structure made popular by Xilinx. This company has grown very rapidly since its beginning in 1984 and is one of the largest suppliers of programmable devices on a chip. Xilinx refers to their chip as a field programmable gate array or FPGA.

The FPGA is laid out in rows and columns of circuit modules with interconnect resources occupying the space between the rows and columns. The circuit modules consist of logic blocks that include combinational logic, flip-flops, and MUXs. Surrounding each logic block are interfacing or I/O blocks that allow interconnection of circuit modules via the interconnect resources.

The FPGA distributes the gates and flip-flops throughout the chip rather than following a gate array with flip-flops followed by another gate array as does the PAL®. The PAL® is sometimes I/O limited since output lines must be coupled back to inputs to create the feedback paths for the state machine. With the FPGA, outputs of state machines can be coupled back to the inputs using interconnections between modules. This interconnection does not decrease the number of available lines for external inputs or outputs.

The various types of programmable devices utilize several different programming methods. PLDs and some gate array chips use fusible links to interconnect the logic elements. In such systems, the circuit configuration, once programmed, is permanent. While the possibility of burning additional fusible links exists, there is no possibility of restoring a link.

In addition to devices that are programmable using fusible link technology, gate arrays are also made with electrically erasable devices [3]. These chips can be configured to realize a state machine in a manner similar to that used for the fusible link device. A programming device creates the proper electrical signals with which to drive the chip lines to configure the internal circuitry. With the electrically erasable CMOS cell technology, a given configuration can be returned to the unprogrammed state and reconfigured several times. The circuit configuration of the electrically erasable devices can be modified several times after the initial programming is completed. Typically, a minimum of 100 erase-write cycles are guaranteed [3]. The erase time required is approximately 50 ms.

Xilinx FPGAs are also reprogrammable, but do not use electrically erasable methods. Instead they are programmed or configured by data stored in an on-board memory that must be loaded externally upon power-up as part of the initialization process. This on-board memory can also be changed at any time using the appropriate command to reconfigure the logic functions of the chip. The on-board memory can activate multiplexers or three-state circuits to connect logic components in the desired configuration.

With the memory-configurable approach of Xilinx, there is no limit on the number of times the circuit configuration is changed. The chip can be reprogrammed electrically without removing the chip from its system. It is programmed in-circuit by providing signals of the prescribed format.

In order to demonstrate some of the chips discussed in the preceding paragraphs, the Signetics PLS and the Xilinx FPGA will be considered in more detail.

8.6.2 THE SIGNETICS PLS

The Signetics PLS is essentially a registered PLA. The succeeding discussion of this chip pertains almost directly to any registered PLA or PAL® device. The PLS chip leads to a very compact method of state machine realization. As mentioned earlier in this section, the PLS includes a PLA along with several edge-triggered flip-flops on a single chip. This family of programmable chips uses fusible link technology and, therefore, can only be programmed one time. An entire state machine can be implemented as a one-chip system. A typical PLS chip configuration is shown in the block diagram of Figure 8.15.

The Signetics PLS155 is specified as a $16 \times 45 \times 12$ device. This chip has a total of 16 input lines consisting of four dedicated inputs plus 12 programmable input/output pins included in the 20-pin package. These 12 lines can be programmed to be either input or output lines. There are 45 AND gates and four edge-triggered JK flip-flops on the chip. A maximum of 12 output lines are available. This device can be used to realize up to 16 states in a state machine. The maximum clock frequency is 15 MHz.

Figure 8.16 represents a simplified schematic of a PLS. The AND gate array can be driven by inputs or their complements and flip-flop outputs or their complements. One section of these AND and OR gates is allocated to the IFL and these gates are driven by inputs and the present state of the flip-flops. A second section of AND and OR gates is allocated to the OFL. Outputs can be unconditional, driven only by the state flip-flops, or conditional, driven by inputs and state flip-flops. Since the

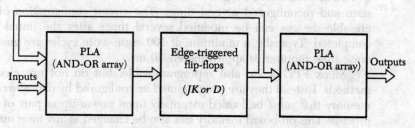

FIGURE 8.15 Block diagram of a programmable logic sequencer.

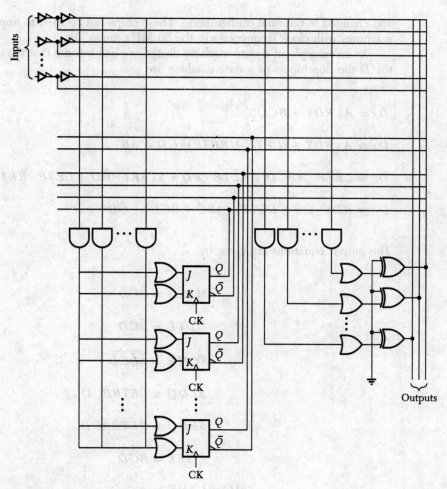

FIGURE 8.16   A simplified PLS schematic.

outputs are coupled back to the inputs of the AND gate array, various delays can be created. There are several flip-flops on some PLS chips that allow an output holding register to be implemented.

Software similar to PALASM® is available from most companies that manufacture PLDs. These programs are used on personal computers to drive programming devices and configure the PLD. In most cases, a development system allows the designed circuit to be simulated before the chip is actually configured. This is particularly important in the case of fusible link devices such as the Signetics PLS. Simulation of the finished system allows debugging to take place before burning the fusible links. For electrically erasable chips, the actual chip can be debugged and changed to the final configuration. These chips can be used to implement state machines with clock frequencies in the 50 MHz range [3].

As an example of a state machine design, we will assume that the equations for the D flip-flop inputs of a state machine are

$$D_A = A \cdot RDY + BC\bar{D}$$

$$D_B = A \cdot RDY + (LETE + RETE) \cdot C\bar{D} + \bar{A}B$$

$$D_C = LETE \cdot \bar{A}B\bar{C}\bar{D} + RETE \cdot BD + START \cdot \bar{B}D + LETE \cdot RETE \cdot \bar{B}C + CD$$

$$D_D = RDY \cdot \bar{C}\bar{B} + LETE \cdot (\bar{A}B\bar{C} + \bar{B}C\bar{D}) + \bar{C}D$$

The output equations are given by

$$MSCG = \bar{B}\bar{C}\bar{D}$$

$$DYL = \bar{B}\bar{C}D$$

$$RCMR = \overline{\bar{A}\bar{B}\bar{C}\bar{D}}$$

$$SLDQ = LETB\bar{B}CD$$

$$SRDQ = RETB\bar{B}CD$$

$$SEL = BC\bar{D}$$

$$UNBLANK = A$$

With the equations for the flip-flop inputs along with the output equations, a suitable PLS chip can be selected. There are four external inputs to the IFL ($LETE$, $RETE$, RDY, and $START$) plus four state flip-flops (A, B, C, and D) that can drive this logic. There are two additional inputs ($LETB$ and $RETB$) that drive the OFL. Thus, there are six external inputs for this system. Seven outputs must also be available. The Signetics PLS155 will satisfy this design.

The PLS155 has four dedicated input lines, eight programmable I/O lines, and four flip-flops. Two I/O lines will be used to increase the number of external inputs to six for this circuit. The seven outputs can be implemented with remaining output lines. The finished circuit is shown in Figure 8.17. A programming device would be used to obtain the PLS on-chip connections shown.

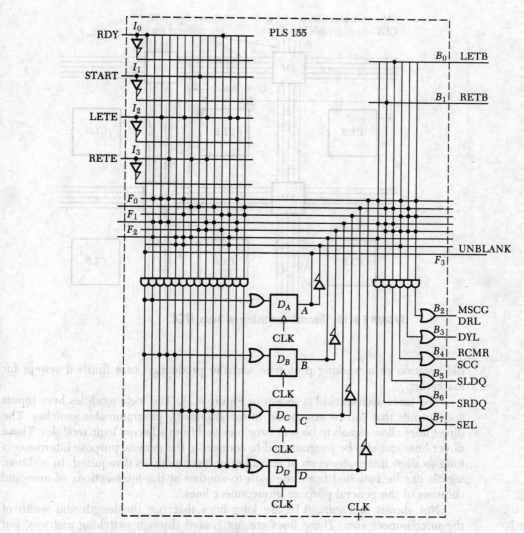

FIGURE 8.17 A PLS state machine controller.

8.6.3 THE XILINX FPGA

Xilinx is a relative newcomer to the programmable logic manufacturing field, having been founded in 1984. This company specializes in high-density programmable gate arrays that include flip-flops within the arrays. With a configuration that can be changed simply by modifying the contents of an on-board RAM, this approach to logic design is becoming very popular. Complex systems can be produced in a short time with the Xilinx chip. This allows a company to minimize the time between

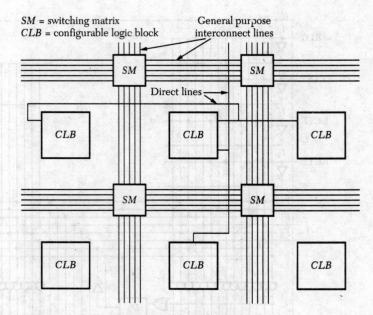

SM = switching matrix
CLB = configurable logic block
General purpose interconnect lines
Direct lines

FIGURE 8.18   General architecture of Xilinx FPGA.

the creation of a working prototype and the production of a finished system for marketing.

The basic architecture is shown in Figure 8.18. The logic modules have inputs and outputs that can be connected to metal lines by programmable switches. The direct lines allow signals to be sent to or received from adjacent logic modules. These direct lines can also be programmed to connect to the general purpose interconnect lines to allow interconnection of nonadjacent logic modules if required. In addition, signals can be switched from one path to another at the intersections of rows and columns of the general purpose interconnect lines.

Not shown in Figure 8.18 are long lines that run the length and width of the interconnect area. These lines are not routed through switching matrices, but can be connected to the logic modules. This allows a common signal to be applied simultaneously to several logic modules with little time skew. Typically, a high-speed clock signal would be applied in this way.

A simplified diagram of the configurable logic block or CLB is shown in Figure 8.19. This block is similar to a two flip-flop, RAM-controlled state machine. The combinational logic is performed by a RAM that can be used as a 32×1 or two 16×1 RAMs. The inputs to the RAM are input variables and the state flip-flop outputs. The switching MUXs can direct the outputs of the combinational logic as well as flip-flop outputs to the inputs of the D flip-flops. The RAM and MUXs make up the IFL for the state flip-flops. A larger state machine can be formed by using additional CLBs in parallel.

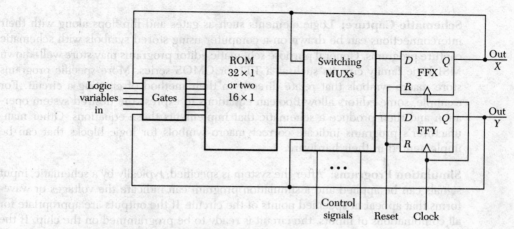

FIGURE 8.19 Simplified diagram of CLB.

Around the periphery of the chip are input/output blocks that provide for connecting from external pins to the internal logic blocks. These I/O blocks include three-state buffering if desired and also include flip-flops or latches to synchronize incoming signals. The configuration memory consists of static memory cells that specify all necessary connections of the components on the chip. This requires a relatively large memory that is volatile with respect to dc power, that is, all data in this memory is lost when the chip loses power. Upon power up, the configuration memory must be loaded before the FPGA is functional. The configuration memory can be loaded from an external memory such as a PROM or from a controlling microprocessor as part of an initialization process. This data can be loaded serially or in an 8-bit parallel format.

An example of a Xilinx FPGA is the XC3042 [5]. This chip includes 3000 gates, 480 flip-flops, and 96 I/O blocks. The configuration memory stores 30,784 bits when fully programmed. The I/O blocks can be programmed to be either TTL or CMOS compatible as they interface with the external pins. The flip-flops can be driven at toggle rates as high as 125 MHz.

8.6.4 TOOLS FOR CREATING STATE MACHINES ON A CHIP

The complex systems that can now be placed on a chip require development and debugging prior to programming the chip. This is particularly critical in fusible link devices that cannot be modified. The electrically erasable device or the Xilinx FPGA can be reprogrammed, but it is more efficient to debug these devices prior to programming also. Various types of tools have been developed to simplify the programming of these complex devices. Most development programs such as PALASM® are now geared to work with personal computers. A few of these tools will be considered in the following paragraphs.

Schematic Capture: Logic elements such as gates and flip-flops along with their interconnections can be drawn on a computer using stored symbols with schematic capture programs. General purpose schematic editor programs may store well-known MSI logic family circuits such as a TTL or CMOS series. More specific programs store macro-symbols that relate directly to their method of creating a circuit. For example, some editors allow Boolean equations to express the desired system operation and then produce a schematic that implements these equations. Other manufacturer's programs indicate correct macro-symbols for logic blocks that can be implemented in their hardware.

Simulation Programs: After the system is specified, typically by a schematic, input signals can be applied and a simulation program can indicate the voltages or waveforms that appear at specified points of the circuit. If the outputs are appropriate for all combinations of inputs, the circuit is ready to be programmed on the chip. If the outputs are not correct, waveforms at other points of the circuit can be examined to locate problems that may exist in the design.

The simulation program accounts for the operating characteristics of the particular type of logic family used. If TTL is specified, voltage and current ratings along with applicable timing parameters will be used in the simulation.

Once the circuit is designed to produce the correct outputs, a design implementation system is used to produce the finished system.

Design Implementation System: The design implementation system is tailored to be specific to the hardware used. If MSI/LSI design on printed circuit boards is the desired finished product, the implementation system might produce the layout of conductors on the PC board. A wiring list will be produced by a system designed for such an output. If the circuit is to be implemented on a chip, the schematic must be translated into a program that creates the correct circuit in the chip to be used. If it is a Signetics PLS or an AMD PAL®, the information is translated into a program that will configure the PLD in terms of AND-OR arrays and flip-flops to create the design. If the chip to be used is a Xilinx FPGA, the schematic information is translated to a program that will be loaded into the on-board RAM as the chip is initialized. Automatic placement of components and routing of connections is done by the program to maximize performance.

Chip Programmer: Each chip is programmed by a device that creates the correct signals for the device used. The fusible link chips must be driven with the proper current for the proper time to remove each fusible link that is to be eradicated. If a ROM is to be used with a Xilinx FPGA, a ROM programmer must be used to program this chip.

Several companies supply powerful schematic capture and simulation programs. Several chip manufacturers make their chips compatible with existing programs in these areas. Mentor Graphics, OrCAD, and VIEWlogic are examples of companies that supply highly capable schematic capture and simulation packages. Data I/O, Logical Devices, and Advantest are examples of companies that supply chip programmers for fusible link and electrically erasable devices.

SUMMARY
* * * * * * * * * * * * * *

1. Programmable logic chips can be used to implement combinational logic on a single chip. Such devices are called programmable logic devices or PLDs.
2. PLDs offer the possibility of changing state machine sequences by changing the program in a single chip.
3. Registered PLDs allow an entire state machine to be implemented on a single chip. Gate array circuits also have this capability.

CHAPTER 8 PROBLEMS
* * * * * * * * * * * * * *

● **Sec. 8.2**

*8.1 Show the contents of all locations of a 16×4 ROM that will convert each input number to the sum of this input number and 1. Assume that the input number 1111 should be converted to 0000.

8.2 Show the contents of all locations of a 16×4 ROM that will produce $F1$, $F2$, $F3$, and $F4$ of Figure 3.24.

8.3 A 32×5 ROM is used as a table look-up multiplier. The input numbers range from binary 0 to binary 15; the output should equal the input number multiplied by 2. Show the contents of all locations.

*8.4 A 6-bit code is to be converted to a second 6-bit code. If a 128×8 ROM is used for this conversion, answer the following:
   a. How many locations will never be accessed?
   b. How many output lines will be unused?
   c. How many bits need not be programmed?

8.5 Realize the expressions $F1 = A\bar{B}C + B\bar{C}\bar{D} + \bar{A}C\bar{D}$, $F2 = B\bar{C}D + ABCD + \bar{A}\bar{C}$, and $F3 = A\bar{D} + B\bar{C}D + ACD$ with MUXs, decoders and gates, and a ROM. Minimize each circuit as much as possible.

● **Sec. 8.3**

8.6 Show the connections of a $6 \times 32 \times 6$ PLA that will accomplish the code conversion of Example 3.2. Use a diagram similar to that of Figure 8.5.

8.7 Show the connections of a $6 \times 32 \times 6$ PLA that will produce $F1$, $F2$, $F3$, and $F4$ of Figure 3.24. Use a diagram similar to that of Figure 8.5.

● **Sec. 8.4**

8.8 Repeat Prob. 8.6 using the 16L8 PAL®.

8.9 Show the connections of the section of a 16L8 PAL® required to implement the functions

$$Y = ABCD + A\bar{B}D + AB\bar{C}\bar{D} + \bar{B}CD$$

and

$$Z = A\bar{B} + \bar{C}\bar{D} + A\bar{C}D$$

8.10 Repeat Prob. 8.9 for the functions

$$Y = A\bar{B}CD + \bar{A}B\bar{C}D + A\bar{B}\bar{C}D + \bar{A}BCD + ABC\bar{D}$$

and

$$Z = \bar{A}BCD + A\bar{B}\bar{C}D + ABCD + \bar{A}BC\bar{D}$$

● **Sec. 8.5**

8.11 Work Prob. 6.14 using a ROM-controlled state machine.

°**8.12** Work Prob. 6.20 using a ROM-controlled state machine.

8.13 Work Prob. 7.11 using a ROM-controlled state machine.

8.14 Work Prob. 6.14 using a 16L8 PAL® chip.

8.15 Work Prob. 6.20 using a 16L8 PAL® chip.

● **Sec. 8.6**

8.16 Work Prob. 6.14 using a PLS similar to that of Figure 8.17.

8.17 Work Prob. 6.20 using a PLS similar to that of Figure 8.17.

REFERENCES AND SUGGESTED READING
. . . .

1. J. S. Schmidt, "Integrated MOS Random-Access Memory," *Solid State Design*, Vol. 6 (1965), 21–25.

2. Engineering Staff, *PAL® Device Data Book*. Sunnyvale, CA: Advanced Micro Devices, Inc., 1992.

3. Engineering Staff, *GAL® Data Book*. Portland, OR: Lattice Semiconductor Corp., 1992.

4. Engineering Staff, *Programmable Logic Devices Data Handbook*. Sunnyvale, CA: Signetics Company, 1992.

5. Engineering Staff, *The Programmable Gate Array Data Book*. San Jose, CA: XILINX, 1992.

• • • • • • • • • Chapter 9

Digital Computing

C urrently, the digital computer is one of the most significant electronic systems in existence. This device has contributed to many of the scientific and engineering developments of the late 20th century and has become part of the everyday routine of many people. It is becoming common for young children to use the digital computer in elementary school or at home. Very few people in industrial countries are not affected by the computer in some way.

This textbook is directed toward the development of the ability to design digital circuits. These circuits may be used for many applications besides the digital computer and thus, the computer has not been treated. However, because the computer industry is a major consumer of the circuits and techniques previously discussed, this chapter will consider the application of digital circuits to the field of digital computers.

Although discussions are limited to basic concepts, a thorough understanding of these fundamentals will allow an easy transition to the more advanced topics treated in a modern computer systems textbook.

The first section includes an overview of the digital computer and its basic organization. This section provides a framework for more specific discussions on computer subsystems such as arithmetic circuits, memory systems, and control circuits that follow in later sections. The second section considers the methods available to perform arithmetic using binary numbers. These methods are then applied to simple arithmetic circuits and a simple arithmetic unit is introduced in the third section. The fourth section considers some brief concepts relative to memory organization. The final section discusses the control unit of a simple computer and describes how this unit coordinates the operation of the various computer subsystems to perform simple calculations. Register transfer notation, which is useful in discussing computer operation, is also introduced in this section.

9.1

THE DIGITAL COMPUTER [2]

SECTION OVERVIEW This section considers the components that comprise a digital computer. The organization or interconnection of these components in the computer is also discussed.

The computer is a collection of electronic circuits and mechanical devices. It is designed to accomplish three tasks:

1. To accept and store instructions and data,
2. To manipulate this data as directed by the instructions to produce and store results, and
3. To present output information as specified by the instructions.

The digital computer consists of five basic sections: the *control* unit, the *arithmetic* unit, the *memory*, the *input* unit, and the *output* unit (Figure 9.1).

The input unit could be one of several devices: a typewriter, a keyboard, a card reader, a paper-tape reader, or an analog to digital converter. This device converts

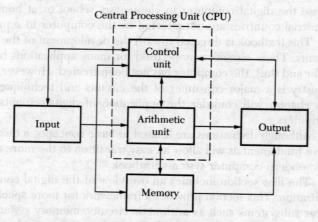

FIGURE 9.1   Block diagram of a computer.

the signals generated to appropriately coded electrical signals. A keyboard requires conversion from mechanical movement of the key to a corresponding digital code that can be recognized by the computer. Card or paper-tape readers sense hole positions on the card or tape and convert this information into the required code. For digital process control, the raw input signal may be the analog voltage of a measurement circuit. This signal, which might represent the present position of some mechanical apparatus, is then converted to digital form for use by the computer. It is possible to have more than one input device to a computer. Because of personal computers and workstations, the keyboard is the most popular form of input device at the present time.

There are also several possible output devices that can be used. Among these are the cathode-ray tube (CRT) terminal or video monitor, the line printer, the dot matrix or laser printer, the card punch, the paper-tape punch, or the digital to analog converter. The output unit accepts the digital code from the computer and converts it to the proper form of output information. Several output devices may be driven by a single computer. For personal computers and workstations, the video monitor and the printer are common output devices.

Every computer has a main memory in which to store binary information. The main memory was constructed primarily from small magnetic cores in the 1960s and 1970s. As indicated in Chapter 8, the semiconductor memory has displaced the magnetic core in memory storage applications. The main memory section of the computer is quite fast but expensive for each bit of storage capacity compared to bulk memory storage. A very large main memory system may have a storage capacity of several million bytes of binary data. Each byte consists of eight binary bits. A medium-sized system may store a few million bytes.

In many computer installations, it is necessary to store several millions or even billions of bits for long periods of time. In such instances, semiconductor storage would be prohibitively expensive. As a result, slower, less expensive bulk storage is used. This is called file memory and may consist of magnetic disks, magnetic tape, or magnetic drums.

The arithmetic unit consists of circuitry for performing arithmetic operations. Some early microprocessors implemented arithmetic units that performed only addition. Subtraction, multiplication, and division had to be performed by writing programs specifically for these operations. Most modern microprocessors now utilize arithmetic units that perform each basic arithmetic operation in response to a single instruction. Several logical operations can also be carried out by the arithmetic unit.

The arithmetic unit and the memory along with all input and output devices operate under the direction of the control unit. This section receives instructions from a stored program and directs the other units to carry out these instructions.

The arithmetic unit and the control unit together make up the *central processing unit* or CPU of the computer. Historically, the first implementation of a CPU on a single chip was appropriately named a microprocessor. Many modern chips include memory and I/O devices on the chip also, but are still called microprocessors.

9.2

· ·

BINARY ARITHMETIC

SECTION OVERVIEW The 1's and 2's complement are important in arithmetic circuits. This section defines complements, shows how to form complements, and indicates how complements can be used in subtraction. A discussion of binary multiplication and division concludes the section.

Since the early days of computing, the digital computer has been known for its advanced computational efficiency. Because binary numbers are used to represent numerical information to the computer, special methods of binary arithmetic have been developed for digital computer usage. Some of these methods are foreign to manual calculation techniques, but are necessary to allow digital circuits to carry out mathematical operations. Formation and use of complements exemplify a technique that is rarely useful in manual calculation, but is very important in computer arithmetic.

The concept of complement arithmetic is also related to the architecture of a digital system. A major use of complements is to create an arithmetic unit that uses adders to generate either sums or differences of two binary numbers. While a separate adder and subtractor could be included in an arithmetic unit, the use of complements requires less circuitry. The next section will consider the formation and application of number complements in digital systems.

9.2.1 NUMBER COMPLEMENTS

· ·

There are two types of complement of concern to the logic designer. One is called the radix complement; the other is the diminished radix complement. The radix refers to the base of the number system being used. If we are dealing with decimal numbers, the radix is 10 and we speak of the 10's complement. The diminished radix complement for decimal numbers is the $(10-1)$'s or 9's complement. In binary, the radix complement is the 2's complement while the diminished radix complement is the 1's complement.

By definition, a complement is a quantity needed to make a thing complete. In number systems, the "thing" we are attempting to complete is itself a number. The size of this number, however, depends on the number of columns with which we are dealing. For the 10's complement, a 4-column number leads to a "thing" equal to 10^4 or 10,000. Given a decimal number, for example, 6784, the 10's complement equals that number that adds to 6784 resulting in 10,000. In this case, the 10's complement is 3216. This information can be expressed algebraically as

$$N_n + \overline{N}_n(10) = 10^n$$

where n is the maximum number of columns dealt with. Generally, we know the number N_n along with the value of n and our problem is to find the complement $\overline{N}_n(10)$. Thus, the more useful form of the previous equation is

$$\overline{N}_n(10) = 10^n - N_n \qquad (\text{10s complement}) \tag{9.1}$$

The 9's complement of a decimal number is formed by finding the number that adds to the given number to result in a sum of $10^n - 1$. The 9's complement of 6784 is 3215. Here the equation is

$$N_n + \overline{N}_n(9) = 10^n - 1$$

Solving for $\overline{N}_n(9)$ gives

$$\overline{N}_n(9) = 10^n - 1 - N_n \qquad (\text{9s complement}) \tag{9.2}$$

Equation (9.2) suggests that we first calculate 10^n, subtract 1 from it, then subtract N_n. There is another way to form the 9's complement of a number that is quite easy. We can form this same value by considering each column individually, subtracting each number from 9. The 9's complement of 6784 can be formed by subtracting 4 from 9, 8 from 9, 7 from 9, and 6 from 9, giving 3215. If we were dealing with a 6-column number and wanted the complement of 6784, we simply add leading zeros to this number to result in 6 total columns. The 9's complement here would be $\overline{N}_n(9) = 993215$.

The 10's complement can be formed in the same manner as the 9's complement, by subtracting each single column digit from 9. After this is done, the result, which is the 9's complement, has 1 added to it to form the 10's complement. The 10's complement of 006784 is $\overline{N}_n(10) = 993216$.

The 2's and 1's complements are formed in much the same way as the 10's and 9's complements. For 2's complement, the number and its complement must add to give 2^n, where n is the number of columns used. The equation is

$$N_n + \overline{N}_n(2) = 2^n$$

Arranging this equation to solve for the complement gives

$$\overline{N}_n(2) = 2^n - N_n \qquad (\text{2s complement}) \tag{9.3}$$

The 2's complement of 1101 is 0011. The sum of these numbers is 10000, which equals 2^n.

The 1's complement differs from the 2's complement by 1, giving the equation

$$\overline{N}_n(1) = 2^n - 1 - N_n \qquad (\text{1s complement}) \tag{9.4}$$

The 1's complement of 1101 is 0010. In this case, the sum of N_n and $N_n(1)$ equals 1111, which is equivalent to $10000 - 1$.

The general equations for radix complement and diminished radix complement are

$$\overline{N}_n\ (r) = r^n - N_n \qquad \text{(radix complement)} \tag{9.5}$$

and

$$\overline{N}_n\ (r - 1) = r^n - 1 - N_n \qquad \text{(diminished radix complement)} \tag{9.6}$$

The 1's complement can be formed easily by inverting each bit of the given number. The 1's complement of $N_8 = 10110010$ is formed by replacing each 0 with a 1 and each 1 with a 0. This gives $\overline{N}_8(1) = 01001101$. From a physical standpoint, a set of n output lines containing the binary code for N_n can be followed by a set of inverters to form $\overline{N}_n(1)$.

Another method of forming the 2's complement is simply to form the 1's complement and then add 1 to this result. The 2's complement of $N_8 = 10110010$ is found by adding 1 to the 1's complement, that is, $\overline{N}_8(2) = 01001101 + 1 = 01001110$.

EXAMPLE 9.1

Form the 10's, 9's, 2's, and 1's complements of the decimal numbers 213, 146, 37, and 13.

SOLUTION

From Eq. (9.1) we form the 10's complement:

$$\overline{N}_3(10) = 10^3 - 213 = 787 \qquad \overline{N}_3(10) = 10^3 - 146 = 854$$

$$\overline{N}_2(10) = 10^2 - 37 = 63 \qquad \overline{N}_2(10) = 10^2 - 13 = 87$$

The 9's complement uses Eq. (9.2) or simply subtracts 1 from the 10's complement, giving $\overline{N}_3(9) = 786$, $\overline{N}_3(9) = 853$, $\overline{N}_2(9) = 62$, and $\overline{N}_2(9) = 86$. Note that $213 + 786 = 999$, as does $146 + 853$. Note also that $37 + 62 = 99$ and $13 + 86 = 99$.

In order to form the 2's and 1's complements, we must convert the decimal numbers to binary giving $213_{10} = 11010101_2$, $146_{10} = 10010010_2$, $37_{10} = 100101_2$, and $13_{10} = 1101_2$. We will form the 1's complement first since this can be done by inspection. The four complements are formed by replacing each 0 with 1 and each 1 with 0 to give $\overline{N}(1) = 00101010$, 01101101, 011010, and 0010. The 2's complement simply adds 1 to the 1's complement to result in $\overline{N}(2) = 00101011$, 01101110, 011011, and 0011. ●

● ● ● DRILL PROBLEMS Sec. 9.2.1

1. Form the 1's and 2's complements of the decimal number 193.

2. Repeat Prob. 1 for the binary number 11010001.

3. Form the 9's complement of 2639.

4. Form the 10's complement of 2639.

9.2.2 SUBTRACTING POSITIVE BINARY NUMBERS WITH ADDERS

Now that we have examined methods of complementing numbers, we move to the subject of using complements and adders to form differences. We will deal with binary numbers exclusively here since we are interested in methods of digital system subtraction.

The first method we will consider is used only for positive binary numbers. The basic idea here is to form the difference $A - B$ by forming $\bar{A} + B$. We can see that $A - B = \overline{\bar{A} + B}$ if we carry out the complementing operations. This method can be applied to either 1's or 2's complement systems.

2's Complement

Since the discussion is limited to 2's complement in the following paragraphs, we will drop the parentheses containing the radix for convenience.

The 2's complement of A is

$$\bar{A} = 2^n - A$$

Adding B results in

$$\bar{A} + B = 2^n - A + B$$

Complementing $\bar{A} + B$ gives

$$\overline{\bar{A} + B} = 2^n - (2^n - A + B) = A - B$$

This last operation yields the correct result if A is greater than or equal to B. If B is greater than A, we must replace the last step of the procedure. In this case, the result would be negative, but we can only deal with a positive result. Thus, when B is larger than A, we must detect this fact and form $B - A$ rather than $A - B$.

We note in the procedure that $\bar{A} + B = 2^n - A + B$. When B is greater than A, this result can be written as $\bar{A} + B = 2^n + B - A$. Since this number is greater than 2^n, it will contain a carry bit, that is, a 1 will be present in the $(n + 1)$st column. This amounts to a carry bit from the nth column. If A were larger than B, the result would not contain a 1 in this column since the result would be less than 2^n. We can therefore examine the carry bit to determine if B is larger than A by the presence of a 1. If this bit is 0 then A is larger than B.

When B is larger than A, we do not want to complement $\bar{A} + B$. Instead, we want to simply subtract 2^n from this result to leave us with

$$\bar{A} + B - 2^n = B - A$$

We can effectively subtract 2^n simply by ignoring the $(n + 1)$st bit after it is used to determine that B is larger than A.

The final procedure to form $A - B$ then is as follows:

1. Complement A.

2. Add B to this complement.

3. If the carry bit is 0, complement the result.

4. If the carry bit is 1, the result is already present if we do not consider the carry bit as part of the result.

We will demonstrate this procedure for 8-bit words by forming $A - B$ for two cases: $A = 01011011$, $B = 00111011$ and $A = 01101100$, $B = 11001101$. For the first case using 2's complement,

$$\bar{A} = 10100101$$

and

$$\bar{A} + B = 11100000$$

Since there is no carry bit, we complete the procedure by forming

$$\overline{\bar{A} + B} = A - B = 00100000$$

In the second case,

$$\bar{A} = 10010100$$

and

$$\bar{A} + B = 101100001$$

The carry bit is 1; therefore, we drop the carry bit which is equivalent to subtracting 2^n. The result is negative with a magnitude of 01100001.

I's Complement

Using the 1's complement system requires some modification of the procedure. If A is greater than B, then

$$\overline{\bar{A} + B} = (2^n - 1) - (2^n - 1 - A + B) = A - B$$

When B is larger than A, this result would be negative. Instead of forming the complement of $\bar{A} + B$, this expression must be slightly modified. We see that

$$\bar{A} + B = 2^n - 1 - A + B = B - A + 2^n - 1$$

We need to drop the 2^n and add 1 to this expression to yield the correct result. This can be done by using the carry bit to add to the LSB. This is referred to as end-around-carry.

The procedure for 1's complement subtraction of one positive binary number B from another positive binary number A is as follows:

1. Complement A.

2. Add B to this complement.

3. If the carry bit is zero, complement the result.

4. If the carry bit is 1, add this bit to the LSB of the existing result.

We will again demonstrate the procedure for the two cases $A = 01011011$, $B = 00111011$ and $A = 01101100$, $B = 11001101$. For the first case

$$\bar{A} = 10100100$$

and

$$\bar{A} + B = 11011111$$

The absence of a carry bit dictates that this result be complemented to form the difference. This gives

$$\overline{\bar{A} + B} = A - B = 00100000$$

In the second case

$$\bar{A} = 10010011$$

and

$$\bar{A} + B = 101100000$$

The carry bit is 1 and must be added to the LSB recognizing that the result is the magnitude of a negative number. When this is done, the answer is found to be 01100001.

● ● ●   **DRILL PROBLEMS** Sec. 9.2.2

1. Use the method outlined in this section to subtract 123 from 247 using 2's complement.

2. Repeat Prob. 1 using 1's complement.

9.2.3   ADDING AND SUBTRACTING SIGNED NUMBERS

A plus or minus sign can be included in a binary number by expressing this sign as the leftmost bit. A 1 is used to represent a minus while a 0 represents a plus. Using this method allows us to express +13 as 01101 and −13 as 11101. If the leftmost bit indicates the sign and the remaining bits express the magnitude of the number, the method is called signed magnitude. There are two other methods that are of more value in calculating applications. These are the 1's complement representation and the 2's complement representation of signed numbers. In both these schemes, positive numbers are expressed in the same way as are signed magnitude numbers. Negative numbers continue to use a 1 for the sign bit, but the number is expressed in complement form. Some examples of 1's complement numbers are +5 = 0101, −5 = 1010, +37 = 0100101, −37 = 1011010, +127 = 01111111, and −127 = 10000000. The same numbers in 2's complement form are +5 = 0101, −5 = 1011, +37 = 0100101, −37 = 1011011, +127 = 01111111, and −127 = 10000001.

The 1's complement form leads to some ambiguity since zero can be represented as either +0 or −0, leading to 00000 or 11111 for 5-bit words. Any circuit using 1's complement arithmetic must account for both possible forms. In 2's complement form, only 00000 applies as a representation of zero since negative 0 cannot be represented.

Microprocessors use either 1's or 2's complement arithmetic. In order to demonstrate basic principles, we will limit the following discussion to 8-bit words.

2's Complement Arithmetic

We will mention that the system used for both 1's and 2's complement arithmetic actually adds the sign bit as if it were part of the number. Thus, as we form the following sums, note that the leftmost bit is added even though it represents the sign.

If B is to be subtracted from A, this number is first complemented and then added to A. This results in

$$A + \bar{B} = A + 2^n - B = 2^n + A - B$$

When both A and B are positive and the magnitude of A is greater than that of B, $A - B$ will be positive. In this case, the result will exceed 2^n and therefore will carry a bit into the sign column (the eighth bit). The sign column will contain a 0 and a 1 since A is positive and B is negative. Adding the carry bit results in a 0 for the sign bit and carries a 1 to the ninth column. If we ignore this 1, it is equivalent to subtracting 2^n from the result, leading to the difference $A - B$.

If B has a larger magnitude than that of A, the sum $A + \bar{B} = 2^n - (B - A)$. This result is the 2's complement form of $B - A$, which again is the correct result. Note here that a 1 will not carry into the eighth column.

The adder simply adds the 8 bits and ignores the carry to the ninth column. If the eighth bit contains a 0, the result is positive. A 1 in this position means the result is negative and therefore appearing in 2's complement form. The only exception to this is when we add two positive numbers or two negative numbers and the result is out of range. We will discuss this case as an example in the following paragraphs.

In order to develop a set of rules or an algorithm for doing 2's complement arithmetic, we must consider six different situations. We will add the following six pairs of numbers using 2's complement numbers: $+51$ and $+32$, -51 and -32, $+51$ and -32, -51 and $+32$, $+68$ and $+86$, and -68 and -86.

The first pair gives

$$+51 = 00110011$$

$$+32 = 00100000$$

$$01010011$$

When two positive numbers are added, the result is given directly if the resulting sum is less than 128. In this case the sum is $+83$.

The two negative numbers give

$$-51 = 11001101$$

$$-32 = 11100000$$

$$110101101$$

If the carry to the ninth column is ignored, the result is the correct number, expressed in 2's complement form. This result is converted as follows:

$$10101101 = -1010011 = -83$$

The next example gives

$$+51 = 00110011$$

$$-32 = 11100000$$

$$100010011$$

Ignoring the ninth bit gives a result of

$$00010011 = +19$$

The next pair of numbers are

$$-51 = 11001101$$

$$+32 = 00100000$$

$$11101101$$

This result is the correct negative number in 2's complement form since

$$11101101 = -0010011 = -19$$

The last two cases are numbers that yield out-of-range results. The two sums are

$$+68 = 01000100$$

$$+86 = 01010110$$

$$10011010$$

and

$$-68 = 10111100$$

$$-86 = 10101010$$

$$101100110$$

The actual numbers resulting are unimportant. The significant point is that when two numbers of similar sign are added and the sum exceeds $2^7 - 1$, in this case 127, the resulting sign bit is opposite to that of the two numbers.

Assuming each negative operand is presented in 2's complement form, the algorithm for addition is:

1. Add the two operands.
2. Check the two sign bits of the operands and compare to the sign bit of the result. If the two operand signs are the same, but the resulting sign is different, the result is out of range. In this case, an overflow flag should be asserted. If the two operand signs are different, the result is valid. If the two operand signs and the sign of the result are the same, the result is also valid.

l's Complement Arithmetic

Use of 1's complement arithmetic is very similar to the use of 2's complement. To form the difference $A - B$, the number B is complemented and added to A, giving $A + \bar{B} = A + (2^n - 1) - B$. Just as in the 2's complement case, when the magnitude of A is greater than B, a carry bit propagates through the sign column to the ninth bit. If this bit is added to the least significant bit (end-around-carry), it is equivalent to subtracting 2^n and adding 1. With these modifications, the result is $A - B$.

When the magnitude of B is greater than that of A, the result is $A + \bar{B} = A + (2^n - 1) - B = (2^n - 1) - (B - A)$. This is the 1's complement of $B - A$, which is the correct result.

The significant points relative to complement arithmetic are as follows:

1. Differences of numbers can be formed by complementing and then adding.

2. The adder always performs the same function whether a sum or difference results.

3. The sign bit is treated by the adder in the same manner as all other bits, thus allowing signed or unsigned numbers to be added with this unit.

● ● ● **DRILL PROBLEMS** Sec. 9.2.3

1. Use the method outlined in this section to subtract -27 from $+123$ using 2's complement.

2. Repeat Prob. 1 using 1's complement.

9.2.4 MULTIPLICATION OF BINARY NUMBERS

Binary numbers can be multiplied by addition and repeated shifting as illustrated by the following 4-bit numbers.

1010	(10)	Multiplicand
1101	(13)	Multiplier
1010	1st	Partial product
0000	2nd	Partial product
1010	3rd	Partial product
1010	4th	Partial product
10000010	(130)	Product

This process is similar to the manual method that humans use for multiplication. An actual multiplier circuit uses an adder to sum the partial products; however, actual adder circuits generally accept only two addends as inputs to form a sum. Consequently, the product may be formed as shown below.

1010	(10)	Multiplicand
1101	(13)	Multiplier
1010	1st	Partial product
0000	2nd	Partial product
01010		Sum of two partial products
1010	3rd	Partial product
110010		Sum of three partial products
1010	4th	Partial product
10000010	(130)	Product

We will later consider circuits that will efficiently implement the multiplication procedure demonstrated.

Multiplication of signed numbers can easily be done using the signed magnitude representation for the operands. Only the magnitudes are used in the multiplying circuit to form the result. An EXOR gate is driven by the sign bits of the multiplier and multiplicand to generate the correct sign of the product.

9.2.5   DIVISION OF BINARY NUMBERS

Division is accomplished by repeated subtraction and shifting as indicated on p. 431.

| 101101 | (45) Dividend |
| 110 | (6) Divisor |

	No subtraction due to negative result
101101	Shift dividend to left
110	Subtract divisor—put 1 in partial quotient (1)

10101	Difference
10101	Shift difference to left
110	Subtract divisor—put 1 in partial quotient (11)

1001	Difference
1001	Shift difference to left
110	Subtract divisor—put 1 in partial quotient (111)

| 011 | Remainder (3) |
| | Quotient = 111 (7) with remainder = 011 (3) |

The n-bit divisor is compared to the most significant n bits of the dividend. If the divisor exceeds these n bits in value, no subtraction is done. Instead, the divisor is compared to $n + 1$ bits of the dividend. When the appropriate bits of the dividend equal or exceed the divisor, the divisor is subtracted from these bits of the dividend. A 1 is placed in the partial quotient. The difference is then shifted one position to the left and the process is repeated.

After a difference is shifted to the left, if the divisor exceeds this shifted difference, no subtraction is done and a zero is placed in the partial quotient. When the remainder is less than the divisor, the process terminates and the last partial quotient is taken as the quotient.

9.3

ARITHMETIC CIRCUITS

SECTION OVERVIEW This section discusses the hardware implementation of adder circuits, multiplier circuits, and division circuits. The arithmetic unit is treated to conclude the section.

9.3.1 ARITHMETIC ADDERS

In adding two single-digit binary numbers, we are concerned with four possible combinations. This information is expressed in the truth table of Figure 9.2. When both A and B are equal to 0, the sum output S and carry output C are both equal to 0. If either A or B is equal to 1, but not both, $S = 1$ and $C = 0$. When $A = B = 1$, the sum output is $S = 0$ while $C = 1$. A binary circuit that produces these outputs is called a half-adder and is shown in Figure 9.3.

Inputs		Outputs	
A	B	S	C
0	0	0	0
0	1	1	0
1	0	1	0
1	1	0	1

FIGURE 9.2 Truth table for a single-digit adder.

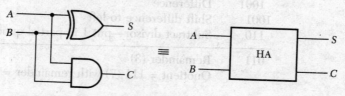

FIGURE 9.3 A half-adder.

Of course, a single-digit adder has little practical value by itself, but it can be used to construct a several-digit adder. In adding a multibit number, only the LSB can be added with a half-adder. All other columns must consider the carry from the column of next least significance in addition to the two inputs. A circuit to consider these three inputs and produce the sum and carry outputs is called a full adder. The truth table for the full adder is shown in Figure 9.4. The full adder can be implemented with two half-adders and an OR gate as demonstrated in Figure 9.5.

The truth table yields the two equations

$$S = \bar{A}\bar{B}C_{in} + \bar{A}B\bar{C}_{in} + A\bar{B}\bar{C}_{in} + ABC_{in}$$

and

$$C_{out} = \bar{A}BC_{in} + A\bar{B}C_{in} + AB\bar{C}_{in} + ABC_{in}$$

Representing each half-adder by its equivalent gates as shown in Figure 9.5 allows us to verify that these equations are implemented by this circuit. The equation for S can be written from inspection of the circuit as

$$S = C_{in} \oplus (A \oplus B) = C_{in} \oplus (A\bar{B} + \bar{A}B)$$

$$= (A\bar{B} + \bar{A}B)\bar{C}_{in} + (\overline{A\bar{B} + \bar{A}B})C_{in}$$

$$= A\bar{B}\bar{C}_{in} + \bar{A}B\bar{C}_{in} + \bar{A}\bar{B}C_{in} + ABC_{in}$$

A	B	C_{in}	S	C_{out}
0	0	0	0	0
0	0	1	1	0
0	1	0	1	0
0	1	1	0	1
1	0	0	1	0
1	0	1	0	1
1	1	0	0	1
1	1	1	1	1

FIGURE 9.4 Truth table for full adder.

This circuit can also be used to find that

$$C_{out} = C_{in}(A \oplus B) + AB = A\bar{B}C_{in} + \bar{A}BC_{in} + AB\bar{C}_{in} + ABC_{in}$$

These equations are equal to the sum and carry outputs generated by the truth table. Thus, two half-adders and an OR gate can be used to construct the full adder.

For slower speed adders, an n-bit adder can be constructed from n full adders. For higher speed systems, the propagation delay of the circuit can become excessive. For the "worst case" of two numbers containing all 1s, the carry propagates from the LSB through the MSB. This requires one propagation delay time for each bit of the numbers being added. This is called the ripple carry method and is demonstrated in Figure 9.6 for a 4-bit adder.

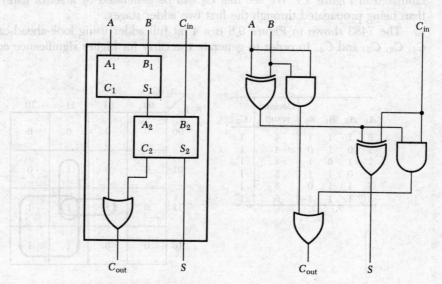

FIGURE 9.5 The full adder.

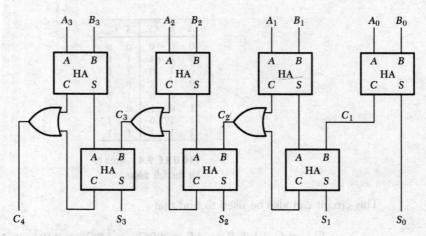

FIGURE 9.6   A four-bit binary adder.

The problem of carry propagation time can be minimized by using a look-ahead-carry circuit. In this method, the carry signals are generated by gates that look for those combinations of inputs that result in carrys. For example, the carry from the second column C_2 of a binary number should be zero if the first two columns add to 0, 1, 2, or 3. If the sum is 4, 5, or 6, a carry of $C_2 = 1$ should be generated. The sums 4, 5, and 6 can occur in several different ways. These possibilities are tabulated in Figure 9.7. We see that C_2 can be generated by a set of gates rather than being propagated through the first two adder stages.

The 7483 shown in Figure 9.8 is a 4-bit full adder using look-ahead-carry on C_1, C_2, C_3, and C_4. In order to generate the carry for higher significance columns

A_1	A_0	B_1	B_0	Decimal result	C_2
0	1	1	1	4	1
1	0	1	0	4	1
1	1	0	1	4	1
1	0	1	1	5	1
1	1	1	0	5	1
1	1	1	1	6	1

(a)

$B_1 B_0$ \ $A_1 A_0$	00	01	11	10
00	0	0	0	0
01	0	0	1	0
11	0	1	1	1
10	0	0	1	1

(b)

FIGURE 9.7   (a) Inputs that lead to $C_2 = 1$. (b) Truth table for C_2.

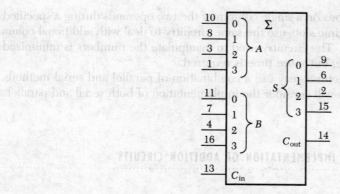

FIGURE 9.8 Block diagram for the 7483 4-bit adder.

(C_3 and C_4), more gates are required. This does not present a serious problem for MSI circuits since many gates can be fabricated on each chip. The 7483 chip can be used as a building block to create larger adders such as 8-, 12-, or 16-bit units. We note that for expansion purposes a carry input to the LSB is required.

The full adder forms the basis for many of the hardware implementations discussed in the following paragraphs.

● ● ● **DRILL PROBLEMS Sec. 9.3.1**

1. Show how to use two 74LS83 chips to construct a 6-bit adder.

2. Repeat Prob. 1 for a 12-bit adder.

● **9.3.2 FLIP-FLOPS IN ARITHMETIC CIRCUITS**

Two factors that strongly influence the configuration or architecture of arithmetic circuits are speed and cost. For large mainframe computers as well as high-performance personal computers, computation speed is often very important. As a result, the computer may carry out calculations on two binary numbers using a large amount of parallel circuitry. These large arithmetic systems manipulate several columns of the two binary operands simultaneously to produce a result in minimum time. The arrangement of several identical or near-identical circuits to perform a similar function during a single time slot is referred to as parallel architecture. While this scheme results in rapid calculations, a great deal of circuitry is required.

For digital control or instrumentation applications, speed of operation may not be as significant as system cost. This opposite extreme of slow speed and minimum cost leads to a serial architecture. With such an arrangement, the system may perform calculations on a single column of the two operands during a specified time slot. Succeeding time slots use the same circuitry to deal with additional columns of the two numbers. The circuitry used to manipulate the numbers is minimized with serial architecture, but more time is required.

Some architectures may use a combination of parallel and serial methods. The following sections will consider the implementation of both serial and parallel arithmetic circuits.

9.3.3 HARDWARE IMPLEMENTATION OF ADDITION CIRCUITS

Serial Adder

From a hardware standpoint, the serial adder is the simplest implementation of an adder. This system requires two shift registers to hold the operands and present them serially to the adder. Figure 9.9 shows the implementation of a 4-bit serial adder.

The operands are first loaded into the two shift registers with a 0 in the leftmost flip-flop of the 5-bit register. These registers can be loaded in a parallel or serial mode. When the add operation begins, the Add Enable signal is asserted high and a gated clock input is applied. These signals along with other pertinent waveforms are shown in Figure 9.10 for the operands 0101 and 1101. The 5-bit register acts as the accumulator and holds the sum after the fifth negative clock transition.

When the Add Enable is deasserted, the 5-bit register contains the result 10010. The carry flip-flop always contains the carry from the previous bit and applies this signal to the carry input of the full adder, C_{in}. This flip-flop is held in the reset state

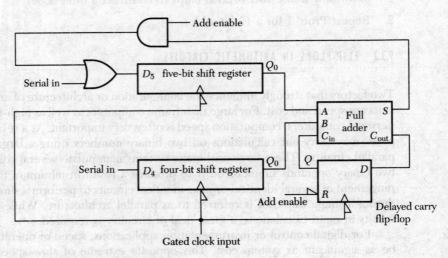

FIGURE 9.9 A 4-bit serial adder.

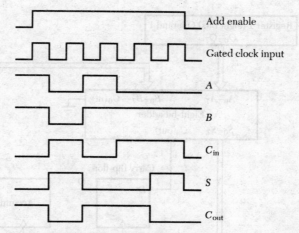

FIGURE 9.10 Timing diagram for serial adder.

by the low level of the Add Enable signal when this signal is not asserted. When Add Enable is asserted, the direct reset of this flip-flop is driven high, allowing clocked operation to take place. Table 9.1 shows the contents of the registers and the C_{in}, S, and C_{out} lines at each clock period.

It is possible to eliminate one flip-flop from the 5-bit register by allowing the delayed carry flip-flop to contain the MSB of the addition after four negative clock transitions have occurred. In this case, the flip-flop must not be reset at the end of the Add Enable gate. It must, however, be reset prior to the first negative transition of the gated clock input.

The serial adder requires $n + 1$ clock transitions to complete an addition where n is the number of bits of the operands. Because of the slow speed of operation, this adder is used only in low-speed applications that require a low-cost system. For high-speed systems parallel adders are used. These systems apply a full adder for each bit of the operand and use look-ahead-carry circuits for improved speed performance.

TABLE 9.1 Register Contents of Serial Adder.

No. of Neg. Clock Trans.	5-Bit Reg.	4-Bit Reg.	C_{in}	S	C_{out}
0	00101	1101	0	0	1
1	00010	0110	1	1	0
2	10001	0011	0	0	1
3	01000	0001	1	0	1
4	00100	0000	1	1	0
5	10010	0000	0	0	0

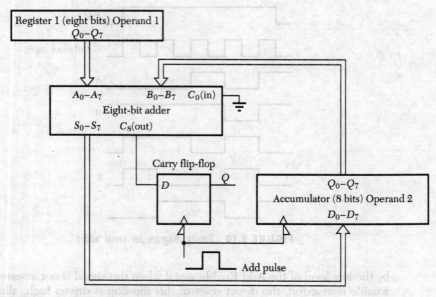

FIGURE 9.11 An 8-bit parallel adder with accumulator.

Parallel Adder

An 8-bit parallel adder with an accumulator is shown in Figure 9.11. The accumulator holds one of the operands prior to the occurrence of the Add Pulse after which the result is contained in the accumulator. Operand 1 and operand 2 must be present long enough to allow the sum outputs and carry output to settle prior to the negative transition of the Add Pulse. The sum bits are clocked into the accumulator on the negative transition of the Add Pulse. The carry bit also enters the carry flip-flop at this time. Although operand 2 is lost at this time, resulting in new, incorrect values at the sum outputs of the adder, the correct values have already been stored.

9.3.4   HARDWARE IMPLEMENTATION OF SUBTRACTION CIRCUITS

The two situations of interest have been discussed in Section 9.2, that is (1) subtraction of one positive number from another positive number and (2) adding or subtracting numbers that can be either positive or negative. The latter case is the most general, but some systems may implement only the first situation. We will consider a circuit for each situation beginning with the case of subtraction of one positive number from another positive number.

A practical implementation of a parallel 1's complement system is shown in Figure 9.12. The LSB must have a carry input to utilize this scheme. Thus, a full adder is used for each of the eight positions. A carry on C_8 will cause the flip-flop

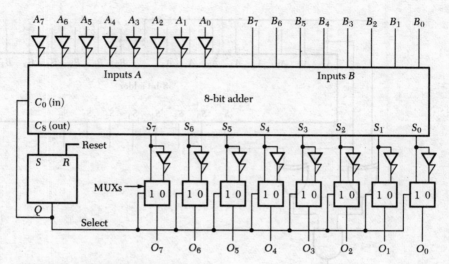

FIGURE 9.12 A subtracting circuit.

output to equal 1. This value will remain even if C_8 returns to a 0 value. The flip-flop will return to a 0 output value only when the reset input is asserted.

The process starts when the numbers A and B are presented to the input lines. The input inverters form the complement of A. The flip-flop must be reset prior to this time. If C_8 is not asserted, Q is not asserted and the 8 inverted sum bits are gated through the MUXs to the output. This produces the complement of the adder output as the result. When a carry bit is generated from the process, C_8 sets the flip-flop to $Q = 1$. This output drives the C_0 input to add 1 to the LSB. In addition, Q drives the select line of the MUXs to present the uncomplemented adder output values as the final results. The fact that $Q = 1$ also indicates that the result represents the magnitude of a negative number.

When implemented in a computer, only an adder and carry flag are required for this method. The number A is held in a complementing register, often the accumulator. This number can be complemented by applying a pulse to the complementing input. If the result is stored in the accumulator, it will be complemented as dictated by the contents of the carry flag.

The second system to be discussed is one that can add or subtract numbers that can be either positive or negative. The theory behind this method has been discussed earlier in Section 9.2.3. A parallel system to implement these ideas is shown in Figure 9.13. The NEXOR and EXOR gates check for out-of-range answers, generating an overflow flag if such occurs.

In reviewing the operation of this 2's complement arithmetic circuit, we note that the 8-bit operands consist of a 7-bit number plus a sign bit. A negative number is represented in 2's complement form. If the number B is to be subtracted from A, B is first complemented and then added to A.

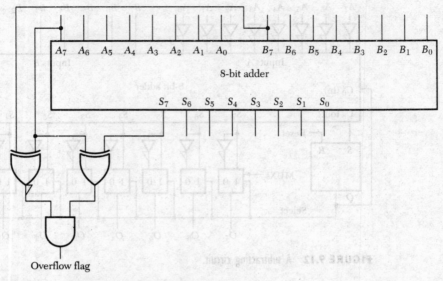

FIGURE 9.13 A 2's complement arithmetic circuit.

It is also possible to construct subtraction circuits based on full subtractors rather than full adders. While this is done in some applications, more hardware is required to implement the full arithmetic unit.

● ● ● **DRILL PROBLEMS Sec. 9.3.4**

1. If $A = 10010110$ and $B = 01110011$, determine the values of Q, S_0 to S_7, and O_0 to O_7 for the circuit of Figure 9.12.

2. Repeat Prob. 1 if $A = 01110011$ and $B = 10010110$.

● **9.3.5 A MULTIPLICATION CIRCUIT**

A simple multiplication circuit is shown in Figure 9.14 [3]. We will consider only the multiplication of two positive 4-bit numbers, since the extension to other cases is straightforward. The multiplier operand is placed in the 4-bit extension of the accumulator. The multiplicand is stored in a 4-bit storage register with parallel outputs that drive one set of inputs to an adder. The LSB of the multiplier operand determines whether the multiplicand is initially added to the accumulator. If this bit is a 1, the control circuit drives the adder to add the multiplicand to the accumulator. After this addition, a shift pulse is applied to the 9-bit shift register consisting of the multiplier register, the accumulator, and the carry bit. The accumulated sum shifts to the right as also does the multiplier. The LSB of the multiplier is lost and

TABLE 9.2 Contents of 9-Bit Register During Multiplication.

No. of Shift Pulses	Multiplicand	9-Bit Reg. Before Add.	9-Bit Reg. After Add.
0	1010	0 0000 1101	0 1010 1101
1	1010	0 0101 0110	0 0101 0110
2	1010	0 0010 1011	0 1100 1011
3	1010	0 0110 0101	1 0000 0101
4	1010	0 1000 0010	

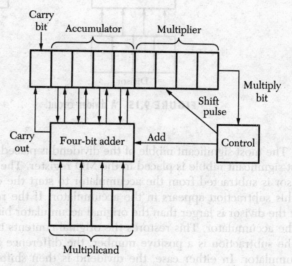

FIGURE 9.14 A binary multiplier.

the LSB of the result has now moved to the 4th bit of the multiplier register. The rightmost bit of the multiplier register determines whether or not the multiplicand is added to the shifted result. A 0 bit in this position indicates that no addition is to be done. After three more shift pulses, the result is held in the 9-bit register as shown in Table 9.2.

9.3.6   A DIVISION CIRCUIT

As indicated in Section 9.2.5, binary division is accomplished by repeated subtraction and shifting. The same register arrangement used for multiplication can also be used for division, but a different control circuit is required. In this case, the 8-bit dividend is placed in the accumulator and the multiplier register. The multiplier register now becomes the quotient register. This register is often called the MQ register to indicate its dual function in multiplication and division. Figure 9.15 shows the arrangement of a divider.

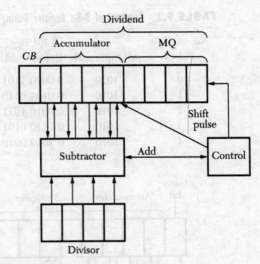

FIGURE 9.15 A divider circuit.

The most significant nibble of the dividend is placed in the accumulator and the least significant nibble is placed in the MQ register. The CB flip-flop is zeroed. The divisor is subtracted from the accumulator to start the division process. The result of this subtraction appears in the accumulator. If the result is negative, indicating that the divisor is larger than the original accumulator bits, the divisor is added back to the accumulator. This restores the original contents to this register. If the result of the subtraction is a positive number, the difference is allowed to remain in the accumulator. In either case, the dividend is then shifted to the left one bit. The rightmost flip-flop of the MQ register is now available to store the first bit of the result. This bit will be a 1 if the result of the previous subtraction was positive and a 0 if the result was negative. This procedure is repeated until the MQ register is filled with the quotient. The remainder is contained in the accumulator. Table 9.3 demonstrates this procedure for a dividend of 137_{10} or 10001001 and a divisor of 10_{10} or 1010. The resulting quotient is 1101 or 13_{10}, while the remainder is 0111 or 7_{10}.

There are several details that lead to additional considerations in the division procedure, but the basic method is demonstrated in the preceding paragraphs. One such consideration is that of determining if the MQ register is large enough to hold the result. It can be seen that, if the first subtraction leads to a positive result, the MQ register will not hold the result. For example, if 198 is divided by 10, the result should be 19 with a remainder of 8. The number 19 requires 5 binary bits and will then exceed the capacity of the MQ register. This condition can be detected prior to starting the process by comparing the divisor to the accumulator. If the divisor is larger than the initial contents of the accumulator, no overflow will occur; otherwise the overflow condition will be indicated. For 198 divided by 10, the initial contents of A and MQ are [A] = 1100 and [MQ] = 0110. The divisor is 1010, which is smaller than [A], thus an overflow flag must be generated.

TABLE 9.3 The Process of Division.

CB	A	MQ	
0	1000	1001	
	1010		Subtract divisor
1	1110	1001	CB = 1 indicates neg. result
0	1000	1001	Original contents restored
1	0001	0010	Shift left 1 bit
	1010		Subtract divisor
0	0111	0011	CB = 0 pos. result-enter 1 in MQ
0	1110	0110	Shift left 1 bit
	1010		Subtract divisor
0	0100	0111	CB = 0 pos. result-enter 1 in MQ
0	1000	1110	Shift left 1 bit
	1010		Subtract divisor
1	1110	1110	CB = 1 neg. result
0	1000	1110	Original contents restored
1	0001	1100	Shift left 1 bit
	1010		Subtract divisor
0	0111	1101	CB = 0 pos. result-enter 1 in MQ
	rem.	quot.	

● **9.3.7 THE ARITHMETIC UNIT**

This section of a computer carries out all arithmetic and logic operations of which the computer is capable. From Figure 9.16 it is seen that an accumulator, a flag register, and an arithmetic logic unit (ALU) make up the basic arithmetic unit [2].

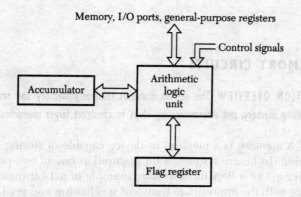

FIGURE 9.16 A basic arithmetic unit.

The accumulator is a very flexible register that allows several operations to be performed on its contents. The contents of the accumulator can be shifted right or shifted left as required. Working in conjunction with the ALU, the contents of the accumulator can be replaced with the 1's or 2's complement of the original number.

The ALU receives instructions in the form of signals from the control unit indicating the operations to be performed. To load the accumulator, a LOAD AC-CUMULATOR instruction from the control unit causes the ALU to gate a word to the accumulator. This word may come from memory, an I/O port, or a register. The appropriate source will also be specified by the control unit.

In a two-operand operation such as addition, the accumulator may provide one operand and then store the result after the operation is carried out. The ALU considers the accumulator operand as one input while the other operand again comes from memory, an I/O port, or a register. The output of the ALU is then loaded into the accumulator, replacing the original operand with the result.

The flag register stores information relative to the results of each operation. For example, a carry flag (flip-flop) is set to 1 when two numbers are added and the result is larger than the maximum number that can be stored by the accumulator. Some flags are loaded prior to an operation to indicate how the operands are to be treated. The flag register may hold information to be used in converting a binary sum into a binary-coded decimal number. The control unit or ALU uses flag data to complete the operation correctly.

The capability of the arithmetic unit in each computer will vary greatly. A simple unit might perform addition, comparison, ANDing, ORing, exclusive ORing, and other simple functions. A more complex unit might include subtraction, multiplication, and division in this list of capabilities. Although multiplication can be carried out with an arithmetic unit that does not include a hardware multiplier, it is an involved process. The control unit must supply a long series of more basic instructions to accomplish the multiplication. This series of instructions is stored in memory and is called on by the control unit when needed. Not only is overall speed much faster for a computer that includes a multiplying arithmetic unit, but also less storage space is required to carry out the operation.

9.4

. . . .

MEMORY CIRCUITS

SECTION OVERVIEW This section discusses read-only memory and read-write memory. Methods of accessing memory and interconnecting chips to construct larger memories are also considered.

A *memory* is a medium or device capable of storing one or more bits of information. In binary systems, a bit is stored as one of two possible states, representing either a 1 or a 0. A flip-flop is an example of a 1-bit memory, and a magnetic tape, along with the appropriate transport mechanism and read-write circuitry, represents the other extreme of a large memory, often with a storage capacity of over a billion bits.

As mentioned previously, the memory of a computer is divided into a main memory and a file memory. The main memory is composed of high-speed semi-conductor memory while the file memory is typically a disk drive system operating considerably slower than the main memory. Instruction or data words can be stored or retrieved from the main memory in tens of nanoseconds while file memory access might require hundreds of microseconds to several milliseconds. The following paragraphs will discuss the main memory of the computer.

There are two broad classifications of semiconductor memories: the read-only memory (ROM) and the read-write memory. The read-write memory is also called a RAM to indicate that this unit is a *random-access memory*. Random access means that any storage location within memory can be accessed in approximately the same time taken to access any other location. This would not be the case for a magnetic tape file memory, which is a serial access memory. A word stored on tape that happens to be located within six inches of the read head can be retrieved much faster than can a word stored 200 feet further down the tape. Both semiconductor ROM and RAM are random-access devices, and thus applying the acronym RAM to only the read-write memory seems inappropriate. Unfortunately, this usage of the acronym RAM is so well established that we understand that it refers to read-write memory.

9.4.1 READ-ONLY MEMORIES

The ROM is used in applications requiring that information be stored over a long period of time with no change. Each location contained in the memory can be accessed to read out information, but the contents cannot be changed. Of course, the correct information must be initially created in the memory before it can be used. This is referred to as programming the ROM. As indicated in Chapter 8, one method of programming, called mask programming, is done by the manufacturer. Using appropriate photographic masks during the semiconductor fabrication process, the desired data are programmed in the ROM. The contents of this type of ROM can never be modified after initial programming.

A major advantage of the ROM is its *nonvolatility*. This refers to the fact that the stored information does not change even if power to the system is lost. Upon reapplication of the power supply voltage, the previous contents are again available.

9.4.2 READ-WRITE MEMORIES (RAM)

The RAM is used in systems that require the contents of the memory to be changed by the system. These devices allow a location to be accessed and written into or read from in fractions of a microsecond. Unfortunately, the basic RAM is a volatile device that loses its contents when power is removed from the system. A common situation finds a RAM being loaded from the file memory with instructions or data. A program to be run on the computer is often read from a floppy or hard disk to the RAM where instructions can be accessed more rapidly than they could if accessed directly from the disk. Furthermore, the system can be powered down and, while the RAM contents become invalid, the disk still contains the program for reloading to the RAM.

Two types of RAM are presently popular in computer systems. These are the *static* RAM and the *dynamic* RAM. The static RAM has individual cells that, once written to, will preserve the stored bit indefinitely, unless a loss of power occurs. The dynamic RAM contains cells that must be initially set to the desired bit, then needs rewriting every few milliseconds to maintain the stored bit. The static RAM is simpler, but requires more transistors for each stored bit than does the dynamic RAM. The dynamic RAM can include more storage cells in a given volume of silicon and requires less current from the power supply for each cell, but additional circuitry is required for the constant rewriting.

The Static RAM

Figure 9.17 shows a single cell of a static RAM chip plus some read-write circuitry common to an entire column of cells. A typical RAM memory chip will contain an array of such cells, for example, a 32-row by 32-column configuration.

MOSFET transistors are used for most large memories because these devices can include more circuitry per chip than TTL or ECL circuits. Some smaller, high-speed memories are fabricated with bipolar junction transistor technology. Transistors T_1 to T_4 make up a bistable circuit. T_1 and T_2 are the amplifying elements while T_3 and T_4 form the active loads for these devices. Less space is required to fabricate a transistor than is required for a resistor; thus, the active loads conserve chip space. Transistors T_5 through T_8 act as gate switches. When off, an open circuit is approximated between the transistor's source and drain. When on, the transistor approximates a short circuit.

An input address to the RAM chip consists of some bits that identify the column number and some bits that identify the row number of the cell to be accessed. The column bits of the address are decoded and drive the proper column select line. This line turns on T_7 and T_8 for each cell of the entire column. All cells of the column are now connected to the read-write circuitry via the bit line and the bit not line, which contains the complement of the data on the bit line. If a data bit is to be entered or written into memory, this bit drives the bit and bit not lines through the read-write circuit. The row select line corresponding to the decoded row address determines which cell of the column will accept the input data. The row select line turns on transistors T_5 and T_6 of all cells of the row. The bistable circuit of the single cell that has both column and row select lines active is driven to the appropriate state by the input data bit.

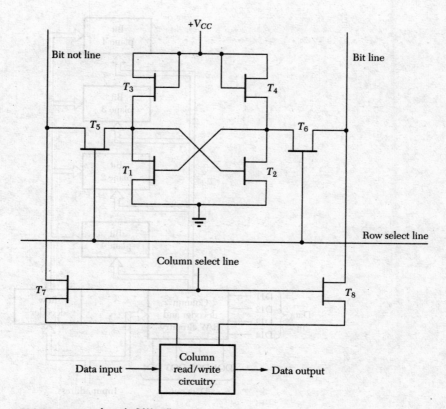

FIGURE 9.17 A static RAM cell.

The read operation addresses the cell in exactly the same way, but in this case the data input is disabled and not allowed to drive the bit lines. Information from the bistable cell is passed over the bit line to an amplifier that drives the data output line.

This matrix of cells, arranged in a row-column configuration, is the basic structure of the static semiconductor RAM. Each matrix can activate only one cell for each input address. Thus, only one bit of data can be written or read from a memory matrix at a given time. It is, however, possible to include several identical matrices on a single chip. Borrowing from core storage technology, each matrix can be referred to as a *bit plane*. If four bit planes of 2^{10} or 1024 locations per plane are created on a chip, the organization would appear as in Figure 9.18. Five bits of each input address is used to access one row and the remaining five address bits are used to address one column of four different bit planes. When writing information to the memory, the four bits of input data are each stored in a different bit plane, but at the same location of each plane. A read operation accesses the same location on each of the four planes and these four bits are presented on the output data lines.

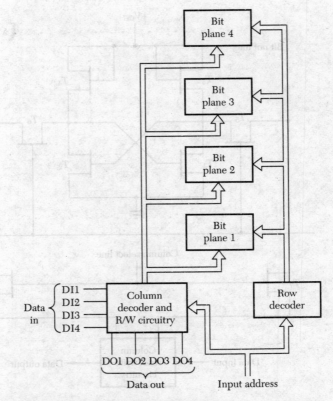

FIGURE 9.18 A memory with four bit planes.

RAM Sizes

The term *byte* is a word coined in the 1960s that refers to 8 bits of a digital code. A 32-bit code would then consist of 4 bytes. In most scientific fields, the symbols k and M stand for factors of 1000 and 1,000,000 respectively. In memory systems, these symbols have slightly different meanings. Locations of memory are accessed with binary input addresses. A 10-bit address will access 2^{10} or 1024 different locations. A k in memory systems refers to this number, that is

$$1k = 2^{10} = 1024$$

Likewise, the symbol M refers to a number equal to 2^{20}. Therefore,

$$1M = 2^{20} = 1,048,576$$

The term 1 kbyte of memory then refers to 1024 bytes of memory and 1 Mbyte refers to 1,048,576 bytes.

There are many different sizes and configurations for RAM chips. Very small chips may be 256 × 1. Single-bit plane chips go up to several megabits of storage. Other popular arrangements consist of 4- or 8-bit plane chips. In the early 1990s, 4 M × 8 and 8 M × 8 memory chips became available. Larger memories continue to be developed each year.

The number of address lines needed on a memory chip is related to the number of locations in the bit plane by

$$2^{n_a} = n$$

where n_a is the number of address lines and n is the number of storage locations. Solving for n_a gives

$$n_a = \log_2 n \tag{9.7}$$

A 1 M × 1 or a 1 M × 8 memory chip requires 20 address lines.

9.4.3 ACCESSING MEMORY

Figure 9.19 shows the block diagram of a 1 k × 4 RAM. Ten address lines are required to access each of the 1024 storage cells.

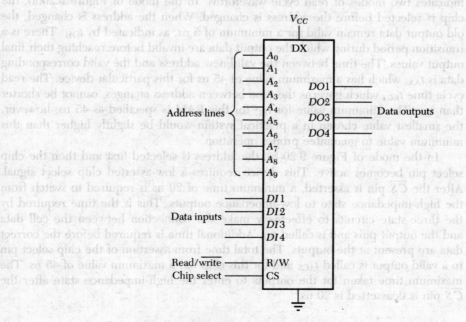

FIGURE 9.19 A 1 k × 4 RAM.

The chip select input is similar to an on-off switch. When this input is not asserted, the chip is disabled using three-state circuits and all output lines are in the high-impedance state. This state approximates an open switch in series with each output data pin. When the chip is disabled, these output pins are effectively disconnected from any circuitry that follows. Not only can data not be read if the chip select pin is not asserted, data cannot be written into the RAM if this input is not asserted.

When the chip is made active by asserting the chip select pin and an address is presented to pins A_0 to A_9, a high level on the $R/\overline{W}$ pin causes the contents of the addressed location to appear on the four output lines. This corresponds to the read mode. If the $R/\overline{W}$ pin is dropped to the low level, the write operation takes place. The data appearing on the input data pins are written into the location being addressed. Any previous information existing at this location is written over by the new data.

The RAM chip of Figure 9.19 has separate data input and data output pins. Other chips reduce the total pin count by combining these pins as DI/DO pins. When the chip is enabled, the output data are present when $R/\overline{W}$ is high and data can be input to the RAM when this pin goes low.

Reading from the RAM

There are certain timing requirements for accessing a memory chip. Figure 9.20 indicates two modes of read cycle waveforms. In the mode of Figure 9.20(a), the chip is selected before the address is changed. When the address is changed, the old output data remain valid for a minimum of 5 ns, as indicated by t_{OH}. There is a transition period during which the output data are invalid before reaching their final output values. The time between the valid new address and the valid corresponding data is t_{AA}, which has a maximum value of 45 ns for this particular device. The read cycle time t_{RC}, which governs the time between address changes, cannot be shorter than t_{AA}. The minimum value for t_{RC} for this RAM is specified as 45 ns; however, the smallest value chosen in a practical system would be slightly higher than this minimum value to guarantee proper operation.

In the mode of Figure 9.20(b), the address is selected first and then the chip select pin becomes active. This device requires a low-asserted chip select signal. After the $\overline{CS}$ pin is asserted, a minimum time of 20 ns is required to switch from the high-impedance state to low-impedance outputs. This is the time required by the three-state circuits to effectively make the connection between the cell data and the output pins and is called t_{LZ}. Additional time is required before the correct data are present at the outputs. The total time from assertion of the chip select pin to a valid output is called t_{ACS} and for this chip has a maximum value of 45 ns. The maximum time taken for the outputs to enter the high-impedance state after the $\overline{CS}$ pin is deasserted is 20 ns.

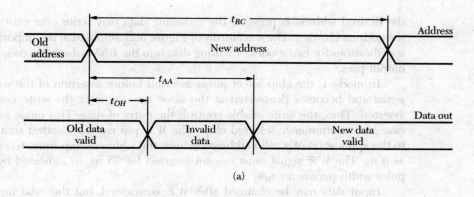

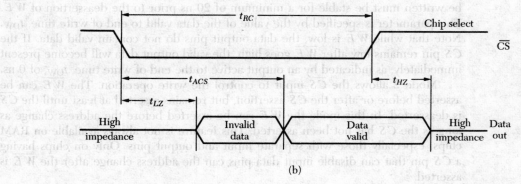

Symbol	Parameter	Time – ns	
		Min	Max
t_{RC}	Read-cycle time	45	
t_{AA}	Address-access time		45
t_{ACS}	Chip-select access time		45
t_{OH}	Output hold from address time	5	
t_{LZ}	Chip select to output in low Z	20	
t_{HZ}	Chip deselect to output in high Z		20

(c)

FIGURE 9.20 Memory read waveforms: (a) for device continuously selected; (b) for address valid before or coincident with chip select; (c) typical times for high-speed RAM.

Writing to the RAM

When writing to the RAM, care must be taken to ensure that the correct address is present before assertion of the write enable. If the write enable is asserted before the desired address is present, the incoming data may write over existing data as the address changes. The waveforms of Figure 9.21 summarize the important timing specifications for two modes of writing data into the RAM which has common input-output pins.

In mode 1, the chip select pin is asserted before assertion of the write enable signal and becomes deasserted at the same time or after the write enable is de-asserted. Thus, the write enable controls the entry of data. This mode includes the case of a continuously selected chip. The $\overline{WE}$ pin can be asserted simultaneously to the application of a valid address because the address setup time, t_{AS}, is specified as 0 ns. The $\overline{WE}$ signal must remain asserted for 35 ns, as indicated by the write pulse width parameter t_{WP}.

Input data may be changed after $\overline{WE}$ is asserted, but the valid input data to be written must be stable for a minimum of 20 ns prior to the deassertion of $\overline{WE}$. This parameter is specified by the value of the data valid to end of write time, t_{DW}. Note that while $\overline{WE}$ is low, the data output pins do not contain valid data. If the $\overline{CS}$ pin remains low after $\overline{WE}$ goes high, the valid output data will become present immediately, as indicated by an output active to the end of write time, t_{OW}, of 0 ns.

Mode 2 allows the $\overline{CS}$ input to control the write operation. The $\overline{WE}$ can be asserted before or after the $\overline{CS}$ assertion, but remains asserted at least until the $\overline{CS}$ is deasserted. In this mode the $\overline{WE}$ can be asserted before the address change as long as the $\overline{CS}$ has not been asserted. This feature is not always available on RAM chips, especially those with separate input and output pins. Only on chips having a $\overline{CS}$ pin that can disable input data pins can the address change after the $\overline{WE}$ is asserted.

The type of timing information given by the waveforms of Figures 9.20 and 9.21 must be carefully considered each time a RAM chip is included in a digital system. The designer must determine which of the specifications are pertinent to the application and design the controlling circuits to meet or exceed these specifications.

9.4.4　MEMORY ORGANIZATION

Individual memory chips can be connected together to form a memory system that is larger than the size of the individual chips. In some instances the word size may be expanded. In other situations, the number of locations may be expanded. Still other situations may call for both word size and the number of locations to be increased.

Expansion of Word Size

Figure 9.22 indicates the organization of a 1 k or 1024 × 8 memory based on 1 k × 1 chips. The number of storage locations in the overall memory is not increased above the number of locations in each chip, but eight bits of data can be stored in each location of the memory.

Waveforms
Write cycle No. 1 ($\overline{\text{WE}}$ controlled)

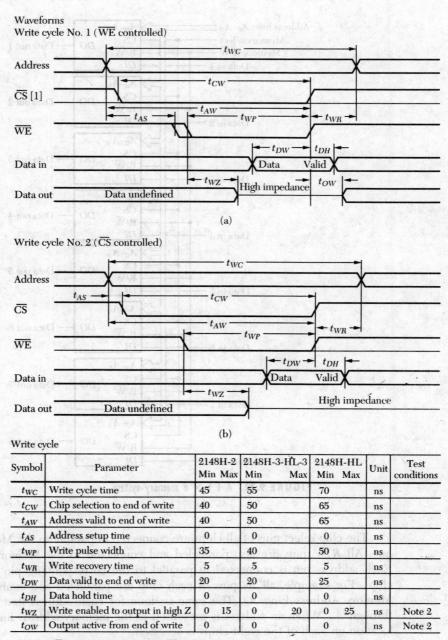

(a)

Write cycle No. 2 ($\overline{\text{CS}}$ controlled)

(b)

Write cycle

Symbol	Parameter	2148H-2 Min Max	2148H-3-HL-3 Min Max	2148H-HL Min Max	Unit	Test conditions
t_{WC}	Write cycle time	45	55	70	ns	
t_{CW}	Chip selection to end of write	40	50	65	ns	
t_{AW}	Address valid to end of write	40	50	65	ns	
t_{AS}	Address setup time	0	0	0	ns	
t_{WP}	Write pulse width	35	40	50	ns	
t_{WR}	Write recovery time	5	5	5	ns	
t_{DW}	Data valid to end of write	20	20	25	ns	
t_{DH}	Data hold time	0	0	0	ns	
t_{WZ}	Write enabled to output in high Z	0 15	0 20	0 25	ns	Note 2
t_{OW}	Output active from end of write	0	0	0	ns	Note 2

Notes: 1. If $\overline{\text{CS}}$ goes high simultaneously with $\overline{\text{WE}}$ high, the output will remain in a high-impedance state.
2. Transition is measured ±500mV from high impedance voltage with Load B. This parameter is sampled and not 100% tested.

FIGURE 9.21 Memory write waveforms: (a) for device continuously selected; (b) for address valid before or coincident with chip select; (c) typical times for high-speed RAM.

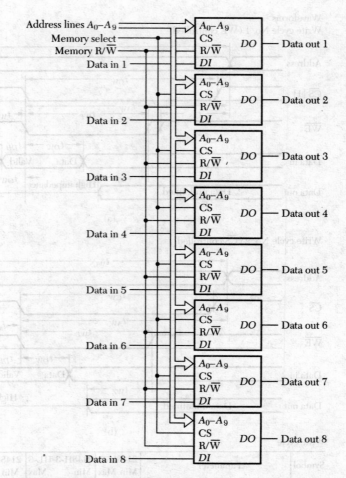

FIGURE 9.22 A l k × 8 memory system.

The chip select pins of all chips are connected in parallel to the Memory Select line. All R/W pins are also paralleled and connected to the Memory R/W line. Each address pin is connected in parallel to the corresponding pin on all other chips. For example, all A_0 pins of each chip are paralleled and connected to the Memory Address Line A_0. The same is done for each address line of the chips through A_9. When an address is presented to the memory address lines, the same location on each of the eight chips is accessed.

The data input and data output pins are not connected in parallel. An 8-bit input word is stored with one bit of the word stored in each chip at the location accessed by the input address. An output word is also made up of 1 bit from each chip, all coming from the corresponding location or address of each chip. This memory has separate data input and data output pins.

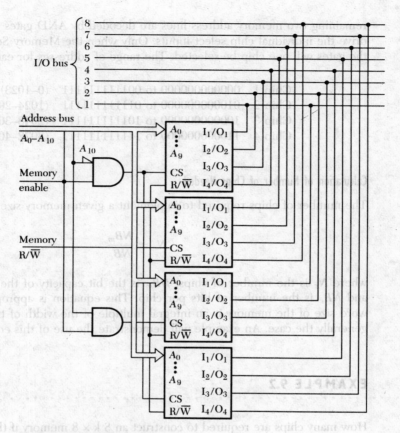

FIGURE 9.23   A 2 k × 8 memory.

Expansion of Number of Locations and Word Size

Figure 9.23 shows a 2 k × 8 memory constructed from 1 k × 4 chips. This memory requires 11 address lines, while each chip has only 10 address pins. Lines A_0 through A_9 connect to the corresponding address pins of the chips as in Figure 9.22. The address line A_{10} is used with the Memory Select line to determine which pair of chips are to be enabled. The two AND gates serve as decoders for this purpose. The upper pair of chips will be active for addresses 00000000000 to 01111111111 (0 to 1023), and the lower pair of chips are active for addresses 10000000000 to 11111111111 (1024 to 2047). These particular chips have common I/O lines that connect to the I/O bus. Two 4-bit wide chips are connected to create 8-bit wide input and output words.

Expansion of the Number of Locations

The 4 k × 8 memory of Figure 9.24 is made up of 1 k × 8 chips. Twelve address lines are required to access 4 k locations, but each chip has only 10 address pins. The remaining two memory address lines are decoded by AND gates (or a decoder) to drive the individual chip select inputs. Only when the Memory Select line enables the gates will one chip be selected. The range of addresses for each chip are

Chip 1	000000000000 to 001111111111	(0–1023)
Chip 2	010000000000 to 011111111111	(1024–2047)
Chip 3	100000000000 to 101111111111	(2048–3071)
Chip 4	110000000000 to 111111111111	(3072–4095)

Calculation of Number of Chips Needed

The number of chips required to implement a given memory size is found from

$$N_c = \frac{NB_m}{NB_c} \tag{9.8}$$

where N_c is the number of chips, NB_m is the bit capacity of the overall memory, and NB_c is the number of bits per chip. This equation is appropriate only if the word size of the memory is an integral multiple of the width of the chip, which is generally the case. An example will demonstrate the use of this equation.

EXAMPLE 9.2

How many chips are required to construct an 8 k × 8 memory if the chips used are those of (a) Figure 9.22, (b) Figure 9.23, or (c) Figure 9.24.

SOLUTION

The total bit capacity of the memory is

$$NB_m = 8 \text{ kbytes} \times \frac{8 \text{ bits}}{\text{byte}} = 64 \text{ kbits}$$

a. The number of 1 k × 1 chips required is

$$N_c = \frac{64 \text{ kbits}}{1 \text{ kbits}} = 64 \text{ chips}$$

b. The number of 1 k × 4 chips required is

$$N_c = \frac{64 \text{ kbits}}{4 \text{ kbits}} = 16 \text{ chips}$$

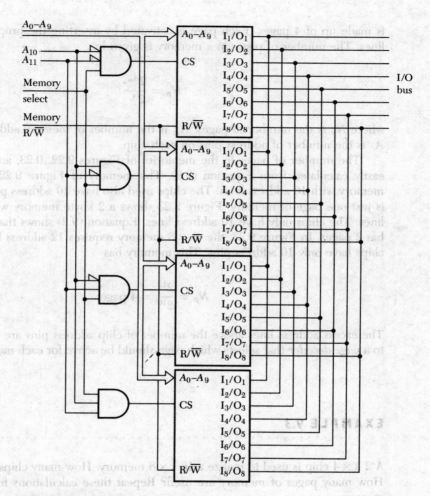

FIGURE 9.24   A 4 k × 8 memory.

c. The number of 1 k × 8 chips required is

$$N_c = \frac{64 \text{ kbits}}{8 \text{ kbits}} = 8 \text{ chips}$$

Often, a memory is expanded in sections called *pages*. Although there are several definitions of a page of memory, we will use the following:

⬤   A page of memory consists of the maximum number of locations that can be accessed directly by the chip address lines.

For example, if the memory is made up of chips having 10 address lines, a page contains 2^{10} or 1k locations. If the entire memory contains 4k locations, the memory is made up of 4 pages. Each page is activated by asserting the proper chip select lines. The number of pages in a memory is given by

$$N_p = \frac{2^{A_m}}{2^{A_c}} \qquad (9.9)$$

where N_p is the number of pages, A_m is the number of memory address lines, and A_c is the number of address pins on each chip.

The number of pages in the memories of Figures 9.22, 9.23, and 9.24 can be easily calculated from Equation (9.9). The memory of Figure 9.22 is a 1 k × 8 memory with 10 address lines. The chips used also have 10 address pins, thus there is just one page of memory. Figure 2.23 shows a 2 kbyte memory with 11 address lines. The chips only have 10 address lines. Equation (9.2) shows that this memory has 2 pages. In Figure 9.23, the 4k × 8 memory requires 12 address lines while the chips have only 10 address pins. This memory has

$$N_p = \frac{2^{12}}{2^{10}} = 4 \text{ pages}$$

The excess address lines above the number of chip address pins are used as inputs to a *page decoder* that selects which page should be active for each memory address.

EXAMPLE 9.3

· ·

A 2 k × 4 chip is used to realize a 16 k × 8 memory. How many chips are required? How many pages of memory are used? Repeat these calculations for a 16 k × 16 memory.

SOLUTION

The number of bits in the memory is

$$NB_m = 16 \text{ kbytes} \times \frac{8 \text{ bits}}{\text{byte}} = 128 \text{ kbits}$$

Each chip stores 8 kbits. From Eq. (9.8) the number of chips is

$$N_c = \frac{128 \text{ kbits}}{8 \text{ kbits}} = 16 \text{ chips}$$

A 2 k × 4 chip has a number of address pins given by $A_c = \log_2 2048 = 11$. The entire memory requires $A_m = \log_2 16,384 = 14$ address lines. The number of pages required is then

$$N_p = \frac{2^{14}}{2^{11}} = 2^3 = 8 \text{ pages}$$

Each page consists of two chips. The page decoder must be a 3-line to 8-line device. For the 16 k × 16 memory, the number of chips required is

$$N_c = \frac{16 \text{ k} \times 16}{2 \text{ k} \times 4} = 32 \text{ chips}$$

The number of total memory locations does not change, and therefore 8 pages are again required. In this case, each page consists of four chips rather than two.   ●

We summarize our discussion of memory organization by stating the following generalizations. The smallest unit of a memory is a chip. Chips are arranged to create the appropriate word size, and these chips make up one page of memory. All pages are connected in parallel—except for the chip select pins—to make up the memory. The page decoder then activates the proper page as determined by the input address.

● **9.4.5 DYNAMIC RAM**
....

The dynamic RAM uses an FET transistor and a capacitor for storage. Figure 9.25 shows a simple representation of a DRAM memory architecture. This figure omits the details of the control and refresh circuitry arrangement, but will suffice to demonstrate the basic operation of one bit plane of a dynamic memory.

The transistor of the cell is nothing more than a gate. The bit is stored as charge on the capacitor. Typically, the 1 bit is stored by charging the capacitor while the 0 bit is represented by an uncharged capacitor. All cell gates in an entire row are opened when the row select line is asserted. Each row capacitor drives a bit-line sense amplifier that also includes a bistable storage cell. With this arrangement, the assertion of a row select line transfers and stores data from all cells of the row to the corresponding sense amplifiers. A particular sense amplifier bit can be gated to the output by assertion of the appropriate column select line.

There are two problems with the dynamic cell that are not encountered in the static cell. The first is that the sense operation draws current from the cell and modifies the capacitor's charge. Enough charge is drawn from the cells that store 1 bits that the operation results in a *destructive readout*. The stored information is modified by reading the cells. The second problem is capacitor leakage that occurs even when the gate transistor remains off. Because a typical cell capacitance is

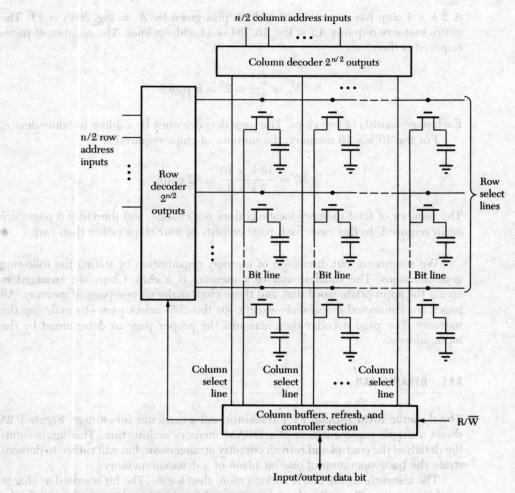

$n/2$ column address inputs

Column decoder $2^{n/2}$ outputs

$n/2$ row address inputs

Row decoder $2^{n/2}$ outputs

Row select lines

Bit line   Bit line   Bit line

Column select line   Column select line   Column select line

Column buffers, refresh, and controller section   ← $R/\overline{W}$

Input/output data bit

FIGURE 9.25   **DRAM memory organization.**

0.4 pF, even a very large leakage resistance to ground results in a low time constant. In order for the cell to maintain enough charge to result in accurately sensed data, a refresh operation must be performed before significant charge is lost. A common refresh time for practical DRAMs is 2 ms. The refresh operation must restore all lost charge on the charged capacitors every 2 ms or less.

The read operation requires two steps. First, the data is read from all cells of a row to the corresponding sense amplifier bistable storage circuits. Second, the data from these sense amplifier circuits are written back to the cells to restore the lost charge to the cells that contained a 1 bit. The data remains in the sense amplifier bistables even after being written back into the cells of the row. The column select line now selects the single row bit of the sense amplifier that is to be read to the output pin. Rewriting information back into the row cells from the sense amplifiers solves the problem of destructive readout of the cells.

The problem of loss of charge on the cells due to leakage is solved by an external circuit that addresses all row select lines once during the required refresh time. This operation is similar to the read operation except that the column select line is not asserted to read information from the sense amplifier. Accessing each row writes bits from the cells to the sense amplifiers, which is then rewritten back to the row to refresh the cell contents of the entire row. The sequential addressing of all rows can be uniformly distributed over the entire refresh time or can be accomplished in a burst mode. In this mode, all rows are rapidly addressed in a small fraction of the refresh time and the process is repeated again after one refresh time has passed.

The total refresh current required for a DRAM is rather small. For example, we can calculate the refresh current needed by a 64-kbit DRAM if we make a few assumptions. If we assume a refresh time of 2 ms and assume an average change of 0.5 V on every capacitor in the system, we can calculate a loss of charge of

$$\Delta Q = \text{number of cells} \times \text{capacitance per cell} \times \text{voltage change}$$

$$= 64 \times 1024 \times 0.4 \times 10^{-12} \times 0.5 = 1.31 \times 10^{-8} \text{ Coulomb}$$

The refresh current is then found from

$$I = \frac{\Delta Q}{\Delta t} = \frac{1.31 \times 10^{-8}}{2 \times 10^{-3}} = 6.55 \ \mu A$$

This is a *worst-case* value that assumes every cell in the memory has a 1-bit stored. Even this worst-case value is only a small fraction of the total current required for the chip's control circuitry.

The column select line determines which of the sense amplifiers connects to the I/O pin. The $R/\overline{W}$ line determines whether the output of the amplifier is passed on to the I/O pin (read) or the external data driving the I/O pin is to be written into the sense amplifier (write). When a data bit is written into a sense amplifier, all other amplifiers retain data that was read from the row cells earlier. Only the amplifier addressed by the column select line is modified. All sense amplifier data are then transmitted back to the row cells, with only the one cell being modified.

The actual design of the DRAM chip is more complex than the conceptual model discussed, but this model demonstrates the major concepts of DRAM operation. In order to minimize the number of pins to the chip, the row address and the column address are applied to the same set of pins at different times. The row address is first applied, and then a row address strobe, $\overline{RAS}$, is asserted. This strobe sets the row address into an internal row address latch. Next the column address is applied to the pins and a column address strobe, $\overline{CAS}$, is asserted to set the column address into a column address latch. The two latches on the chip now hold the row address and the column address simultaneously.

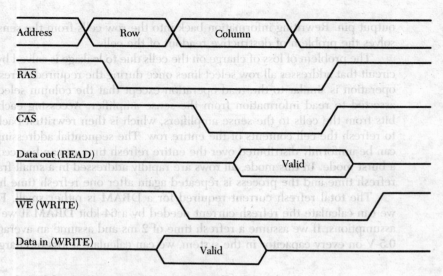

FIGURE 9.26 Timing diagram for a DRAM.

Figure 9.26 shows a timing diagram of read and write cycles for a DRAM. Methods of DRAM refreshing will not be considered. The write operation can be accomplished by first asserting $\overline{CAS}$ followed by $\overline{WE}$ or by first asserting $\overline{WE}$ followed by $\overline{CAS}$. Data are not accepted until the leading edge of $\overline{CAS}$ or $\overline{WE}$, whichever occurs last.

The number of bit planes included on a chip can easily be increased to extend the word length of each location just as in the static RAM. Although the DRAM has provided larger memories for computers than has the static RAM for years, advances in integrated circuit fabrication technology now allow static RAMs to offer several megabytes of storage.

9.5

THE CONTROL UNIT

SECTION OVERVIEW The control unit directs the operations carried out by the computer. This section introduces a basic control unit and considers the means of generating the necessary control signals. These control signals initiate the required operations in the proper sequence to carry out instructions. A simple program is considered in order to demonstrate the operation of the control unit. Practical modifications to the basic unit are then considered followed by a discussion of instruction pipelining.

Because of the popularity of the microprocessor chip (μP), the discussion of this section will focus on the basic operation of such a chip rather than on a large mainframe computer. There are various arrangements of the systems that make up

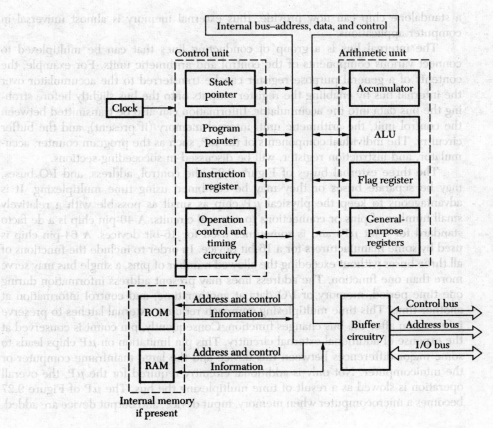

FIGURE 9.27 Basic μP organization.

the μP chip. A basic arrangement that can introduce the major concepts of μP operation is shown in Figure 9.27 [2].

The central processing unit (CPU) of a computer consists of the control unit and the arithmetic unit. A μP chip includes these two units as a minimum. Some chips contain clock circuitry for timing purposes while external clock circuitry is required for other μP chips. If an on-chip clock is implemented by the manufacturer, an external crystal or tuned circuit is used to accurately control the clock frequency.

Some μPs include memory on chip while others require external memory chips. Those devices containing on-chip memory can also utilize external memory chips if necessary. Generally, these μP chips contain a somewhat limited amount of memory, for example, 1024 (1 k) or 2048 (2 k) bytes of ROM along with 128 bytes of RAM. These μPs are often used in control and instrumentation applications that do not require external memory. This type of device is called a standalone μP, a one-chip μP, or a microcontroller and leads to a minimal chip-count system. Most general purpose microcomputer systems require a greater amount of memory than a standalone chip can now provide, thus external memory is almost universal in computer applications.

The internal bus is a group of conducting lines that can be multiplexed to connect various components of the control and arithmetic units. For example, the contents of a general purpose register can be transferred to the accumulator over the internal bus by enabling the register outputs onto the bus slightly before strobing the bus data into the accumulator. Information can also be transmitted between the control unit, the arithmetic unit, internal memory (if present), and the buffer circuitry. The individual components of the μP, such as the program counter, accumulator, and instruction register, will be discussed in succeeding sections.

The three external buses of Figure 9.27, the control, address, and I/O buses, may be separate buses or they may be combined using time multiplexing. It is advantageous to keep the physical μP chip as small as possible with a relatively small number of pins or connections to external circuits. A 40-pin chip is a de facto standard for 8-bit μPs and is sometimes used for 16-bit devices. A 64-pin chip is used by some manufacturers for a 16-bit device. In order to include the functions of all three buses without exceeding the allowed number of pins, a single bus may serve more than one function. The address lines may present address information during one time period, memory or I/O data at another time, and control information at another time. This time multiplexing approach requires external latches to preserve information after the bus changes function. Consequently, pin count is conserved at the expense of additional external circuitry. This pin limitation on μP chips leads to some major differences between this device and the large mainframe computer or the minicomputer. Not only is additional circuitry required for the μP, the overall operation is slowed as a result of time multiplexing the bus. The μP of Figure 9.27 becomes a microcomputer when memory, input device, and output device are added.

9.5.1 REGISTER TRANSFER LANGUAGE

In dealing with registers and memory storage locations that contain several bits, it is convenient to use *register transfer language* or RTL [1]. This notation has not been standardized, but there are enough similarities among the many proposed RTLs that it is easy to convert from one to another.

When dealing with registers, each register will have a name or a letter designation. The accumulator may be designated A and the program counter PC. Individual bits are numbered with numbers in square brackets. The leftmost number is the most significant bit and is numbered $n - 1$. If A is a 4-bit register, the individual bits are

$$A = A[3], A[2], A[1], A[0]$$

where the commas mean concatenation. Bits 0 through 2 of the A register can be referred to as $A[2:0]$.

The contents of a memory location can be written as M(01100), which refers to the word stored at the memory location specified by the number in parentheses (01100).

A typical RTL statement is

$$A \leftarrow R$$

which means that the contents of register R are written to register A. It is to be noted that whatever was contained in register A will be overwritten with the contents of register R while register R will maintain its contents. After execution of this statement, both registers have the same contents.

Another example of an RTL statement is

$$A \leftarrow A + B$$

This statement means that the contents of A, prior to the execution of the statement, will be added to the contents of B and will replace the contents of A with this sum. Register A contains one of the two operands prior to execution then stores the result when the statement is executed.

A rotation of bits in a register to the right would be indicated by

$$A \leftarrow A[0], A[3:1]$$

This operation is equivalent to

$$A[3] \leftarrow A[0]; A[2] \leftarrow A[3]; A[1] \leftarrow A[2]; A[0] \leftarrow A[1].$$

Typically, each RTL operation is performed during a clock cycle of the control unit as will be demonstrated in the next subsection. RTL affords an efficient notation for the specification of rather complex operations.

9.5.2   A SIMPLE CPU

Most μPs are based on a von Neumann architecture with slight variations. This architecture is named after John von Neumann who was born and educated in Hungary. In the early 1930s, von Neumann was appointed to the professorial staff of the School of Mathematics at Princeton University. He held this position until his death in 1957. While several others contributed to the concept now known as the von Neumann machine, von Neumann's significant contributions were such that this type of system bears his name.

One idea central to the von Neumann machine is that of the stored program. Not only are the instructions governing program execution stored in memory, so also are all data values to be used in the program. A second important characteristic of this machine is based on the assumption that a program can be carried out by executing a series of predominantly sequential instructions. These instructions are then placed in adjacent locations in memory. A program counter (PC) can be used to specify the address or location in memory from which to retrieve the next instruction to be executed. The program counter is automatically incremented after an instruction has been retrieved and thus identifies the location of the next instruction to be retrieved. This method of instruction retrieval eliminates the need for more complex addressing schemes that would be required if the instructions were not stored in sequence. Often a program will require an instruction to be retrieved that is not the next instruction in sequence. This is referred to as branching and can be handled by modifying the program counter in a nonincremental way. Although this modification slows the program execution, the overall speed is not degraded seriously as long as the branching instructions are comparatively few.

The basic ideas of control unit operation can be introduced by considering a very simple machine using 8-bit words. This machine will be capable of executing only 8 instructions and includes a memory with 32 total locations [2]. While the smallest μP is far more capable than this system, the basic operation of a von Neumann machine can be described easily by means of this simple model.

Each instruction to be included in an instruction set must be identifiable by a decoder in the control unit and, therefore, must be represented by a unique binary code. The number of instructions I that can be represented by an n bit code is

$$I = 2^n \tag{9.10}$$

Given a number of instructions, n can be found from

$$n = \log_2 I \tag{9.11}$$

where n is always rounded to the next higher integer. Equations (9.10) and (9.11) apply to instructions having a fixed length. As we will later see, many modern μPs use a variable length code to represent the instruction.

The same type of calculation also applies to memory addressing. If L is the number of locations in memory and m is the number of address bits, these numbers can be related by

$$L = 2^m \tag{9.12}$$

or

$$m = \log_2 L \tag{9.13}$$

We will assume a fixed length instruction code in this simple system. In order to represent 8 instructions, the required number of bits is

$$n = \log_2 8 = 3$$

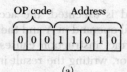

(a)

OP code	Operation
000	Add
001	Subtract
010	Shift right
011	Load
100	Store
101	Branch unconditionally
110	Branch on zero
111	Halt

(b)

FIGURE 9.28 (a) Instruction word. (b) Op code.

In order to access 32 locations the number of address bits is

$$m = \log_2 32 = 5$$

An instruction word is generally composed of two parts, the n op code or instruction bits and the m memory address bits as shown in Figure 9.28. The total length of the instruction word is $n + m$, which is 8 bits for this system. Figure 9.28 also defines the assigned meaning of the op codes. The portion of an 8-bit instruction register that holds the op code is referred to as op(IR), while the portion that contains the address is referred to as ad(IR).

In this 8-bit instruction word, the address designates the location of a word or operand that is important to the instruction specified by the op code. Data is also stored in the form of eight-bit words. We will apply the popular single-address system to our model. In such a system, an operation requiring two operands assumes that one operand is in the accumulator and the other is at a location specified by the instruction word. For example, the instruction word of Figure 9.28(a) indicates that the number stored in location 26(11010) is to be added to the number in the accumulator. An instruction such as the shift right also assumes that the operand is in the accumulator. An address is unnecessary in this case and only the 3-bit op code need be specified.

The instruction set for this control unit consists of the eight commands listed in Figure 9.28(b). The add instruction causes a number in memory to be added to the contents of the accumulator. In RTL this is expressed as

$$A \leftarrow A + M(Addr)$$

After this command is executed, the accumulator will contain the result of the addition of the initial operand in A and the number stored in memory at the address indicated. The subtract instruction causes a number in memory to be subtracted from the accumulator, writing the result into the accumulator, or

$$A \leftarrow A - M(Addr)$$

The shift right instruction shifts each bit of the accumulator one position to the right. It is conventional that the most significant bit (MSB) of any number stored in the accumulator is the leftmost bit. The least significant bit (LSB) is the rightmost bit. A shift right instruction drops the original LSB from the register. The new MSB becomes a zero. This operation can be expressed as

$$A \leftarrow 0, A[7:1]$$

The load instruction loads the accumulator with a word from memory. The RTL expression is

$$A \leftarrow M(Addr)$$

Store moves a word from the accumulator to memory or

$$M(Addr) \leftarrow A$$

The unconditional branch instruction moves the program to a specified point when nonsequential instruction execution is desired. The specified point is contained in ad(IR). The RTL expression for this operation is

$$PC \leftarrow ad(IR)$$

where PC is the program counter.

The branch on zero instruction will force the program to jump to a nonsequential instruction only if the accumulator contains the number 0 when this instruction is executed. The RTL statements for this operation are

$$\text{If } A = 0 \text{ then } PC \leftarrow ad(IR)$$

$$\text{If } A \neq 0 \text{ then do nothing}$$

If A contains a zero, the PC will be filled with the new number. If A is nonzero, no operation is carried out.

The halt command stops the execution of further instructions.

To solve a mathematical problem with this computer, the problem must be broken down into a series of small operations, each of which corresponds to one of the instructions of the set. These instructions are then stored sequentially in memory. All operands or numbers to be operated on are also stored in memory in locations that differ from those of the instructions. The solution is carried out by executing the first stored instruction, then executing each successive instruction in sequence, unless instructed to depart from the sequence by a branch instruction.

Program execution begins by loading the first instruction from its memory location to the instruction register. This process of loading the instruction register from memory is called fetching the instruction.

An operation network decodes or identifies the instruction to be executed and the control unit executes this instruction. In this simple control unit, the first three bits of the instruction word identify the instruction while the last five bits identify the operand. Each instruction is fetched then executed until the program has been completed.

As an example, let us consider the simple problem of calculating the sum of consecutive numbers from 1 to some input number [2]; that is, if the input number is 5, the program should calculate $5 + 4 + 3 + 2 + 1 = 15$. We will assume a maximum input number of 20. We reserve the first 20 memory locations (00000 to 10011) for instruction words and use the last 12 locations (10100 to 11111) for data. One approach to accomplish this might be as follows, assuming the input number is stored in memory location 10100 and the number 1 is stored in location 10101 (for decrementing purposes):

1. Load the accumulator with the contents of location 10100 (input number).
2. Store the input number in location 10110. This location will contain the sum.
3. Subtract the contents of location 10101 from the accumulator. This decrements the accumulator.
4. Store the result in location 10100.
5. If accumulator is zero, go to the halt instruction (branch to the last step). If not, proceed to the next instruction.
6. Add the contents of location 10110 to the accumulator.
7. Store the result in location 10110.
8. Load the accumulator with the contents of location 10100.
9. Return to step 3 of the procedure.
10. Halt.

When this procedure is completed by reaching the HALT instruction, the sum will be stored in location 10110.

This program can be implemented in terms of machine language (binary code for instruction words) by using the sequence of Figure 9.29. This program assumes that the required information has been initially stored in memory.

The program counter determines which memory location is to be accessed to fetch an instruction word. When a program is to be executed the program counter is reset to zero and the contents of location zero (00000) are placed in the instruction register. The op code is identified and carried out (executed), then the program counter is incremented by one and the next instruction is fetched. The

Memory location	Contents (machine language)		Meaning[*]
	Op code	Address	
00000	011	10100	Load A with contents of location 10100. This is the input number.
00001	100	10110	Store A contents in location 10110. This location stores the sum as it is being formed.
00010	001	10101	Subtract contents of location 10101 from number in A. This decrements A.
00011	100	10100	Store A contents in location 10100. This is the decremented number.
00100	110	01001	If contents of A is zero (decremented number), the program will be terminated by the HALT command.
00101	000	10110	Add contents of location 10110 to A. This forms the updated sum.
00110	100	10110	Store new sum in location 10110.
00111	011	10100	Load A with contents of location 10100.
01000	101	00010	Program returns to instruction at location 00010 and proceeds from that point.
01001	111	xxxxx	Halts program.
⋮	⋮	⋮	⋮
10100	00010000		Initially this is input number (16). Final value will be zero (00000000).
10101	00000001		Size of decrement = 1.
10110	xxxxxxxx		Nothing originally at this location. Sum is stored during program. Final value = 136 (10001000).

[*]A = Accumulator

x = Don't care

FIGURE 9.29 Machine language program and memory contents that form the sum of consecutive numbers from 1 through 16.

program counter is always incremented unless a branch instruction occurs. If a branch instruction does occur, the location specified by the instruction is placed in the program counter. The counter is then used to specify the location from which the instruction is fetched. After this operation, the counter is incremented to specify the location of the successive instruction. The HALT instruction stops the program counter and the execution of instructions is suspended until the counter is reset to begin execution again.

In order to execute each instruction, the instruction word must be transferred from memory to the instruction register after which the indicated operation is carried out. The time taken to transfer the instruction is called the instruction fetch time or simply fetch time. For a simple machine such as the one under discussion, fetch time is constant regardless of which memory location is accessed.

The time required to carry out the operation indicated by the instruction word is called the execution time and will require different amounts of time, depending on the complexity of the operation. A shift-right instruction would require less time than an add instruction since the latter must access the operand from memory before adding to the accumulator A. The shift-right instruction requires only that a pulse be applied to the parallel clock inputs of A, along with a gating signal to determine the direction of shift. The total time for fetching and executing is called the program cycle time or instruction cycle time.

The master clock is used to create required time slots to carry out each step of the program cycle. The clock may be followed by a counter to divide the clock frequency to create appropriate time slots.

Let us consider the various operations that must be carried out to fetch and execute each of the eight instructions of this computer. The process of fetching and executing an instruction is broken down into a series of small operations that can be performed by the control unit. These operations are called microoperations, and the instructions that specify each of these individual operations are referred to as microinstructions. Note that all computers, not just μPs, use microinstructions and microoperations to carry out basic instructions. These terms were used prior to the invention of the μP.

Figure 9.30 lists the microoperations that must be performed during each time slot to fetch and execute the eight instructions of our simple computer. The first three microinstructions of each instruction cycle are the same. Performing these microoperations results in fetching an instruction from memory and inserting it into the IR. The next two through five microinstructions are different depending on which instruction has been loaded into the IR.

	ADD Add	SUB Subtract	SHR Shift right	LOD Load	STO Store	BRU Branch	BRZ Branch if [A] = 0		HLT Halt
T_1	[PC] → MAR	[PC] → MAR	[PC] → MAR	[PC] → MAR	[PC] → MAR	[PC] → MAR	[PC] → MAR		[PC] → MAR
T_2	[M(MAR)] → MBR	[M(MAR)] → MBR	[M(MAR)] → MBR	[M(MAR)] → MBR	[M(MAR)] → MBR	[M(MAR)] → MBR	[M(MAR)] → MBR		[M(MAR)] → MBR
T_3	[MBR] → IR	[MBR] → IR	[MBR] → IR	[MBR] → IR	[MBR] → IR	[MBR] → IR	[MBR] → IR		[MBR] → IR
T_4	ad [IR] → MAR	ad [IR] → MAR	SR[A] → A	ad [IR] → MAR	ad [IR] → MAR	ad [IR] → PC	If [A] ≠ 0 nothing	If [A] = 0 ad [IR]→PC	Reset state machine and disable
T_5	[M(MAR)] → MBR	[M(MAR)] → MBR	[PC] + 1 → PC	[M(MAR)] → MBR	[A] → MBR		[PC] + 1 → PC		
T_6	[A] + [MAR] → A	[A] + [MAR] → A		[MBR] → A	[MBR] → M(MAR)				
T_7	[PC] + 1 → PC	[PC] + 1 → PC		[PC] + 1 → PC	[PC] + 1 → PC				

FIGURE 9.30 Microoperations to carry out each instruction.

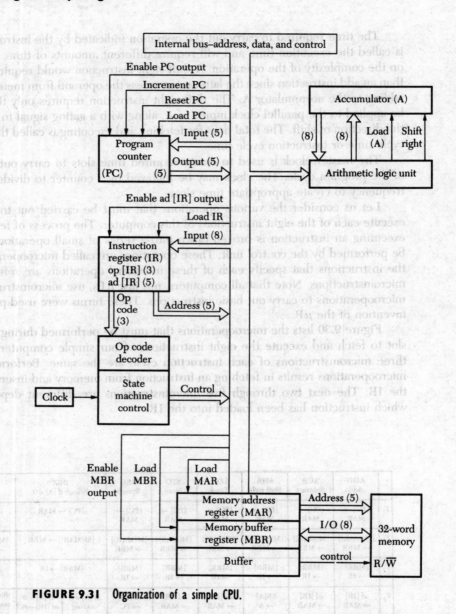

FIGURE 9.31 Organization of a simple CPU.

The control unit must create seven time slots to carry out the longer instructions of the set. During each time slot, the proper gating and data transfers must take place. Figure 9.31 shows a CPU along with a main memory that will accomplish the desired functions.

A state machine is used to create the time slots along with the proper control signals. The program counter is a 5-bit binary counter that can be reset, incremented, or loaded from the last 5 bits of the IR via the internal bus. We note that the program counter must be able to specify any possible memory location; thus this counter must contain 5 bits. The memory address register (MAR) can be filled in parallel from one of two possible sources: the PC or the last five bits of the IR or ad(IR). When accessing an instruction word from memory, the PC will fill the MAR at the proper time. When data is to be written to or read from memory, the MAR is filled from ad(IR).

The external memory can be written into or read from respectively by taking the $R/\overline{W}$ line low or by driving this line high. The location accessed in memory depends on the address present on the address bus. When memory is accessed the 8-bit word is transferred between memory and memory buffer register (MBR). The MBR can receive data from the accumulator or transmit to the IR or ALU via the internal bus. The direction of transfer is determined by the state machine output signals. These, in turn, depend on the instruction present in the IR and the time slot or state in which the state machine resides.

We will demonstrate the generation of the various control signals necessary to execute the instructions of the set by using the program of Figure 9.29. We will assume that the PC has been reset to 0 and the controlling state machine is reset to start execution.

The first task to be accomplished is the fetching of the instruction from memory location 00000. The microoperations to be performed at each time slot can be seen in Figure 9.30. The fetching operation begins at T_1 by transferring an address of 00000 to the MAR. At T_2, the memory is enabled and the contents of location 00000 are transferred to the MBR via the I/O bus. The memory $R/\overline{W}$ line must be high at this time. For this program, the word 01110100 is loaded into the MBR. The contents of the MBR are transferred to the IR during T_3. This is accomplished by enabling the MBR output and activating the load IR line. This completes the instruction fetch time for the system. These first three microoperations are always the same, regardless of the instruction fetched. The RTL statements corresponding to these operations are

$$\text{MAR} \leftarrow \text{PC} \qquad (\text{during } T_1)$$

$$\text{MBR} \leftarrow \text{M(MAR)} \qquad (\text{during } T_2)$$

$$\text{IR} \leftarrow \text{MBR} \qquad (\text{during } T_3)$$

The op code decoder now identifies the instruction to be executed by decoding the first three bits of the IR. The 011 corresponds to the LOD instruction. This information is used as input information by the state machine, which will control the microoperations necessary to load the accumulator from memory. At T_4, the state machine enables ad[IR] onto the internal bus and loads the MAR with this information (10100) or

$$\text{MAR} \leftarrow \text{ad(IR)} \qquad (\text{during } T_4)$$

TABLE 9.4   Outputs to Be Generated during T_4.

Instruction in IR	Outputs during T_4
ADD + SUB + LOD + STO	Enable ad(IR) output and load MAR
SHR	Shift right
BRZ $\cdot \bar{A}_0\bar{A}_1\bar{A}_2\bar{A}_3\bar{A}_4\bar{A}_5\bar{A}_6\bar{A}_7$ + BRU	Enable ad(IR) output and load PC

This same operation should also be carried out during T_4 if the instruction in the IR happens to be one of the four instructions ADD, SUB, LOD, or STO.

If op[IR] had contained the code for SHR, the shift-right control of the accumulator would be enabled during T_4. If BRU were the instruction, then ad(IR) would be transferred to the PC during T_4 by enabling ad(IR) onto the bus and generating a load PC signal. This same transfer would also be the case if BRZ were present in the IR and the accumulator contained zero.

We summarize the requirements on the controlling state machine during T_4 in Table 9.4. The HLT instruction should reset the PC and the state machine and disable the state machine clock. During T_5, the memory should be accessed, transferring data from the location specified by the MAR to the MBR. This transfer would be carried out when the LOD, ADD, or SUB instructions are being executed. In the example program, the LOD instruction is executed and transfers the contents of memory location 10100 to the MBR. When the STO instruction is in the IR, the accumulator contents are transferred to the MBR during T_5. Figure 9.30 shows that the SHR instruction or the BRZ instruction (when $A \neq 0$) should result in the PC being incremented. If BRU or BRZ (when $A = 0$) is present in IR, the microoperation sequence has been completed and the fetching sequence is repeated. Table 9.5 summarizes the actions taken during T_5.

TABLE 9.5   Outputs or Actions Taken during T_5.

Instruction in IR	Outputs during T_5
ADD + SUB + LOD	Enable memory, $R/\overline{W}$ high, load MBR
STO	Enable A, load MBR
BRZ $\cdot (A_0 + A_1 + A_2 + A_3 + A_4 + A_5 + A_6 + A_7)$ + SHR	PC $\leftarrow$ PC + 1
BRZ $\cdot (\bar{A}_0\bar{A}_1\bar{A}_2\bar{A}_3\bar{A}_4\bar{A}_5\bar{A}_6\bar{A}_7)$ + BRU	Reset state machine to fetch state

TABLE 9.6   Outputs or Actions Taken During T_6.

Instruction in IR	Outputs or Action Taken
ADD	Enable MBR onto bus, activate ADD input on ALU
SUB	Enable MBR onto bus, activate SUB input on ALU
LOD	Enable MBR onto bus, activate LOAD A input on ALU
STO	Enable MBR onto external I/O bus, enable memory, take $R/\overline{W}$ low

Figure 9.30 shows that only four instructions require microoperations during T_6. All four of these instructions result in the enabling of the MBR onto the internal bus. The ADD instruction then instructs the ALU to add the data on the internal bus to the contents of A. The SUB instruction subtracts the bus data from A. The LOD instruction transfers the bus data to A and the STO instruction transfers this data to memory with location specified by the MAR. Table 9.6 summarizes these actions. For these same four operations, the PC is incremented at the end of T_6 and the state machine is reset to the fetch state.

We see that the simple CPU operates in a cyclical manner. The first three time slots are reserved for instruction fetching. The proper control signals are provided by the state machine controller during these time slots to fetch an instruction. The contents of the PC determines the address of the fetched instruction. After the instruction is placed in the IR, the op code decoder identifies the operation to be executed. This information is passed on to the state machine controller, which generates the proper control signals during time slots T_4 through T_7 to execute the instruction.

●●●   **DRILL PROBLEMS** Sec. 9.5.2

1.   Using the instruction set of Figure 9.28, write a program to form $Z = 4X + Y/2$ where X is an integer between 1 and 32 while Y is an even integer between 2 and 128. Assume X is stored in memory location 28 and Y is in location 29. Store Z in location 30.

2.   Write a logical expression for the load MAR signal of Figure 9.31 in terms of the time slots and instructions of the system.

9.5.3 A PRACTICAL CPU

The CPU of Figure 9.31 demonstrates the essential features of the μP, but is too limited to perform as a practical device. A typical μP is capable of executing between 50 and 200 instructions. Not only is the control circuitry more complex, larger instruction and data words must be handled. The fixed length instruction word of the simple CPU gives way to a variable length instruction word for practical systems. An 8-bit μP will use from 1 to perhaps 6 bytes for the instruction word. This large number of bits allows a reasonable number of op codes, address bits, and data bits to be represented. The number of bytes used per instruction word depends on the nature of the instruction itself.

The size of the address space or maximum number of unique memory locations is a function of the op code length and the maximum length of the instruction word.

EXAMPLE 9.4

An 8-bit μP uses both 1- and 2-byte instruction words. The processor executes 58 different instructions using a fixed length op code. What is the maximum memory size that can be used with this system?

SOLUTION

Equation (9.10) is used to calculate the number of bits required for the op code. This gives

$$n = \log_2 58 = 5.86$$

Six bits are required for the op code. Since the largest instruction word consists of 2 bytes or 16 bits, only 10 bits are available for the memory address. For $m = 10$, Eq. (9.11) gives

$$L = 2^{10} = 1024 \text{ locations}$$

The address space consists of only 1 kbyte in this case. The maximum size can then be 1024×8 or 8 kbits.

In order to conserve instruction word size, μP designers have implemented several schemes to pack more capability into a limited instruction word size. One method that increases the addressing capability is to organize the memory in banks. For example, four banks, each consisting of 1 k locations, can comprise the memory. A single byte instruction is used to select the bank to be addressed. Once the bank selection instruction has been executed, the internal circuitry generates the corresponding 2-bit bank number. These two bits are combined with the 10-bit address from the instruction word to create a 12-bit address to access one of 4096 locations. When a bank has been selected, all memory accesses are made from this bank until a different bank is selected. In most programs, memory accesses can be confined to a particular bank with only an occasional need to cross the boundary into another bank. The bank selection instruction is then required very rarely.

If the number of bits generated by the μP to identify the memory bank is b, the total number of memory locations is given by

$$LB = 2^{b+m} \tag{9.14}$$

Again, m is the number of address bits of the instruction word.

Another method of compacting information in the instruction word is to use a variable length op code. In order to demonstrate this idea, we might consider a μP with 45 instructions. A fixed length op code would require 6 bits. We note that if the codes representing the 45 instructions were chosen to be 000000 through 101100 consecutively, the codes from 101101 to 111111 would not be used. It would be possible to use the first four bits of some of the latter codes to represent an instruction. For example, the code 1111 might represent the command to move data from memory to the accumulator. The code 1101 might instruct the μP to move data from the accumulator to memory. The first two bits of these instructions are different from all other instructions and can then identify a move instruction. The next two bits indicate the direction of the data transfer. The remaining bits can be used for addressing. In this instance, a 2-byte instruction would allow 12 bits for the address rather than 10. The variable length op code allows more addressing capability and is used in virtually all modern μPs.

● ● ● **DRILL PROBLEM Sec. 9.5.3**

If a computer has 226 separate instructions and 4 kwords of storage in main memory, how many bits are required for the instruction word in a single-address system? Assume a fixed op code length.

● **9.5.4 MANIPULATING LARGER DATA WORDS**

It is possible, although inefficient, to handle data words having a number of bits that exceeds the number for which the μP was designed. Sixteen-bit arithmetic can be performed by an 8-bit μP. Since all instructions in an 8-bit processor are directed toward 8-bit words, a series of instructions is required to handle each 16-bit word manipulation. If two 8-bit words from memory are to be added without considering the carry, the sequence of instructions might consist of the following:

1. Transfer word A to accumulator.
2. Add word B to accumulator.
3. Store accumulator as word C (result).

For a 16-bit operation having word 1 stored in locations $A1$(LSByte) and $A2$ and word 2 stored in locations $B1$(LSByte) and $B2$, the sequence of instructions might appear as

1. Transfer byte $A1$ to accumulator.
2. Add byte $B1$ to accumulator.
3. Store accumulator in location $C1$ (LSByte of result).
4. Transfer byte $A2$ to accumulator.
5. Add carry flag to accumulator.
6. Add byte $B2$ to accumulator (generally steps 5 and 6 are combined).
7. Store accumulator in location $C2$ (MSByte of result).

More than twice as many instructions are required for the add operation on the 16-bit word than for the 8-bit word. A 16-bit or 32-bit μP can be used if 16-bit or 32-bit manipulation must be done quickly. In many instrumentation or control applications, the program execution time is not critical; thus, an 8-bit μP may result in a less expensive system.

9.5.5 INSTRUCTION PIPELINING

The serial nature of the microoperations in the simple computer limits the speed of instruction execution. When an instruction is being executed, the fetching procedure is inoperative, a result of at least two factors. The first factor is the internal bus usage conflict. Because operand information or data are transferred via this bus during instruction execution, it cannot be used simultaneously for fetching instructions. The second factor is the control unit's serial design. It generates serial control signals and carries out serial microoperations. All early computers and all early μPs utilized this serial mode of operation.

In the early 1960s, the concept of *instruction look-ahead* or *pipelining* was introduced. In pipelined architecture, the internal bus may be broken into two paths, allowing instruction fetches to take place simultaneously with the internal transfer of data. During the execution of one instruction, the next instruction to be executed can be fetched and stored in the CPU. When the execution of the first instruction is completed, the execution of the next instruction can be started almost immediately without waiting for a fetch cycle. In a practical system, several instructions are prefetched and stored in a high-speed queue or *first-in-first-out memory* (FIFO) to be transferred to the instruction register when appropriate.

This FIFO is referred to as a *pipeline*, as it channels instructions sequentially to the IR.

The speed advantage of pipelining is lost when a branch instruction is encountered. If a branch instruction dictates that a nonsequential instruction be executed next, all information in the queue will be worthless. For example, when the program of Figure 9.29 is being executed and the IR contains the instruction fetched from memory location 00100, a pipeline would contain the instruction of memory location 00101 as the next instruction to be executed. If the accumulator happens to contain zero, the BRZ instruction must send the program to the HLT instruction contained in memory location 01001. In this situation, the information in the queue is of no value. When a program branches to a nonsequential instruction, the queue must be refilled with those sequential instructions that succeed this nonsequential one. Simple pipelining is more effective when used in a program that has relatively few branching instructions.

There are various advanced methods used to improve the concept of pipelining. Some processors use two pipelines, filling one with sequential instructions and another with instructions that succeed the branch instruction. When it is determined whether the branch takes place or not, the appropriate pipeline is used to fill the IR. Another method uses a section of the control unit to pre-evaluate the branch conditions, before this instruction reaches the IR. If knowledge of the upcoming branch is known in advance, the pipeline can begin filling with the correct instructions.

While the practical CPU is more complex than the simple one of Figure 9.31, the basic concepts involved in digital computers are embodied in this simple processor.

SUMMARY

• • • • • • • • • • • • • • • • •

1. The digital computer consists of a control unit, an arithmetic unit, a memory, an input unit, and an output unit.
2. Binary arithmetic uses number complements for many operations.
3. Memory chips can be arranged in many different configurations to implement a memory system.
4. A μP consists of at least a control unit and an arithmetic unit on a single chip. On some μPs, memory is also included.

CHAPTER 9 PROBLEMS

• • • • • • • • • • •

● Sec. 9.2.1

°**9.1** Find the 10's complement of 2436.

9.2 Find the 1's complement of 34_{10} for an 8-bit number.

9.3 Find the 2's complement of 123_{10} for an 8-bit number.

9.4 If $A = 10010010_2$ and $B = 00110010_2$ use 2's complement arithmetic to form $A - B$. Assume that A and B are positive, unsigned numbers.

9.5 Repeat Prob. 9.4 using 1's complement numbers.

°9.6 Repeat Prob. 9.4 if both A and B are signed 2's complement numbers.

9.7 Repeat Prob. 9.4 using 1's complement arithmetic if both A and B are signed numbers.

9.8 Use addition and repeated shifting to multiply the unsigned binary numbers 0110 and 1011.

9.9 Repeat Prob. 9.8 for the signed 1's complement numbers 10110 and 01011.

9.10 Use repeated subtraction and shifting to divide 0101_2 into 10110010_2.

°9.11 Write the expression for C_3 of a 4-bit look-ahead-carry adder. C_3 is the carry from the third to the fourth column.

9.12 Implement the serial adder of Figure 9.9 using standard TTL chips. Show all pin connections. *Hint:* The 7494 and 7496 may be used for the shift registers.

9.13 Implement the parallel adder of Figure 9.11 with standard TTL chips. Show all pin connections.

9.14 Implement the subtracting circuit of Figure 9.12 with standard TTL circuits. Show all pin connections.

9.15 If $A = 10010001$ and $B = 00011011$ in the circuit of Figure 9.13, indicate the values of S_0 to S_7 and the overflow flag.

°9.16 Repeat Prob. 9.15 for $A = 00011011$ and $B = 10010001$.

● **Sec. 9.3.5**

9.17 Implement the binary multiplier of Figure 9.14 with standard TTL circuits. Show all pin connections.

● **Sec. 9.3.6**

9.18 Implement the divider circuit of Figure 9.15 with standard TTL circuits. Show all pin connections.

● **Sec. 9.4.2**

9.19 How many bits can be stored in a 1 k × 4 memory? How many address lines are required to address all of the 4-bit words?

°9.20 A memory chip has 8 bit planes and uses 12 address lines. How many total storage locations does the memory chip contain? What is the specification on memory size for this chip?

● **Sec. 9.4.3**

9.21 If the address of the RAM described in Figure 9.20 is always set up before the chip is selected, what is the upper limit on the number of read operations that can be completed in one second?

9.22 If the address of the 2148H-2 RAM chip described in Figure 9.21 is set up, then the chip is selected within a maximum time of 10 ns from the point at which the address becomes stable, what is the upper limit on the number of write operations that can be completed in one second?

● **Sec. 9.4.4**

9.23 Show how to use the 1 k × 4 chips of Figure 9.23 to construct a 1 k × 16 bit memory.

9.24 Repeat Prob. 9.23 for a 4 k × 8 memory.

9.25 Show how to use the 1 k × 8 chips of Figure 9.24 to construct a 1 k × 16 bit memory.

9.26 Repeat Prob. 9.25 for a 2 k × 16 memory.

Sec. 9.5.1

9.27 Write the RTL for a series of operations that
(1) loads the 8-bit accumulator A from register C,
(2) rotates A one position to the left with the leftmost bit circulating to the rightmost position,
(3) stores the result into memory location 18.

9.28 Write the RTL for a series of operations that
(1) zeros the contents of memory location 16,
(2) loads the accumulator A from memory location 12,
(3) adds the contents of memory location 14 to A,
(4) increments the contents of memory location 16,
(5) checks the contents of memory location 16,
(6) if this number is less than four, repeat steps 3 through 5, otherwise write the accumulated sum to memory location 24.

°9.29 What is the maximum number of instructions that can be included in a 16-bit system that addresses 4 kwords of storage?

9.30 A system uses 8 memory banks and reserves 11 bits of the instruction word for addressing purposes. What is the maximum number of locations that can be accessed by this system?

°9.31 Add an operation called AVG to the simple CPU of Sec. 9.5. This operation fetches an operand from memory, adds it to the operand in the accumulator, and divides the result by 2. The fractional part of the result is neglected. Indicate how many time slots are needed and show the microoperation taking place during each time slot.

9.32 Repeat Prob. 9.31 for an instruction called BRM that retrieves an operand from memory and branches if this operand is nonzero. If the operand is zero, the next instruction of the program will be fetched. *Hint:* More than 7 time slots may be required.

9.33 Write a program showing contents of every storage location to
(1) read in a value of X and Y from the I/O register.
(2) calculate Z from

$$Z = \frac{4X + Y}{4} \text{ if } X < Y$$

$$Z = \frac{2Y + X}{2} \text{ if } X > Y$$

(3) read Z out to the I/O register.

(4) repeat steps 1, 2, and 3 until a negative value of X is read in.
A 10-bit instruction word is used along with an op code as follows:

0000	ADD (to accumulator A)
0001	SUBTRACT (from A)
0010	SHIFT RIGHT (contents of A)
0011	SHIFT LEFT (contents of A)
0100	BRANCH UNCONDITIONALLY
0101	BRANCH IF A POSITIVE OR ZERO
0110	LOAD A (from memory)
0111	STORE A (to memory)
1000	READ (from I/O to memory)
1001	WRITE (from memory to I/O)
1010	BRANCH IF A IS NEGATIVE
1011	HALT

REFERENCES AND SUGGESTED READING

1. K. J. Breeding, *Digital Design Fundamentals,* 2nd ed. Englewood Cliffs, NJ: Prentice-Hall, 1992.
2. D. J. Comer, *Microprocessor-Based System Design.* New York: Holt, Rinehart and Winston, 1986.
3. C. H. Roth, Jr., *Fundamentals of Logic Design,* 4th ed. St. Paul, MN: West, 1993.

Asynchronous State Machines

T o this point in the text, we have considered only those systems with clock-driven state changes. A major advantage of such a system is that all signals are stable when the state flip-flops change. If there is any conflict with unstable signals, certain variables can be delayed with respect to the clock signal to resolve this problem. There are two prices paid for the advantage of having this time reference. The first is in additional circuitry required to generate and control the clock signal. The second is that state changes must wait until a clock transition occurs even though the signals determining the state change may have stabilized long before the change takes place. The speed of operation is limited when the system is driven by a clock.

It is possible to design systems, called asynchronous state machines, to avoid these problems. Although the advantages of asynchronous over synchronous systems are significant, there is one great disadvantage that leads to the continued popularity of the synchronous machine. It is considerably more difficult to design reliable asynchronous systems than to design reliable synchronous systems. For this reason, synchronous state machines abound while asynchronous systems are used much less often. This chapter will discuss some basic principles relative to the design of simple asynchronous state machines.

10.1

THE FUNDAMENTAL-MODE MODEL

SECTION OVERVIEW The fundamental-mode model is introduced in this section. This model aids in the analysis of asynchronous digital circuits.

An asynchronous system uses feedback to produce memory elements as does the synchronous state machine. The synchronous system utilizes clocked flip-flops driven by the system clock. The asynchronous machine generally uses gates rather than flip-flops. Inputs to these gates are composed of system inputs and feedback signals from gate outputs, which drive the system to change states. Figure 10.1 demonstrates a simple asynchronous circuit.

X and Y are the system inputs while Z is the system output. The signal Z is fed back, however, to a gate input and in this way helps determine its own value. One difference between Z as an input signal and the system inputs X and Y is that there is always a delay from change in X or Y to change in Z. When an X or Y change dictates a change in Z, this change occurs only after the cumulative propagation delay time through the gates. We cannot be sure of the value of Z until the effect of this variable as an input has taken place.

It is characteristic of asynchronous circuits that the feedback variables along with system inputs determine the values of these same feedback variables. It is possible to force a change in feedback variable Z, by changing a system input, then Z may change a second time after a small delay time as feedback variables drive the input gates.

An idealized model has been proposed to reflect this behavior [2]. This fundamental-mode model is shown in Figure 10.2 for the circuit under discussion. The gates are considered to have no delay in this model, while the delay element has an output that follows its input after a delay of Δt.

The variable at the input of the delay element is called the excitation variable, while the feedback variable appears at the output of the delay element. This model produces the delay of the feedback variable but predicts no delay between input and output. Although this is not completely accurate, the overall behavior of the actual circuit can be predicted rather accurately applying this particular model.

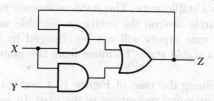

FIGURE 10.1 An asynchronous circuit.

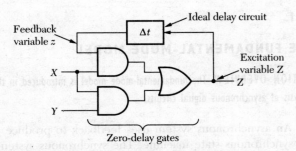

FIGURE 10.2 Fundamental-mode model for the circuit of Figure 10.1.

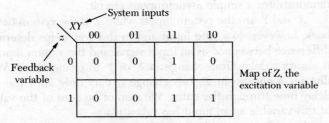

FIGURE 10.3 Excitation map for the circuit of Figure 10.1.

In order to characterize the behavior of a circuit, we plot a map of excitation variable as a function of gate inputs. Of course, gate inputs consist of both system inputs and feedback variables. Figure 10.3 shows an excitation map for the circuit of Figure 10.1. The system inputs are always plotted horizontally across the map, while the feedback variables are plotted vertically.

It is important to realize that the value Z takes on will also be the value assumed by z after a delay of Δt. Thus, the information depicted by the map represents a dynamic situation. This can be demonstrated by supposing the system inputs are $X = Y = 1$ and $z = 1$, which leads to $Z = 1$. This is called a stable state since $z = Z$. If X is then changed to 0, the output Z changes to 0 as indicated by the map location corresponding to $X = 0$, $Y = 1$, and $z = 1$. This condition will persist for only Δt since z will assume a value of 0 at this time, moving the system to the $X = 0$, $Y = 1$, and $z = 0$ location. The location 011 is a transient state, while 010 is a stable state. The stable states are normally identified on the map by drawing a circle around the excitation variable such as in Figure 10.4. Whenever $z = Z$, the gate inputs will not be changed by the feedback variable; therefore, the state is a stable state. To move from this state requires that a system input be changed.

In examining the map of Figure 10.4, we note that system input changes lead to a direct horizontal movement in the map. In an effort to simplify the succeeding

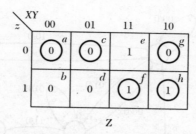

FIGURE 10.4 An excitation map showing stable states.

theory, we impose the restriction that only one input can change at a given time. We may change from $XY = 10$ to 11 or 00, but never to 01. With this restriction, the system can only move from $XY = 10$ to 01 by moving first to either 11 or 00, stabilizing, and then proceeding to $XY = 01$. While some practical systems allow simultaneous changes of more than one input, the treatment of such systems is beyond the scope of this chapter.

Although system input changes force the direct horizontal movement in the map, vertical movement is indirectly influenced. For example, if the system of Figure 10.4 exists in the stable state corresponding to $XYz = 010$, a change of X to 1 forces the system to move horizontally to location 110. This is a transient state causing Z to change from 0 to 1. After a delay of Δt, z will also change from 0 to 1, leading to a stable state located at $XYz = 111$.

We refer to every cell or location in the map as a state. In effect, the state is determined by the values of X, Y, and z. A primitive state diagram containing a number of states equal to the number of cells can be drawn from the map. Figure 10.5 contains the diagram for the map of Figure 10.4. The format specifies all gate inputs, dividing the system inputs from the feedback variables by a slash. There are several points reflected by the primitive state diagram that deserve emphasis.

1. Only in a stable state can a system input be changed. All transient states have the same value of system inputs entering and exiting from the state. Transient states are shaded in the state diagram.
2. In order to leave a stable state, a system input must be changed.
3. In order to cause a feedback variable change, the system must pass through a transient state.
4. Only one input can be changed in moving from one stable state to another.

It is not always possible to move from one stable state directly to all other stable states. To move from state a to state h, the system may move through states c, e, and f before reaching h. One input would change to exit from each stable state. The feedback variable changes only as a transient state is traversed.

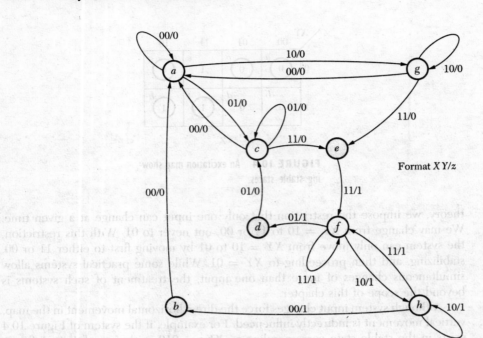

Format XY/z

FIGURE 10.5 Primitive state diagram for map of Figure 10.4.

EXAMPLE 10.1

Use the fundamental-mode model to develop the excitation map for the circuit of Figure 10.6. Identify all stable states.

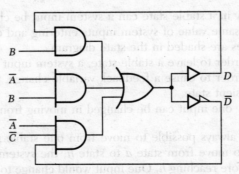

FIGURE 10.6 Circuit for Example 10.1.

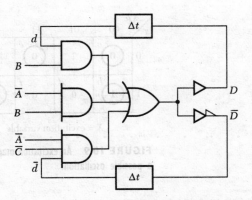

FIGURE 10.7 Fundamental-mode model for the circuit of Figure 10.6.

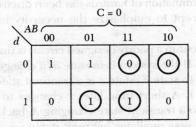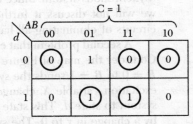

D = excitation variable

FIGURE 10.8 Excitation map for the circuit of Figure 10.6.

SOLUTION

The fundamental-mode model is shown in Figure 10.7. The expression for the excitation variable D is

$$D = \bar{A}B + \bar{A}\bar{C}\bar{d} + Bd$$

This expression is plotted in the map of Figure 10.8. Those states with $D = d$ are circled to identify the stable states. ●

10.2

PROBLEMS OF ASYNCHRONOUS CIRCUITS

SECTION OVERVIEW This section considers three problems in asynchronous circuits that must be avoided for reliable operation. These three problems are hazards, oscillations, and critical races.

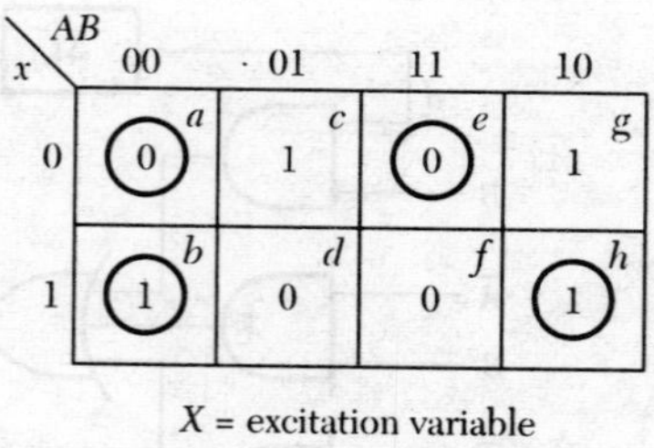

X = excitation variable

FIGURE 10.9 An excitation map with a possible oscillation.

In Chapter 2, we discussed the problem of static and dynamic hazards. Any glitch occurring as a result of a hazard may cause an unwanted change of state in an asynchronous system. Since the elimination of hazards has been discussed previously, we will not discuss it further except to emphasize the necessity in asynchronous circuits of eliminating all hazards.

A second problem that can occur in a poorly designed circuit is that of oscillation. Consider the map of Figure 10.9. If the system is in state a, a change of input from $B = 0$ to $B = 1$ sends the system to state c. State c is a transient state, and thus the excitation variable X changes to 1. A short time later x changes to 1, moving the system to state d. This state is also a transient state changing X back to 0, followed by a change in x to 0. The system now oscillates between states c and d. Of course, this type of situation can be used to advantage in a clock circuit by adding a delaying network to control the delay time Δt to create the desired oscillation frequency. In most systems, the oscillation is unacceptable, and the situation depicted by states c and d of the map must be avoided.

A third problem in asynchronous design is that of critical races. This situation can occur only when two or more feedback variables are present in the system. The excitation map of Figure 10.10 demonstrates the critical race problem. This system has two external inputs, A and B, and two excitation variables, X and Y, that are fed back to the input of the circuit.

One critical race occurs if the system starts in state e and input B changes from 1 to 0. The excitation variables begin to switch from $XY = 00$ toward $XY = 11$. Due to unequal propagation delays, one of the excitation variables will reach a value of 1 while the other has not changed from a value of 0. If the condition $XY = 10$ is reached, the system moves to stable state d. If the condition $XY = 01$ is reached rather than 10, the system moves to stable state b. The final stable state reached from this input condition depends on the relative switching speeds of variables X and Y. This situation is referred to as a critical race. Although a particular circuit may always reach the same final state when AB switches from 01 to 00, an identical circuit may reach different states for these input values. In fact, if a gate requires replacing, this may result in a different circuit behavior. Critical races obviously must be avoided.

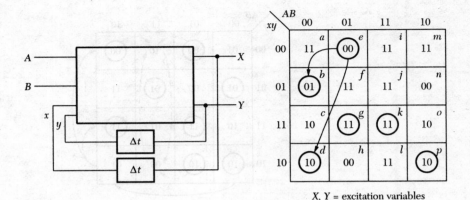

FIGURE 10.10 An excitation map with critical races.

Noncritical races can be tolerated, as demonstrated by changing the input conditions from $AB = 01$ to 11 when the system starts in state e. The system again moves toward $XY = 11$, but will reach either 01 or 10, depending on gate speed. The system will move to either state j or state l, then to a final state of k. Regardless of which excitation variable wins the race, the same final state is reached.

Another critical race exists when the system starts in state b and the input changes to $AB = 10$. The system switches to transient state n with $XY = 00$. After a short period of time, xy changes to 00, moving the system to transient state m. A race situation again exists as the system moves toward $XY = 11$. If XY reaches 10, the system stops in stable state p. If XY reaches 01, however, the system moves back to state n and proceeds to oscillate between states m and n. Although a final state is never reached, we will also call this situation a critical race.

In designing asynchronous circuits, hazards, oscillations, and critical races must be avoided.

EXAMPLE 10.2

Locate any critical races and oscillation problems that may exist in the excitation map of Figure 10.11. Indicate the starting state, the change in AB required to initiate the problem, and the ending states for each problem.

SOLUTION

Each stable state must be examined to see if a change of A or B leads to a critical race or oscillation. The initial state of $ABxy = 0010$ can move to a critical race if A is changed to 1. The system moves to the transient state of 1010, which produces an output of $XY = 01$ after some delay. Since a value of $XY = 10$ will change

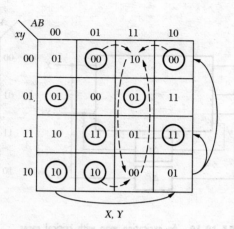

FIGURE 10.11 Excitation map for Example 10.2.

to 01 for this situation, the value of $XY = 11$ or $XY = 00$ might be generated at the system output, depending upon whether X changes slightly faster than Y or Y changes faster than X. If Y is faster, the ending state will be 1011, whereas a faster X change will lead to a final state of 1000. This critical race is indicated by the solid lines in Figure 10.11.

An oscillation problem exists when a starting state of 0110 is changed by asserting input A. The system then moves to the transient state 1110, which produces $XY = 00$ after a delay. This moves the system to transient state 1100, which, after a delay, returns the system to 1110. Oscillation between these transient states continues until a change in A or B occurs. Two other starting states that allow entry into this oscillating mode are 0100 and 1000 as shown by the dashed lines of Figure 10.11.   ●

10.3

. . . .

BASIC DESIGN PRINCIPLES

SECTION OVERVIEW This section discusses methods of avoiding critical race and oscillation problems when designing asynchronous systems.

When given an excitation map to analyze, all stable and transient states are determined. The system behavior for any combination of inputs can be precisely found. When we must design a system, however, we are given only the system behavior. As usual in a design problem, there may be several different circuits that can satisfy the specifications. In particular, we generally can only define the stable

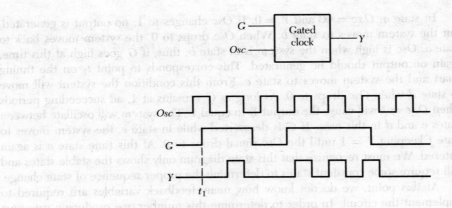

FIGURE 10.12   Gated clock system.

states in terms of the specs. We then must add transient or cycle states to cause the system to sequence through the proper states without introducing any critical races.

We will introduce a simple problem as a vehicle to discuss some design principles used for asynchronous circuits. Figure 10.12 shows the block diagram and timing chart of the circuit to be constructed.

The timing chart indicates that the input G gates the Osc input to the output. Additionally, if Osc is high when G is asserted, the output does not go high until the beginning of the next Osc cycle. If G is deasserted when Osc is high, the output Y remains high until Osc drops to a low level. We see that there must be a memory function included in the system to avoid cutting the length of output pulse when G changes level. We therefore can use a state machine to satisfy the specifications. A state diagram of this system appears in Figure 10.13.

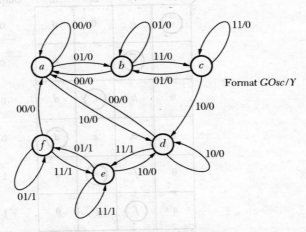

FIGURE 10.13   State diagram for gated clock.

In state a, $Osc = 00$ and $Y = 0$. If Osc changes to 1, no output is generated, but the system moves to state b. When Osc drops to 0, the system moves back to state a. Osc is high when the system is in state b; thus, if G goes high at this time, again no output should be generated. This corresponds to point t_1 on the timing chart and the system moves to state c. From this condition the system will move to state d when Osc drops to 0. As long as G remains at 1, all succeeding periods when $Osc = 1$ will cause the output Y to equal 1. The system will oscillate between states e and d in this case. If G is deasserted while in state e, the system moves to state f keeping $Y = 1$ until the Osc signal drops to 0. At this time state a is again entered. We must recognize that this state diagram only shows the stable states and will require some transient states to determine the proper sequence of state change.

At this point, we do not know how many feedback variables are required to implement the circuit. In order to determine this number, we produce a primitive excitation table having only one stable state per row. With two inputs and six stable states, a 4-column, 6-row map is required as shown in Figure 10.14.

This map is constructed by first inserting one stable state in each row. Then transient states are added to cause the proper movement between stable states as indicated by the state diagram. The requirement that inputs must not be changed simultaneously leads to the "don't care" conditions. Any location that requires a simultaneous change of two variables to move from the stable state contains a "don't care" symbol.

It would be possible to synthesize a circuit from the primitive excitation map, but the circuit would not be minimal. This map has a great deal of redundancy that can be removed. Each row of an excitation map corresponds to a different combination of feedback variables. The primitive map has six rows and would require three

$GOsc$				
00	01	11	10	Output Y
ⓐ	b	θ	d	0
a	ⓑ	c	θ	0
θ	b	ⓒ	d	0
a	θ	e	ⓓ	0
θ	f	ⓔ	d	1
a	ⓕ	e	θ	1

FIGURE 10.14 Primitive excitation map.

$GOsc$	00	01	11	10	Output Y
	(a)	(b)	(c)	d	0
	a	θ	c	(d)	0
	a	(f)	(e)	d	1

FIGURE 10.15 Excitation map for diagram of Figure 10.13.

feedback variables. By combining rows that contain the same states in corresponding locations, the number of feedback variables can be reduced. This process is called merging and we realize that a "don't care" condition can be taken as any state to allow a row to be merged. Before merging we must note which stable states produce an output and assign an output to each row. This is shown in the primitive state map of Figure 10.14. As we merge rows, we will not merge similar rows if outputs are different. The elimination of redundant states proceeds in the same way that equivalent states are removed from a synchronous system. We continue to limit the scope of our discussion to simpler asynchronous systems and will not consider more complex reduction methods.

Following the rules mentioned, the excitation map of Figure 10.15 is constructed. Comparing the state diagram with this map, we see that the same system behavior is predicted. With three rows, two feedback variables must be used resulting in the map of Figure 10.16. The assignment of feedback variable value is arbitrary.

$GOsc$ xz	00	01	11	10
00	(00)	(00)	(00)	01
01	00	11	11	(01)
11	00	(11)	(11)	01
10	00	θθ	θθ	θθ

X, Z = excitation variables

FIGURE 10.16 Excitation map with feedback variables.

Once the feedback variables are assigned, the excitation variables for the stable states are also determined. The transient states are not fixed at this point and can be selected to eliminate races or oscillations. To determine potential problems, we start in each stable state and then move to logically adjacent input combinations. If excitation variables remain the same or are logically adjacent to those of the stable state, critical races cannot occur. In this instance, there are no problems and the assignment in Figure 10.16 is appropriate.

The next step is to implement the excitation variables X and Z. The separate K-maps of Figure 10.17 are useful for this purpose. All hazards must be covered in this step. The final circuit is shown in Figure 10.18.

In order to demonstrate the use of transient state assignment, we will consider Example 10.3.

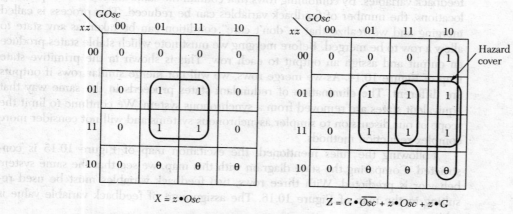

$X = z \cdot Osc$

$Z = G \cdot \overline{Osc} + z \cdot Osc + z \cdot G$

FIGURE 10.17   K-maps for excitation variables.

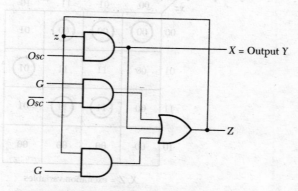

FIGURE 10.18   Gated clock circuit.

EXAMPLE 10.3

The asynchronous circuit in Figure 10.19 exhibits unreliable behavior. Redesign the circuit to produce reliable operation.

SOLUTION

The expressions for the excitation variables, X and Y, are developed in terms of A, B, x, and y. These are given in Figure 10.19, along with the excitation map for X and Y. A critical race problem from an initial state of 1000 is located along with an oscillation problem between states 0011 and 0010. The oscillation mode can be entered from initial states of 0111, 0110, or 1010. These possibilities are indicated on the map of Figure 10.19.

The transient or cycle states causing the critical races can be modified on the map to eliminate the ambiguity of final states. This is done in Figure 10.20. By modifying the two states shown, the oscillation and critical race problems are eliminated from the excitation map. The circuit must now be changed to conform to the information reflected by the modified excitation map. This can be easily done by plotting the excitation maps separately for X and Y as shown in Figure 10.20.

While implementing the race-free maps, we must also avoid any possible hazards. The expressions for X and Y given in Figure 10.20 include hazard covers. The reliable system is realized by the circuit of Figure 10.21. ●

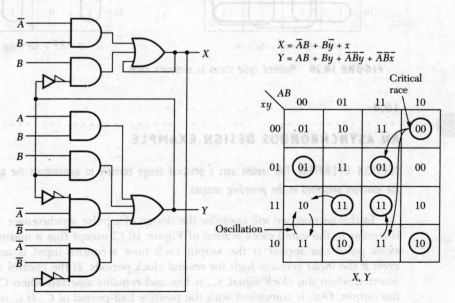

$$X = \overline{A}B + B\overline{y} + x$$
$$Y = AB + By + \overline{A}\overline{B}\overline{y} + \overline{A}\overline{B}x$$

FIGURE 10.19 An asynchronous circuit with race problems.

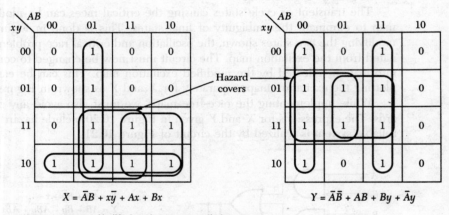

FIGURE 10.20   Modified cycle states to eliminate races.

10.4

. . . .

AN ASYNCHRONOUS DESIGN EXAMPLE

SECTION OVERVIEW This section uses a practical design example to demonstrate the application of the concepts presented in the preceding sections.

In this section, we will consider the design of a pulse synchronizer. This circuit is similar to the gated clock system of Figure 10.12 except that a maximum of one clock pulse can appear at the output each time a control input is asserted high even if the input remains high for several clock periods. If the control signal, W, is asserted when the clock signal, C, is low and remains asserted when C goes high, the output, Out, is coincident with the positive half-period of C. If C is high when W goes high and W remains asserted until C goes low and returns high for a second

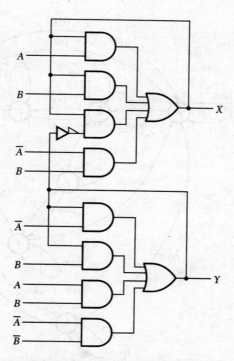

FIGURE 10.21 Asynchronous circuit with no race problems.

time, Out is coincident with this second assertion of C. If C is high when W goes high, but W returns low before the second assertion of C, no output occurs. We also assume that successive assertions of W will be separated by at least one-half clock period. Figure 10.22 shows the block diagram for this system along with a timing chart that reflects the preceding description.

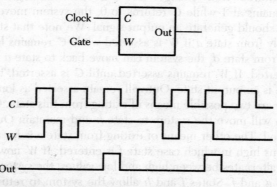

FIGURE 10.22 A pulse synchronizing system.

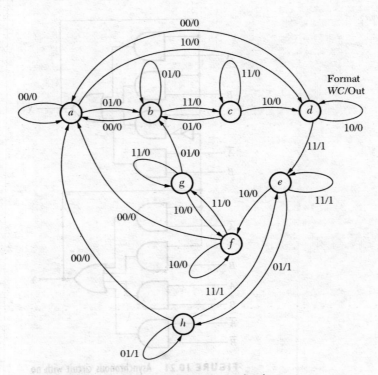

FIGURE 10.23 State diagram of the pulse synchronizer.

The creation of the state diagram must consider all possible combinations of the clock signal, the control signal, and the required output. The state diagram of Figure 10.23 represents the behavior of the pulse synchronizer system. State a exists when both W and C are low. If C changes to 1 while W remains equal to 0, state b is entered. If W changes to 1 after C has changed to 1, state c is reached. If W remains at 1 while C returns to 0, the system moves to state d. None of these states should generate an output signal. We note that state d could also be reached directly from state a if W is asserted while C remains low.

From state d, the system can move back to state a if W is deasserted before C is asserted. If W remains asserted until C is asserted, the system moves to state e, which is an output state. Out will remain asserted as long as the system is in state e. There are two possible means of exiting from this state. One is the deassertion of W, which will move the system to state h and maintain Out = 1 as long as C remains asserted. The other means of exiting from state e is for the clock to go low while W remains high in which case state f is entered. If W now remains asserted while the clock alternates between high and low values, the system moves alternately between states g and f. States f and h allow the system to return to state a if both W and C are low. The system moves from state g to b if W goes low while C remains high.

WC	00	01	11	10	Out
	(a)	b	θ	d	0
	a	(b)	c	θ	0
	θ	b	(c)	d	0
	a	θ	e	(d)	0
	θ	h	(e)	f	1
	a	θ	g	(f)	0
	θ	b	(g)	f	0
	a	(h)	e	θ	1

FIGURE 10.24 Primitive excitation map for pulse synchronizer.

The primitive excitation map, having one stable state per row, is constructed from the state diagram as shown in Figure 10.24. Note that transient states are added to cause the system to move correctly from one stable state to another as indicated by the state diagram.

We now examine the primitive excitation map to find which rows can be merged. In this map it is possible to merge rows 1, 2, and 3 into a single row. Rows 6 and 7 can be merged as can rows 5 and 8. The merged excitation map requires four rows and one version appears in Figure 10.25(a). This map presents a problem with a critical race that cannot be resolved. The noncritical race need not be considered since the same final state of $WCxy = 0000$ is reached, although different paths may be taken from state 1011. The critical race from 1111 to 0100 can be resolved by selecting the correct transient states, but the race existing when the initial state is 1001 leads to a problem. If WC is changed to 11, the final state should be 1110, but depending on the speed of X compared to Y, the system will move to either 0000 or 1111.

This problem can be overcome by exchanging the positions of the third and fourth rows resulting in the map of Figure 10.25(b). This map has no critical races and the expressions for X and Y can be found to be

$$X = Cy + Wx$$

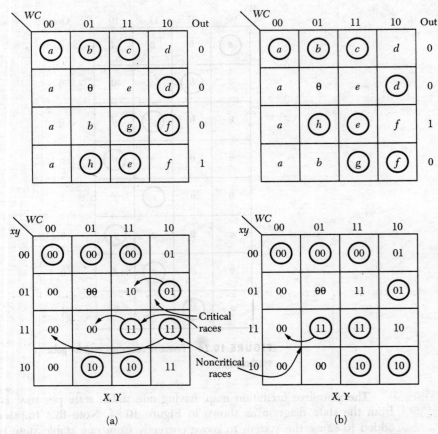

FIGURE 10.25 Excitation maps: (a) map with critical races, (b) map with no critical races.

and

$$Y = Cy + W\bar{C}\bar{x} + W\bar{x}y$$

The last term in the expression for Y is a hazard cover. Since states e and h are to generate an output signal, from Figure 10.25(b) the expression for the output is Out = XY. The implementation of these equations results in the circuit of Figure 10.26.

SUMMARY

• • • • • • • • • • • • • •

1. The asynchronous state machine can operate at higher frequencies than can the synchronous state machine, but requires a more complex design procedure.

2. Hazards, critical races, and oscillations are possible problems that must be eliminated in asynchronous state machines.

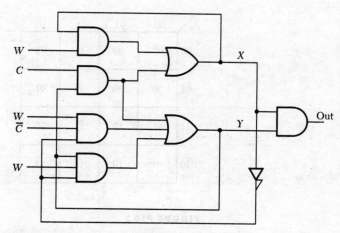

FIGURE 10.26   Pulse synchronizing circuit.

CHAPTER 10 PROBLEMS

● ● ● ● ● ● ● ● ● ● ● ● ● ● ●

● **Secs. 10.1 and 10.2**

°10.1 Given the excitation map shown in Figure P10.1 for an asynchronous system:
   a. Identify all stable states by circling them.
   b. Identify all noncritical races that could occur. Indicate the starting stable state (AB/Y_1Y_2), the change of inputs initiating the race, and each state cycled through.
   c. Identify any oscillations that could occur.
   d. Identify all critical races that could occur.

10.2 Repeat Prob. 10.1 for the excitation map of Figure P10.2.

y_1y_2 \ AB	00	01	11	10
00	11	01	11	00
01	01	10	11	01
11	01	11	10	11
10	11	10	10	00

Y_1, Y_2

FIGURE P10.1

AB xy	00	01	11	10
00	00	00	11	01
01	00	00	10	01
11	11	10	11	10
10	01	10	11	11

X, Y

FIGURE P10.2

AB xy	00	01	11	10
00	10	00	10	00
01	01	00	01	11
11	00	11	01	11
10	10	10	00	01

X, Y

FIGURE P10.3

10.3 Locate a critical race and an oscillation problem in the excitation map of Figure P10.3 by specifying the starting state ($ABxy$) and the necessary change in AB to cause the problem.

10.4 For the asynchronous circuit of Figure P10.4,
    a. Draw the fundamental mode model and clearly label the feedback variables and the excitation variables.
    b. Draw the excitation table for the circuit.
    c. Identify any potential problems.

Secs. 10.3 and 10.4

10.5 Determine if the circuit shown in Figure P10.5 has critical race problems. If so, minimally redesign the circuit to remove any problems.

10.6 Modify the excitation table of Figure P10.4 to eliminate critical races and oscillation problems, while maintaining the state diagram shown in Fig are P10.6.

10.7 Construct an implementation of the state diagram of Figure P10.6.

10.8 Design a circuit to produce the output shown in Figure P10.8.

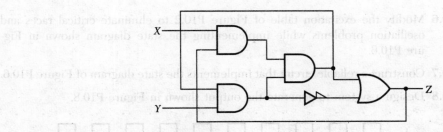

FIGURE PI0.4

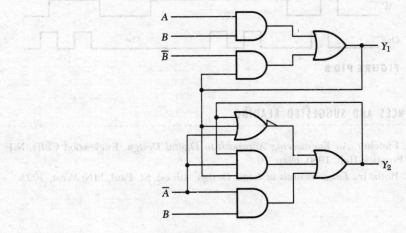

FIGURE PI0.5

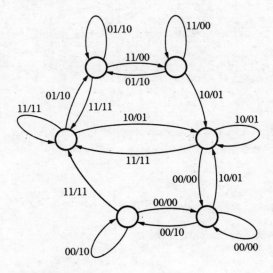

FIGURE PI0.6

°10.6 Modify the excitation table of Figure P10.2 to eliminate critical races and oscillation problems while implementing the state diagram shown in Figure P10.6.

10.7 Construct a reliable circuit that implements the state diagram of Figure P10.6.

10.8 Design a system to generate the output shown in Figure P10.8.

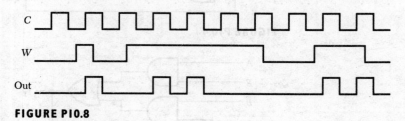

FIGURE P10.8

REFERENCES AND SUGGESTED READING

1. W. I. Fletcher, *An Engineering Approach to Digital Design.* Englewood Cliffs, N.J.: Prentice-Hall, 1980, chap. 10.
2. C. H. Roth, Jr., *Fundamentals of Logic Design,* 4th ed. St. Paul, MN: West, 1993.

Appendix 1

Logic Families

S ince the early days of transistor usage in electronic computers, it was realized that different configurations of logic circuits offered different advantages. It was also apparent that computer design could be expedited if the designer did not have to be concerned with constructing the gates, flip-flops, and other elements with discrete components. With the development of the printed circuit board, the computer manufacturers were able to produce logic circuits on PC boards in the early 1960s. Typically, two to four AND gates would be constructed on one board. One or two flip-flops may appear on a single board. All necessary logic components were constructed on boards that could then be interconnected to form digital systems.

As integrated circuits became available, those circuits occupying a single board were implemented on a single chip. Digital system design methods changed little as computers began to be constructed with integrated circuits. The major difference was in volume occupied by the circuits. These integrated circuits of the mid-1960s are now called small-scale integrated or SSI circuits.

Several types of technology have been developed for IC fabrication in the last two decades. Three very popular families are the TTL, the CMOS, and the ECL technologies. TTL is a mnemonic for transistor-transistor logic. CMOS signifies complementary metal-oxide silicon circuits, and ECL signifies emitter-coupled logic. Some of the outstanding features of these families are tabulated in Table A1.1.

TABLE AI.I Broad Characteristics of Logic Families.

Characteristic	TTL	CMOS	ECL
Power input	Moderate	Low	Moderate-high
Frequency limit	High	Moderate	Very high
Circuit density	Moderate-high	High-very high	Moderate
Circuit types per family	High	High	Moderate

The TTL family has been the most popular for several years because of its great versatility. Many types of logic circuit are available in this technology. Certain subfamilies of TTL have propagation delay times near 1 ns.

The CMOS family has become very significant in the last decade due to the low power dissipation of CMOS circuits. Portable devices such as calculators and high-density circuits such as computer memories must minimize drain on the dc power source. In addition to low power requirements, CMOS circuits require very few, if any, resistors to be fabricated on the IC chip. Consequently, more circuits can be produced in a given volume of silicon with this technology. One disadvantage that has prevented the dominance of the CMOS family is the lower speed of operation. The higher-speed CMOS gates have propagation delays in the 10 ns range, however, this figure continues to improve each year.

For very high-speed logic, the ECL family is a requirement. Propagation delays of 0.4 ns can be achieved with ECL gates. Fewer types of logic circuit are available in ECL than are available in either TTL or CMOS.

AI.I

TRANSISTOR-TRANSISTOR LOGIC (TTL)

The TTL family is based on the multiemitter construction of transistors shown in Figure A1.1. The operation of the input transistor can be visualized with the help of the circuit of Figure A1.2, which shows the bases of the three transistors connected in parallel, as are the collectors, whereas the emitters are separate.

If all emitters are at ground level, the transistors will be saturated by the large base drive. The collector voltage will be only a few tenths of a volt above ground. The base voltage will equal $V_{BE(on)}$, which may be 0.5 V. If one or two of the emitter voltages are raised, the corresponding transistors will shut off. The transistor with an emitter voltage of 0 V will still be saturated, however, and saturation will force the base voltage and collector voltage to remain low. If all three emitters are raised to a higher level, the base and collector voltages will tend to follow this signal.

Returning to the basic gate of Figure A1.1, we see that when the low logic level appears at one or more of the inputs, T_1 will be saturated with a very small voltage appearing at the collector of this stage. Since at least $2V_{BE(on)}$ must appear at the

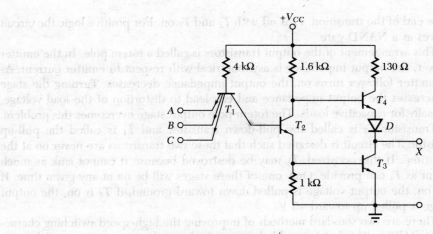

FIGURE AI.I   Basic TTL gate.

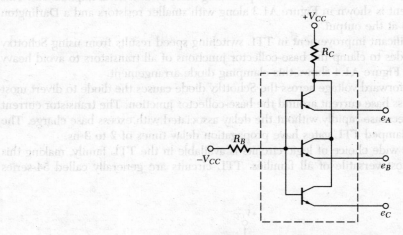

FIGURE AI.2   Discrete circuit equivalent
to the multiemitter transistor.

base of T_2 in order to turn T_2 and T_3 on, we can conclude that these transistors are off. When T_2 is off, the current through the 1.6 kΩ resistance is diverted into the base of T_4, which then drives the load as an emitter follower.

When all inputs are at the high voltage level, the collector of T_1 attempts to rise to this level. This turns T_2 and T_3 on, which clamps the collector of T_1 to a voltage of approximately $2V_{BE(on)}$. The base-collector junction of T_1 appears as a forward-biased diode, whereas in this case the base-emitter junctions are reverse-biased diodes. As T_2 turns on, the base voltage of T_4 drops, decreasing the current through the load. The load current tends to decrease even faster than it would if only T_4 were present, because T_3 is turning on to divert more current from the load.

At the end of the transition, T_4 is off with T_2 and T_3 on. For positive logic the circuit behaves as a NAND gate.

This arrangement of the output transistors is called a totem pole. In the emitter follower, the output impedance is asymmetrical with respect to emitter current. As the emitter follower turns on, the output impedance decreases. Turning the stage off increases the output impedance and can lead to distortion of the load voltage, especially for capacitive loads. The totem-pole output stage overcomes this problem.

Transistor T_3 is called the pull-down transistor and T_4 is called the pull-up transistor. The circuit is designed such that these two transistors are never on at the same time. If this occurred, T_3 may be destroyed because it cannot sink as much current as T_4 can provide. Only one of these stages will be on at any given time. If T_3 is on, the output voltage is pulled down toward ground; if T_4 is on, the output voltage is pulled up toward +5 V.

There are two standard methods of improving the high-speed switching characteristics of TTL. The first is to add clamping diodes to the input emitters of the gate to reduce transmission line effects by providing more symmetrical impedances. This improvement is shown in Figure A1.3 along with smaller resistors and a Darlington connection at the output.

A significant improvement in TTL switching speed results from using Schottky barrier diodes to clamp the base-collector junctions of all transistors to avoid heavy saturation. Figure A1.4 shows this clamping diode arrangement.

A low forward voltage across the Schottky diode causes the diode to divert most of the excess base current around the base-collector junction. The transistor current can then decrease rapidly without the delay associated with excess base charge. The Schottky-clamped TTL gates have propagation delay times of 2 to 3 ns.

A very wide choice of logic circuits is available in the TTL family, making this line the most versatile of all families. TTL circuits are generally called 54-series

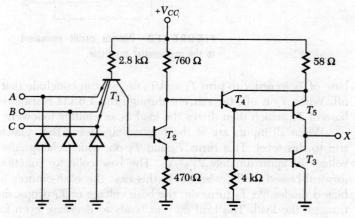

FIGURE A1.3   High-speed TTL gate.

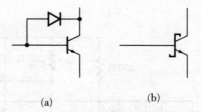

(a)   (b)

FIGURE AI.4   (a) Schottky-clamped transistor. (b) Symbol for a clamp.

or 74-series circuits. For example, a 5400 or 7400 is a quad 2-input NAND gate. The 54 indicates the circuit is designed to meet military requirements by satisfying specified performance over a temperature range of $-55°$ to $125°$C. The 74 series is the commercial series with an operating temperature of $0°$ to $70°$C.

A typical TTL IC may be numbered 74S10. The letter or letters after the 74 indicate TTL type. If there is no letter, the element is conventional TTL, and its average propagation delay is 13 ns, while its average power dissipation is 10 to 15 mW per gate. The letter H designates high-speed TTL, such as the gate shown in Figure A1.3, with propagation times that are approximately half that of conventional TTLs. The letter S indicates Schottky-clamped TTL, which is even faster than the H-type circuit with typical propagation delays of 2 to 3 ns and power dissipation slightly higher than conventional TTLs. Low-power TTL has a designator of L following the first two numbers. These circuits cut down system power dissipation by a factor of approximately 10 when replacing conventional TTL circuits, and their gate delays are increased by a factor of approximately 2.

The low-power, Schottky-clamped TTL is popular at the present time. The power dissipation is about five times lower than conventional while delay times are comparable. An LS follows the first two numbers for low-power, Schottky TTL circuits.

The numbers after the type designator indicate what kind of circuit is contained in the IC. This number ranges from 00 up to a three-digit number in the hundreds. There may be a letter following the number of the IC that indicates package type. Another three- or four-digit number may appear on the package to identify the year and week the IC was manufactured.

The 74S10 is a commercial Schottky TTL, with triple 3-input NAND gates having a propagation delay time of 3 ns and a 19 mW power dissipation. Its V_{IHmin} is specified as 2 V; V_{ILmax} is specified as 0.8 V; V_{OHmin} is specified as 2.7 V; and V_{OLmax} is specified as 0.5 V. The current I_{IHmax} is given as 50 μA when $V_{\mathrm{IH}} = 2.7$ V, I_{ILmax} is specified as -2 mA when $V_{\mathrm{IL}} = 0.5$ V. The short-circuit output current varies from a minimum of -40 mA to a maximum -11 mA.

Advanced Schottky and advanced low power Schottky were introduced around 1985. These circuits will be considered in Section A1.4.

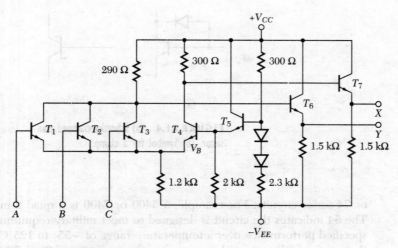

FIGURE A1.5   An ECL gate.

A1.2
. . . .

EMITTER-COUPLED LOGIC (ECL)

One of the older families is the ECL family. For many years this configuration was unrivaled in high-speed switching applications, but it now competes with Advanced Schottky TTL. The good switching characteristics of this family again result from the avoidance of saturation of any transistors within the gate. Figure A1.5 shows an ECL gate with two separate outputs. For positive logic X is the OR output while Y is the NOR output.

Often the positive supply voltage is taken as 0 V and V_{EE} as -5 V. The diodes and emitter follower T_5 establish a base reference voltage for T_4. When inputs A, B, and C are less than the voltage V_B, T_4 conducts while T_1, T_2, and T_3 are cut off. If any one of the inputs is switched to the 1 level, which exceeds V_B, the transistor turns on and pulls the emitter of T_4 positive enough to cut this transistor off. Under this condition output Y goes negative while X goes positive. The relatively large resistor common to the emitters of T_1, T_2, T_3, and T_4 prevents these transistors from saturating. In fact, with nominal logic levels of -1.9 V and -1.1 V, the current through the emitter resistance is approximately equal before and after switching takes place. Thus, only the current path changes as the circuit switches. This type of operation is sometimes called current mode switching. Although the output stages are emitter followers, they conduct reasonable currents for both logic level outputs and, therefore, minimize the asymmetrical output impedance problem.

The ECL family has a disadvantage of requiring more input power than the TTL line. Furthermore, the great variety of logic circuits that can be realized with TTL cannot be duplicated with ECL. Propagation times of 0.4 ns can now be achieved with ECL circuits.

AI.3

MOSFET LOGIC

The p-channel (p-MOS) and n-channel (n-MOS) MOSFET devices offer several advantages over TTL and other bipolar logic families. Three big advantages of the MOS families are (1) the improved packing density because on-chip resistors are not required, (2) the lower power dissipation resulting from high-impedance active loads, and (3) the simplicity of a fabrication process that allows higher yields and the integration of more complex systems. Many microprocessors are based on MOS logic to make them cost effective. The disadvantages of MOS logic compared with TTL are (1) a slower operating speed and (2) a lower drive-current capability.

CMOS or complementary-symmetry MOS is a well-established logic family that includes many SSI and MSI devices. Because both p- and n-channel devices must be fabricated on the same chip for CMOS, LSI systems are generally implemented by the simpler p-MOS or n-MOS methods. For less complex systems, CMOS offers the lowest power dissipation of any logic family and can also operate over a wide range of power supply voltages from a typical low of 3 V to a high of 18 V.

The inverter of Figure A1.6 is the basic block for a CMOS gate. Both p-channel and n-channel devices are enhancement-type MOS transistors. When the input voltage is near ground potential, the n-channel device T_1 is off. The voltage from gate to source of T_2 is approximately $-V_{DD}$ and, therefore, T_2 is on. With T_2 in the high conductance state and T_1 in the low conductance state, the power supply voltage drops across T_1 and appears at the output. When e_{in} increases and approaches V_{DD}, T_2 turns off and T_1 turns on, dropping the power supply voltage across T_2 and leading to zero output voltage. The high impedance of the off transistor results in a negligible current drain on the power supply. The threshold voltage can be controlled during fabrication and is generally designed so that switching occurs at an input voltage of approximately $V_{DD}/2$.

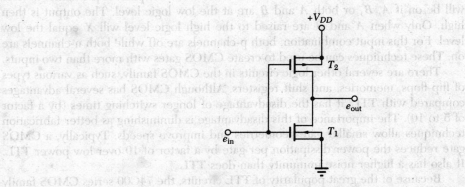

FIGURE AI.6 Basic CMOS inverter.

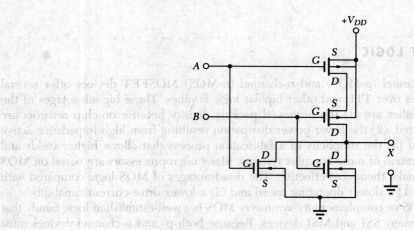

FIGURE A1.7 CMOS NOR gate for positive logic.

The output voltage levels of a gate are typically less than 0.01 V for the 0 state and 4.99 to 5.00 V for the 1 state ($V_{DD} = 5$ V). The required input voltages for this latter case might be 0.0 to 1.5 V for a 0 and 3.5 to 5.0 V for a 1. The noise margins in either state are then approximately 1.5 V.

Figure A1.7 shows a 2-input NOR gate using CMOS with positive logic. If both A and B are held at the 0 logic level, both n-channel devices are off while both p-channel transistors are on. The output is near V_{DD}. If input A moves to the 1 state, the upper p-channel device turns off while the corresponding n-channel turns on. This leads to an output voltage near ground. If input B is at the upper logic level while A is at the 0 level, the lower p-channel device shuts off while the corresponding n-channel device turns on again, resulting in an output 0. When both A and B equal 1, the output is obviously 0.

A positive logic NAND gate is shown in Figure A1.8. In this case, at least one of the series n-channel devices will be off and one of the parallel p-channel devices will be on if A, B, or both A and B are at the low logic level. The output is then high. Only when A and B are raised to the high logic level will X equal the low level. For this input combination, both p-channels are off while both n-channels are on. These techniques can be used to create CMOS gates with more than two inputs.

There are several other logic circuits in the CMOS family, such as various types of flip-flops, memories, and shift registers. Although CMOS has several advantages compared with TTL, it has the disadvantage of longer switching times (by a factor of 5 to 10). The importance of this disadvantage is diminishing as better fabrication techniques allow smaller gate geometries and improve speeds. Typically, a CMOS gate reduces the power dissipation per gate by a factor of 10 over low power TTL. It also has a higher noise immunity than does TTL.

Because of the great popularity of TTL circuits, the 74C00 series CMOS family was designed to be equivalent to the 7400 series TTL family in terms of function

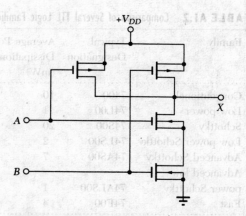

FIGURE A1.8 A CMOS NAND gate for positive logic.

and pin assignments. Although this CMOS family can drive low-power TTL, buffers are required to drive other TTL types.

A1.4

TTL GATES

The conventional or standard TTL family was the first TTL family introduced, appearing in the mid-1960s. Over the years, the performance of this family has improved and several other TTL families have been introduced. Another family introduced in the late 1960s is the low power TTL circuit. This configuration is basically the same as that of the conventional circuit but uses larger resistors to minimize input current requirements. While power dissipation per gate is decreased, so also is switching times and output drive current (see Table A1.2).

The Schottky TTL family became prominent in the early 1970s. This circuit uses Schottky diodes to clamp the base-collector junctions of key transistors to prevent saturation of the transistor. Avoiding saturation of switching transistors decreases switching times, resulting in a higher-speed circuit. Switching speeds are decreased by a factor of two or three over conventional TTL in the Schottky circuit. Unfortunately, the power dissipation increases by this same factor.

Low power Schottky reduces the power dissipation of the Schottky circuit by increasing the chip resistors and slightly modifying the basic current configuration. The power dissipation is about five times less than that of conventional TTL while switching times are comparable.

Advanced Schottky and advanced low power Schottky were introduced around 1985. Advanced Schottky gates have lower switching times than does any other family of TTL circuits. The power dissipation per gate is slightly less than conventional TTL

TABLE A1.2   Comparison of Several TTL Logic Families.

Family	Typical Designation	Average Power Dissipation/Gate (mW)	Typical Propagation Delay (ns)
Conventional	7400	10	10 (Note 1)
Low power	74L00	1	33 (Note 2)
Schottky	74S00	20	3 (Note 3)
Low power Schottky	74LS00	2	10 (Note 4)
Advanced Schottky	74AS00	7	1.5 (Note 5)
Advanced low power Schottky	74ALS00	1	4 (Note 5)
Fast	74F00	4	3 (Note 6)

Note 1: 15pF/400 Ω load
Note 2: 15pF/4k Ω load
Note 3: 15pF/280 Ω load
Note 4: 15pF/2k Ω load
Note 5: 50pF/2k Ω load
Note 6: 50pF/500 Ω load

while switching times are decreased by a factor of six to eight. Advanced low power Schottky decreases power dissipation compared to conventional TTL by a factor of ten while decreasing switching times by a factor of two or three.

In addition to a change of input configuration, the advanced TTL families are fabricated with smaller device geometries to minimize parasitic capacitors that limit switching speeds. As fabrication technology improves, leading to comparable prices between TTL families, the lower performance circuits become obsolete. The advanced Schottky families are becoming very prominent in the design of TTL systems. Another TTL family introduced in recent years is the fast series that combines low power dissipation with high switching speed. Table A1.2 summarizes the characteristics of several TTL families.

A1.4.1   AVAILABLE GATES

Although several thousands of gates can be integrated on a chip, SSI circuits contain only a maximum of three or four gates on each chip. This limitation is imposed by the number of pins allowed on the IC package. These pins are metal contacts connected internally to the gates, extending through the package to allow external connections to be made to the gates. SSI chips generally limit the number of pins to 12, 14, or 16. The power supply requires two connections to the chip (+5 V and ground), and each separate gate requires an output and two (or more) input pins. Figure A1.9 shows several typical TTL chips demonstrating each type of gate discussed and an inverter chip. The required power supply for TTL is 5 ± 0.25 V.

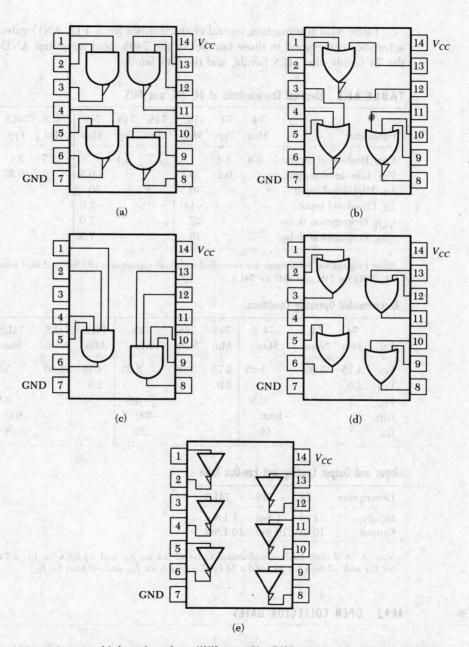

FIGURE AI.9 (a) A quad two-input NAND gate chip (7400). (b) A triple three-input NOR gate chip (7427). (c) A dual four-input AND gate chip (7421). (d) A quad two-input OR gate chip (7432). (e) A hex inverter chip (7404).

Table A1.3 summarizes several characteristics for a TTL AND gate. The characteristics correspond to three families of the 7408 quad two-input AND gate chip: the 74 family, the 74LS family, and the 74S family.

TABLE AI.3 Electrical Characteristics of 74, 74S, and 74LS.

Parameter	74 Min	74 Typ	74 Max	74S Min	74S Typ	74S Max	74LS Min	74LS Typ	74LS Max	Units
V_{OH} High-level output	2.4	3.4		2.7	3.4		2.7	3.4		V
V_{OL} Low-level output		0.2	0.4			0.5		0.35	0.5	V
I_{IH} High-level input			40			50			20	μA
I_{IL} Low-level input			−1.6			−2.0			−0.4	mA
t_{pLH} Propagation delay			27			7.0			15	ns
t_{pHL} Propagation delay			19			7.5			20	ns

Note: Propagation delay times are measured with load capacitance of 15pF and load resistors of 400Ω for 74, 280Ω for 74S, and 2kΩ for 74LS.

Recommended Operating Conditions.

Par.	74 Min	74 Nom	74 Max	74S Min	74S Nom	74S Max	74LS Min	74LS Nom	74LS Max	Units
V_{CC}	4.75	5.00	5.25	4.75	5.00	5.25	4.75	5.00	5.25	V
V_{IH}	2.0			2.0			2.0			V
V_{IL}			0.8			0.8			0.8	V
I_{OH}			−1000			−400			−800	mA
I_{OL}			16			20			8	mA

Input and Output Loading and Fan-Out Table.

Description	74	74S	74LS
Inputs	1 ul	1 Sul	1 LSul
Output	10 ul	10 Sul	10 LSul

Note: A 74 ul (unit load) is understood to be 40 μA for I_{IH} and −1.6mA for I_{IL}, a 74 Sul is 50μA for I_{IH} and −2.0mA for I_{IL}, and a 74 LSul is 20 μA for I_{IH} and −0.4mA for I_{IL}.

AI.4.2 OPEN COLLECTOR GATES

In μP systems, it is not uncommon to drive a single input line with several different gate outputs. Figure A1.10 shows an arrangement with three NAND gates driving line *A* and three NAND gates driving line *B*. With normal TTL gates, this arrangement is inappropriate and generally results in the destruction of some of

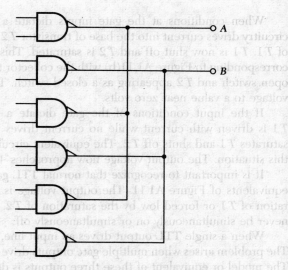

FIGURE A1.10 Multiple gates driving lines *A* and *B*.

the connected gates. While it is not the intent of this text to discuss the circuit configuration of gates, it is important to understand the source of this problem.

Figure A1.11 shows the major features of a typical TTL gate output circuit along with the equivalent circuits for each possible output state. The input circuitry of the gate performs the designated gate function and creates the drive current for the two output transistors. This arrangement of $T1$ and $T2$ is called a totem-pole output and is used in most TTL gates with slight modifications.

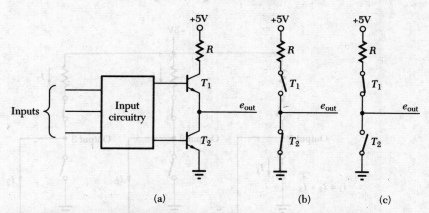

FIGURE A1.11 Output circuits for TTL gates: (a) actual output circuit for low output, (b) equivalent circuit for low output, and (c) equivalent circuit for high output.

When conditions at the gate inputs dictate a low voltage output, the input circuitry drives current into the base of transistor $T2$ and cuts off current to the base of $T1$. $T1$ is now shut off and $T2$ is saturated. This results in an equivalent circuit corresponding to Figure A1.11(b) with the collector to emitter of $T1$ appearing as an open switch and $T2$ appearing as a closed switch. This condition forces the output voltage to a value near zero volts.

If the input conditions of the gate dictate a high level output, the base of $T1$ is driven with current while no current drives the base of $T2$. This condition saturates $T1$ and shuts off $T2$. The equivalent circuit of Figure A1.11(c) represents this situation. The output voltage now approaches +5 V.

It is important to recognize that normal TTL gates allow only the two possible equivalents of Figure A1.11. The output voltage is either forced high by the saturation of $T1$ or forced low by the saturation of $T2$. With this gate, $T1$ and $T2$ can never be simultaneously on or simultaneously off.

When a single TTL output drives an input line, the gate functions as intended. The problem arises when multiple gate outputs drive a line as shown in Figure A1.12. The model or equivalent of these three outputs is depicted in Figure A1.13. In this

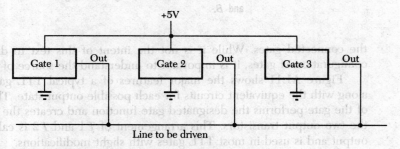

FIGURE A1.12 Three TTL gates driving the same line.

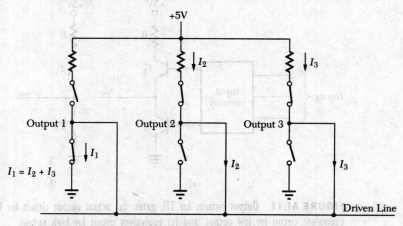

FIGURE A1.13 Model of three TTL gate outputs driving the same line.

figure, it is assumed that gate 1 is attempting to drive the line to ground level while gates 2 and 3 are attempting to pull the line toward +5 V. Since the TTL load resistors are relatively small, rather high currents flow through the upper transistors of outputs 2 and 3. These currents sum together to form the current I_1 if negligible current is drawn by the line to be driven. The current I_1 flows through the lower transistor of output 1. This current is often high enough to cause destruction of the conducting transistor of output 1, especially if more than three circuits are connected to the same line. Typical μP systems may require 20 circuits to connect to a given bus line. Thus, normal TTL gate outputs cannot be used to drive these lines.

The *open collector gate* is used to solve this problem in a small number of applications. In this gate, transistor $T1$ is removed from the normal totem-pole output configuration. If open collector gates are used to implement the system of Figure A1.12, the equivalent circuit of the outputs appears as in Figure A1.14. With $T1$ missing from the circuit, the open collector gate can only pull the driven line to ground level through $T2$. If all output gates have $T2$ shut off, which is equivalent to open switches connected to the outputs, the line is not forced high. In order to solve this problem, a single pull-up resistor is used to cause the voltage to rise when all output transistors are off. When a single output stage is turned on, as in Figure A1.14, the current through this output is limited by the value of the external pull-up resistor. A single pull-up resistor is required for each set of common output lines as shown in Figure A1.15.

The AND symbols overlaying the connection to the pull-up resistors do not represent the physical presence of gates. It is conventional to include this symbol to emphasize that only if all gates are driven to an output high condition will the circuit output go high. If a single gate is driven to an output low condition, the circuit output goes low regardless of other gate conditions. This behavior suggests an AND-type relationship and, hence, the overlaid symbol.

When a gate output is driven low, the total current absorbed by this output is supplied by the resistor rather than from the other gates. This can easily be limited to a safe value by proper selection of the resistance value.

There are upper and lower limits on the value of R. We will assume that point A in Figure A1.15 is connected to a gate input with $I_{IHmax} = 50\ \mu A$, $I_{ILmax} = -2.0$ mA, $V_{IHmin} = 2.7$ V, and $V_{ILmax} = 0.5$ V and that I_{OL} for each open collector gate

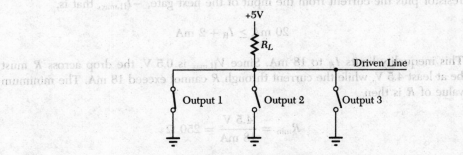

FIGURE AI.14 Open-collector outputs.

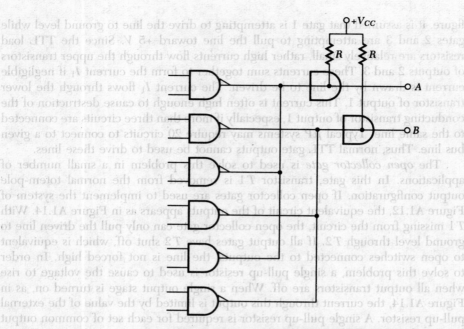

is 20 mA and $I_{OH} = 250$ μA. When the three open collector gates are at the high
level, a maximum of 250 μA leakage current flows into these gates. The resistor
must supply 50 μA to the input of the driven gate plus 750 μA leakage current to
the open collector outputs and the voltage must exceed 2.7 V. Since $V_{CC} = 5$ V, a
2.3 V drop across R is the maximum allowable drop. The upper limit of R is found
to be

$$R_{max} = \frac{2.3 \text{ V}}{800 \text{ }\mu\text{A}} = 2.9 \text{ k}\Omega$$

When one gate output is driven low, it must absorb or sink the current through the
resistor plus the current from the input of the next gate, $-I_{ILmax}$, that is,

$$20 \text{ mA} \geq I_R + 2 \text{ mA}$$

This inequality limits I_R to 18 mA. Since V_{ILmax} is 0.5 V, the drop across R must
be at least 4.5 V, while the current through R cannot exceed 18 mA. The minimum
value of R is then

$$R_{min} = \frac{4.5 \text{ V}}{18 \text{ mA}} = 250 \text{ }\Omega$$

A typical resistance value might be 1 kΩ.

The general equations for R_{max} and R_{min} are given by

$$R_{max} = \frac{V_{CC} - V_{IHmin}}{N_1 I_{OH} + N_2 I_{IH}}$$

and

$$R_{min} = \frac{V_{CC} - V_{ILmax}}{I_{OL} + N_2 I_{IL}}$$

where N_1 is the number of open collector gates and N_2 is the number of gate inputs driven.

The 7403 is an example of a NAND gate with open collectors. This chip includes four 2-input gates.

AI.4.3 THREE-STATE OUTPUTS

Another circuit designed to allow several gate outputs to drive a single input is the *three-state device*. Three-state output circuits are often used with μP circuits and have become very popular as this device has grown in importance. In three-state circuits, an additional input, called the enable, is provided on the chip. The symbol for a three-state buffer is shown in Figure A1.16(a).

When the enable input is asserted low, the buffer functions as a normal TTL circuit. In this case, the output voltage level follows that of the input. When the enable input is deasserted by a high voltage, the output is in a high-impedance state.

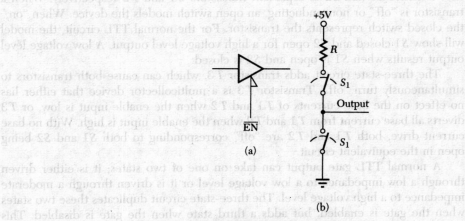

FIGURE AI.16 (a) A three-state buffer symbol. (b) An equivalent circuit for the three-state output.

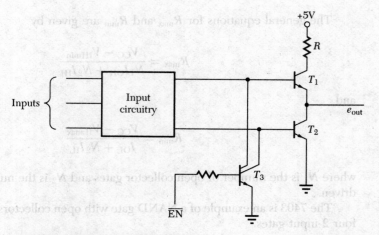

FIGURE A1.17   A three-state output gate.

In this state, the buffer circuit is connected through a high impedance to the output terminal of the chip. This high impedance may exceed 20 MΩ and from a practical standpoint appears as an open circuit. In effect, the buffer output is disconnected from the output pin of the chip. This pin is not influenced by the input voltage level when the circuit is disabled.

Figure A1.17 shows the mechanism used to create the third (high impedance) state. Transistors $T1$ and $T2$ are present in a normal TTL output circuit. These transistors are always driven to opposite conditions: when $T1$ is "on," $T2$ is "off," and when $T1$ is "off," $T2$ is "on." The currents required to cause these conditions are generated in TTL chips by the input circuit. The equivalent output circuit uses switches $S1$ and $S2$ to represent transistors $T1$ and $T2$ respectively. When a transistor is "off" or nonconducting, an open switch models the device. When "on," the closed switch represents the transistor. For the normal TTL circuit, the model will show $S1$ closed and $S2$ open for a high voltage level output. A low voltage level output results when $S1$ is open and $S2$ is closed.

The three-state circuit adds transistor $T3$, which can cause both transistors to simultaneously turn "off." Transistor $T3$ is a multicollector device that either has no effect on the base currents of $T1$ and $T2$ when the enable input is low, or $T3$ diverts all base current from $T1$ and $T2$ when the enable input is high. With no base current drive, both $T1$ and $T2$ are "off" corresponding to both $S1$ and $S2$ being open in the equivalent circuit.

A normal TTL gate output can take on one of two states; it is either driven through a low impedance to a low voltage level or it is driven through a moderate impedance to a high voltage level. The three-state circuit duplicates these two states when the gate is enabled, but adds a third state when the gate is disabled. This third state effectively presents very high impedances from +5 V to the output line and from ground to the output line. The transistors $T1$ and $T2$ are "off," leading

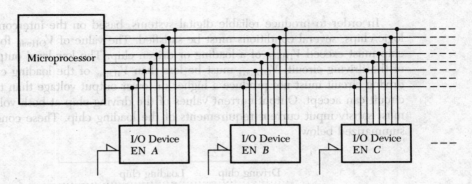

FIGURE AI.18 Microprocessor data bus organization.

to impedances from collector to emitter in the MΩ range. For most applications, the open switches of the equivalent circuits, which imply infinite impedance, can be used with sufficient accuracy. The output line is no longer driven by low impedance devices and this line assumes a voltage determined by other lower impedance circuits connected to this point.

When three-state devices drive the same line, a maximum of one device can be enabled at any given time. This device then drives the line high or low and must be disabled before another gate is enabled. Although this method solves the problem of excessive gate current, an additional input line is required to enable each circuit that drives the line.

Three-state outputs are used for certain latches, memory circuits, buffers, counters, registers, and other circuits that require a third high-impedance state at the outputs. The 74LS244 is an example of an octal, three-state buffer that is useful in many μP applications. This enable/disable feature is available in several MOS logic systems and is often used in connection with μP input/output (I/O) buses.

Figure A1.18 shows a μP with several I/O devices connected to a bidirectional data bus. When data presented by a given device is to be read into the μP, the enable line is taken low so that the device can drive the microprocessor bus. At this time all other devices must be disabled, presenting high impedances to each line of the bus. With this configuration a single set of bus lines can serve a great number of devices that could include memories, input and output registers, and other digital devices.

AI.4.4 LOGIC COMPATIBILITY

When one logic circuit must drive another, certain conditions must be satisfied to guarantee reliable operation. We would not expect proper operation if one gate of the system produces $V_{OH} = 3.2$ V and drives a circuit having $V_{IHmin} = 4.6$ V. The high-level output of the gate is too low to be interpreted by the following circuit as a high level.

In order to produce reliable digital systems, based on the interconnection of logic chips, several conditions must be satisfied. The value of V_{OHmin} for a driving chip must exceed V_{IHmin} of a loading or driven chip. The low-level output voltage of the driving circuit, V_{OLmax}, must be less than V_{ILmax} of the loading circuit. The driving circuit must not produce a higher or lower output voltage than the loading circuit can accept. Output current values of the driving chip at both voltage levels must satisfy input current requirements of the loading chip. These conditions are summarized below.

Driving chip		Loading chip
V_{OHmin}	>	V_{IHmin}
V_{OLmax}	<	V_{ILmax}
V_{OHmax}	<	Maximum input voltage
V_{OLmin}	>	Minimum input voltage
$-I_{OHmax}$	>	I_{IHmax}
I_{OLmax}	>	$-I_{ILmax}$

I_{OHmax} and I_{ILmax} will have negative values, since they flow out of the corresponding terminals. If these conditions are satisfied, the driving chip and the loading chip are said to be compatible. All chips within a logic family are designed to be compatible with other chips of the same family. In fact, most families are designed to allow one circuit to drive several loading circuits.

Most TTL families are basically compatible with one another, but mixing families should be done with care, recognizing that overall circuit performance may be degraded over that of a single-family system. Mixing conventional TTL with ALS leads to slower switching speeds than an ALS system. If noise induced by external sources is a problem, the higher impedance LS family may maximize the noise effects while the AS family would minimize such effects.

If it is necessary to connect one type of logic to another, additional interfacing circuitry may be required. If voltage conditions are satisfied, but current conditions are not, current amplifier or buffer stages must be used. If current requirements are met, but voltage requirements are not, a level shifter with a required current gain near unity must be used. When all conditions are met, no additional circuitry is required. Of course, if the driving circuitry is loaded by several circuits, the source and sink current capability of the driving gate must exceed the sum of the loading current requirements.

Some CMOS families are designed to be compatible with certain TTL families. Figure A1.19 shows the compatibility of the CMOS series 74C with TTL series 74L. Although there are newer families in both TTL and CMOS circuits, these will be used to demonstrate the principle of compatibility. The noise margin for the situation of a 74C gate driving a 74L gate is found to be $0.7 - 0.4 = 0.3$ V for the low level and $2.4 - 2 = 0.4$ V for the high level.

FIGURE AI.19 Demonstration of 74C compatibility with TTL 74L.

When TTL series 74L becomes the driving circuit and CMOS 74C becomes the loading circuit, voltage compatibility does not quite exist: $V_{OHmin} = 2.4$ V, but $V_{IHmin} = 3.25$ V; and $V_{OLmax} = 0.4$ V, while $V_{ILmax} = 0.8$ V. The high-level output voltage of the 74L will not meet the required input value of the 74C. Level shift circuits are required to remedy this situation.

Recent advances in CMOS technology have led to much higher switching speeds. Newer CMOS families are fully compatible with TTL families and have comparable switching speeds. The 74HC and 74HCT CMOS families have propagation delays that are comparable to or better than low power Schottky TTL. Circuits of the 74HCT family are designed to be direct replacements for 74LS TTL circuits. The average power dissipation per CMOS gate is typically 20 to 50 times less than that for the 74LS TTL gate. The output current and fan-out capability of the CMOS circuit, in terms of 74LS TTL inputs, is a factor of 2 less than that for the 74LS TTL circuit.

CMOS 74C $V_{CC} = 5$ V	$\xrightarrow{\ V_{Voltage} = +2.4\ V\ }$ $I_{OH(max)} = -100\ \mu A$ $I_{IH(max)} = -360\ \mu A$	TTL 74L $V_{CC} = 5$ V

TTL 74L $V_{CC} = 5$ V	$\xrightarrow{\ V_{Voltage} = 0.4\ V\ }$ $I_{IL(max)} = 360\ \mu A$ $I_{OL(max)} = 180\ \mu A$	CMOS 74C $V_{CC} = 5$ V

FIGURE A1.19 Demonstration of 74C compatibility with TTL 74L.

When TTL series 74L becomes the driving circuit and CMOS 74C becomes the loading circuit, voltage compatibility does not quite exist. $V_{OH(min)} = 2.4$ V but $V_{IH(min)} = 3.25$ V, and $V_{OL(max)} = 0.4$ V, while $V_{IL(max)} = 0.8$ V. The high level output voltage of the 74L will not meet the required input value of the 74C. Level shift circuits are required to remedy this situation.

Recent advances in CMOS technology have led to much higher switching speeds. Newer CMOS families are fully compatible with TTL families and have comparable switching speeds. The 74HC and 74HCT CMOS families have propagation delays that are comparable to or better than low power Schottky TTL. Circuits of the 74HCT family are designed to be direct replacements for 74LS TTL circuits. The average power dissipation per CMOS gate is typically 20 to 50 times less than that for the 74LS TTL gate. The output current and fan-out capability of the CMOS circuit, in terms of 74LS TTL inputs, is a factor of 2 less than that for the 74LS TTL circuit.

Pulse-Generating Circuits

A2.1

THE MONOSTABLE MULTIVIBRATOR

The monostable multivibrator or one-shot produces an output pulse of precise width, initiated by an input trigger signal. The width of the pulse is determined by a resistor value and a capacitor value; thus, this width can be selected to be almost any desired value.

In the last fifteen to twenty years, very few IC chips have been as popular as the 555 timer. This chip is used in a variety of timing applications. One important use of the 555 is that of a monostable multivibrator; another is that of an astable multivibrator. This section will consider the 555 monostable circuit before proceeding to other chip implementations of the one-shot.

A2.1.1   THE 555 ONE-SHOT

A block diagram of the 555 is shown in Figure A2.1. The three equal values of resistance establish the reference voltages of $V_{CC}/3$ and $2V_{CC}/3$. The trigger comparator

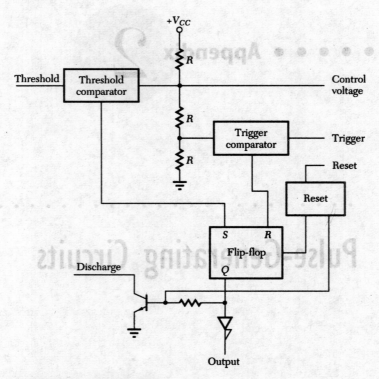

compares the voltage level of the trigger input to the reference voltage $V_{CC}/3$. As long as the input voltage is greater than $V_{CC}/3$, the comparator output is not asserted. When the input signal drops below $V_{CC}/3$, the comparator output is asserted and drives the flip-flop to the $Q = 0$ state.

If the trigger input voltage exceeds $V_{CC}/3$ and the threshold input voltage is below $2V_{CC}/3$, neither comparator is asserted. When the threshold input voltage increases above $2V_{CC}/3$, this comparator output is asserted and sets the flip-flop to the $Q = 1$ state. If both comparators are asserted simultaneously, requiring that the trigger voltage is less than $V_{CC}/3$ and the threshold voltage is greater than $2V_{CC}/3$, the reset input overrides the set and forces the flip-flop to the $Q = 0$ state. This is called a reset-overrides-set flip-flop for obvious reasons.

The output buffer inverts the flip-flop output to cause a high-level output when the flip-flop is reset and a low output when the flip-flop is set. When the flip-flop is set, it provides base drive current to the discharge transistor allowing this transistor to saturate when the collector circuit is completed. We note that the flip-flop can also be set by applying a low voltage to the reset input. This action saturates the discharge transistor and causes the output to be at the low level.

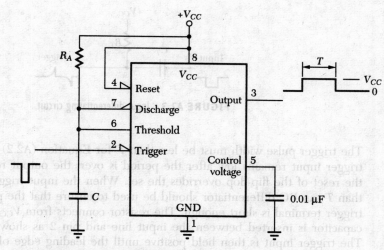

FIGURE A2.2 The 555 timer used as a one-shot.

The circuit of Figure A2.2 shows the connections required to convert the 555 timer into a one-shot. The 0.01 μF capacitor is connected to the control voltage input to hold this voltage constant even if a transient voltage appears on the power supply. The normal state of the flip-flop is the set state that saturates the discharge transistor. This places a short circuit across the capacitor holding the capacitor voltage to 0 V. Since the threshold voltage is tied to this point, the threshold comparator is not asserted. The trigger input is held at a high level until the output pulse is to be initiated.

When the trigger input drops below $V_{CC}/3$, the flip-flop changes state. The output is now at the high level and the discharge transistor is turned off, becoming an open circuit. The capacitor can now charge through R_A toward the voltage V_{CC}. As the capacitor charges, the threshold voltage reaches a value of $2V_{CC}/3$ causing the flip-flop to return to the set state. This ends the period, dropping the output to 0 V, and again saturates the output transistor. The period T is the time taken for the capacitor to charge from 0 V to a voltage of $2V_{CC}/3$ with a target voltage of V_{CC}. This time can be calculated from the general equation for a charging capacitor.

$$v_c = v_i + (v_t - v_i)\left[1 - e^{-t/t_c}\right] \tag{A2.1}$$

Here v_i is the initial voltage, v_t is the target voltage, and t_c is the time constant. For the case under consideration $v_i = 0$, $v_t = V_{CC}$, and $t_c = R_A C$. The period T is then found by solving Equation (A2.1) with these values, giving

$$T = R_A C \ \ln 3 = 1.1 R_A C \tag{A2.2}$$

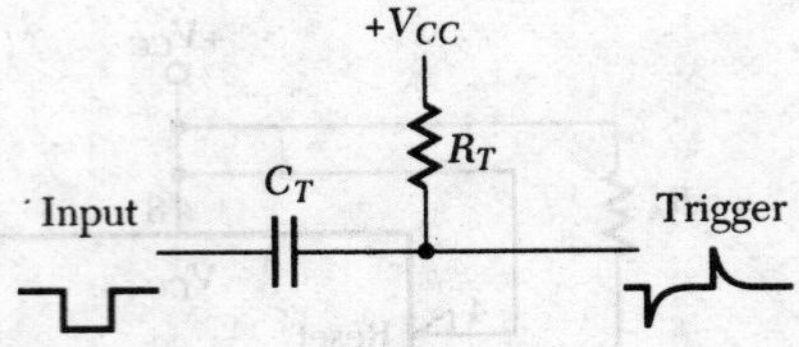

FIGURE A2.3   Input differentiating circuit.

The trigger pulse width must be less than T for Equation (A2.2) to be valid. If the trigger input remains low after the period is over, the output remains high since the reset of the flip-flop overrides the set. When the input trigger pulse is longer than T, an RC differentiator should be used to ensure that the pulse reaching the trigger terminal is short enough. The resistor connects from V_{CC} to pin 2, and the capacitor is inserted between the input line and pin 2 as shown in Figure A2.3. The trigger input is then held positive until the leading edge of the trigger pulse arrives. This transition is coupled through the capacitor to initiate the period. Pin 2 then charges toward V_{CC} with a time constant largely determined by the $R_T C_T$ product of the trigger circuit since the input impedance to the trigger comparator is very high. This $R_T C_T$ product should be considerably smaller than that of the one-shot period.

The period T can be varied from a minimum of approximately 10 μs to a maximum of several hours by varying the elements of R_A and C. Values of $R_A = 10$ kΩ and $C = 0.001$ μF lead to $T = 11$ μs, whereas $R_A = 10$ MΩ and $C = 100$ μF result in $T = 1100$ s. The 555 timer is popular for lower frequency applications, but it cannot be used when smaller rise and fall times are required.

The 555 timer is designed to operate with a power supply voltage ranging from 5 V to 18 V. With $V_{CC} = 5$ V, the circuit is compatible with TTL logic. The magnitude of the output voltage varies directly with the power supply, allowing the 555 timer to be used in a variety of applications.

A2.1.2   TTL ONE-SHOTS

All one-shot circuits can be classified as retriggerable or nonretriggerable. The nonretriggerable type can be triggered only once during the period T. Any subsequent trigger pulses are ignored until the period has ended. The period T is not affected by the input signals that occur after the period is initiated by a single pulse. The 555 is nonretriggerable. A retriggerable one-shot can be triggered several times during the period. Each time a trigger pulse occurs, the period is adjusted as though it had started at the time of the last input pulse. If the period is T, the output signal will last T seconds after the final trigger pulse. Figure A2.4 shows the difference between the retriggerable and nonretriggerable circuit.

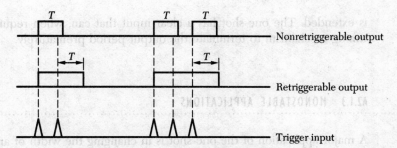

FIGURE A2.4 Outputs of retriggerable and nonretriggerable one-shots.

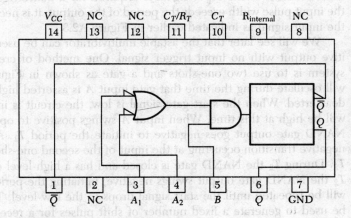

FIGURE A2.5 The 74121 one-shot.

The 74121 of Figure A2.5 is an example of a TTL, nonretriggerable one-shot whose period can be varied from 30 ns to 40 s. If the R_{INT} pin is connected to V_{CC} and the C_T and C_T/R_T pins are left open, period T is typically 30 ns. If the R_{INT} pin is left open and a capacitor is connected from the C_T to the C_T/R_T pin (positive to C_T/R_T) and a resistor R_T is connected from the C_T/R_T pin to V_{CC}, the one-shot will have a period given by

$$T = C_T R_T \ \ln 2 = 0.69 \ C_T R_T \tag{A2.3}$$

The circuit is triggered by holding either A_1 or A_2 low and taking B from a low to high level. For triggering on a negative transition, A_1, A_2, and B are all high; then either A_1 or A_2 is taken to a low level to initiate the period. The input pulse width can be longer or shorter than period T with no effect on the period duration.

The 74123 is an example of a retriggerable one-shot that is similar to the 74121 except that if a second trigger pulse occurs before the period ends, the output pulse is extended. The one-shot has a clear input that can, when required, override the timing mechanism to terminate the output period prematurely.

A2.1.3 MONOSTABLE APPLICATIONS

A major application of the one-shot is in changing the width of an input pulse to a different, but fixed value. For example, a short 1 μs pulse may be used to initiate an 8ms output pulse. If the output period is longer than the input pulse width, no differentiation of the input signal is required for a nonretriggerable circuit. When the input pulse width exceeds the period of the output, it is necessary to differentiate the input signal as indicated earlier in Figure A2.3.

We will see later that the astable multivibrator can be used to generate a repetitive output with no input trigger signal. One method of creating a gated astable system is to use two one-shots and a gate as shown in Figure A2.6. This circuit will oscillate during the time that gate input A is asserted high and stops after A is deasserted. When the start gate signal is low, the circuit is inactive. Gate input B will be high at this time. When input A swings positive to open the start gate, the NAND gate output goes negative to initiate the period T_1. At the end of T_1, the negative transition occurring at the input of the second one-shot initiates the period T_2. During T_2 the NAND gate is closed and has a high-level output. At the end of T_2 the NAND gate output swings negative, initiating the period T_1. This sequence will be repeated until the start signal drops to the low level. This gated circuit can be used to generate a fixed number of shift pulses for a receiving, or transmitting shift register.

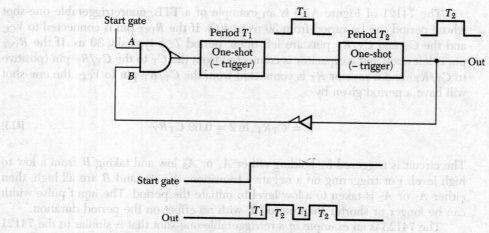

FIGURE A2.6 Generation of repetitive waveform with one-shots.

A2.1.4 RELIABILITY CONSIDERATIONS OF THE ONE-SHOT

Some digital systems must be completely reliable in order to be useful, while others may allow a small percentage of errors to occur and still not detract from the effectiveness of the system. The control unit of a computer must be highly accurate while a multiplexed LED display can occasionally ignore a current pulse to a display with no serious effect. Most control systems must be highly accurate even in the presence of extraneous noise signals that are often present in electronic circuits.

The one-shot circuit is generally avoided in high performance systems. This device does not function well in a noisy environment and also introduces timing problems in systems depending on precise time periods. The edge-triggered flip-flop can be considerably more reliable than the one-shot and is used in the popular state machine introduced in Chapter 5. The state machine is used almost exclusively in applications requiring high reliability.

Noisy Data Lines

The generation of gated clock signals that can be applied to a receiving shift register is one possible application of the one-shot circuit. The incoming data line can be applied to the input of the shift register and also to a gated clock circuit. The edge of the first bit of the data (start bit) triggers the gated clock, which then applies shift pulses to the shift register. These pulses are thus synchronized to the incoming data and shift this information into the register.

If the gated clock is based on a one-shot circuit and the incoming data line is applied to the trigger input, the one-shot can be triggered by a noise pulse on the line. The clock generation circuit treats this noise pulse as a start bit and shifts an erroneous word into the receiving register. Any noise pulse with an amplitude great enough to trigger the one-shot leads to the reception of a false word.

Synchronization of Incoming Bits

In addition to the problem of poor noise immunity, a second problem with a one-shot circuit is that asynchronous input data cause asynchronous outputs. If there is a clock circuit in the receiver, the incoming data will have no fixed time relationship to the clock transitions. If these data are used to trigger a one-shot, the output period is initiated with no fixed relationship to the clock signal. If synchronization is important, the one-shot circuit cannot be used. We learned earlier that an asynchronous input can be synchronized with a clocked flip-flop when synchronization is important.

The third problem is that the period of the one-shot is determined by imprecise elements that vary with temperature. Even if the trigger input is synchronized to the clock signal, the trailing edge of the one-shot output will not be synchronized. If this is an important consideration, a flip-flop circuit must again be used.

Although the last section suggests several applications of the one-shot, care must be taken to determine if the poor reliability can be tolerated in each specific application under consideration.

A2.2

THE ASTABLE MULTIVIBRATOR

The 555 timer can also be used for an astable with the connections shown in Figure A2.7. The capacitor voltage can swing between $V_{CC}/3$ and $2V_{CC}/3$. As the capacitor voltage rises toward $2V_{CC}/3$, it does so with a target voltage of V_{CC} and a time constant of $C(R_A + R_B)$. When it reaches $2V_{CC}/3$, the threshold comparator causes the flip-flop to change state (see Figure A2.1). The discharge transistor saturates, and the capacitor voltage heads toward ground with a time constant of CR_B. When the voltage drops to $V_{CC}/3$, the trigger comparator changes the state of the flip-flop and shuts off the discharge transistor. The capacitor again charges toward V_{CC}. The output and capacitor voltage waveforms of the 555 are shown in Figure A2.8.

The duration of the positive portion of the waveform is called t_1 and is calculated from Equation (A2.1) with $v_i = V_{CC}/3$, $v_t = V_{CC}$, and $t_c = (R_A + R_B)C$. This gives

$$v_c = \frac{2V_{CC}}{3} = \frac{V_{CC}}{3} + \left[V_{CC} - \frac{V_{CC}}{3}\right]\left[1 - e^{-t_1/t_c}\right]$$

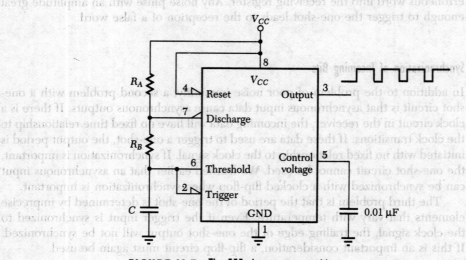

FIGURE A2.7 The 555 timer as an astable.

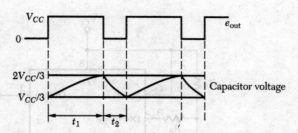

FIGURE A2.8 Timing waveforms.

which can be solved for t_1 to yield

$$t_1 = C(R_A + R_B) \ln 2 = 0.69\, C(R_A + R_B) \qquad \text{(A2.4)}$$

Likewise, the time t_2 can be found to be

$$t_2 = 0.69\, C R_B \qquad \text{(A2.5)}$$

The duty cycle of the astable is defined as the ratio of the duration of the positive portion of the period to the total period, or

$$\text{Duty cycle} = \frac{t_1}{t_1 + t_2} = R_A + \frac{R_B}{R_A + 2R_B}$$

If the value of R_B becomes much greater than R_A, the duty cycle will approach a minimum value of 50%.

Some manufacturers give an alternative definition of duty cycle as the ratio of the time the output transistor is on (low voltage) to the total period. This definition leads to

$$\text{ON duty cycle} = \frac{t_2}{t_1 + t_2} = \frac{R_B}{R_A + 2R_B}$$

Another configuration used to obtain a 50% or even smaller duty cycle, with a smaller spread of timing resistor values, is shown in Figure A2.9. The value of t_1 is found by noting that the capacitor charges through R_A only, rather than $R_A + R_B$. This value is then

$$t_1 = 0.69\, C R_A$$

During t_2, the capacitor discharges from $2V_{CC}/3$ to $V_{CC}/3$. The target voltage and discharge resistance can be calculated from the circuit of Figure A2.10. (Note that the discharge transistor is saturated.)

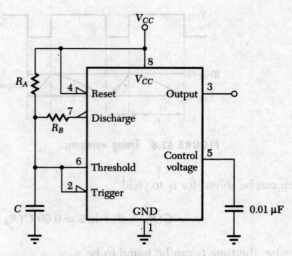

FIGURE A2.9   Astable capable of 50% duty cycle.

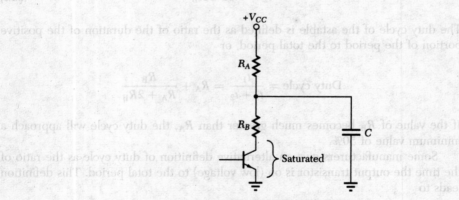

FIGURE A2.10   Equivalent circuit during t_2.

The Thevenin equivalent voltage, which is also the target voltage, is

$$v_1 = \frac{R_B V_{CC}}{R_A + R_B}$$

and the discharge resistance is

$$R_{th} = \frac{R_A R_B}{R_A + R_B}$$

Using Equation (A2.1) with $v_i = 2V_{CC}/3$, $v_t = V_{CC}R_B/(R_A + R_B)$, $t_c = R_{th}C$, and solving for the time for the capacitor voltage to reach $V_{CC}/3$ results in

$$t_2 = R_{th}C \ \ln \frac{2R_A - R_B}{R_A - 2R_B} \tag{A2.6}$$

The resistor R_B can be adjusted to result in a t_2 that equals t_1. In using Equation (A2.6), we must limit the value of R_B to lead to a target voltage of less than $V_{CC}/3$ or the circuit will not oscillate. This imposes an upper limit on R_B of $R_A/2$.

Using Equation (A2.1) with $v_i = 2V_{CC}/3$, $v_f = V_{CC}$, $R_A + R_B$, $t = R_B C$, and solving for the time for the capacitor voltage to reach $V_{CC}/2$ results in

$$t_a = R_B C \ln \frac{2R_A - R_B}{R_A - 2R_B} \qquad (A2.4)$$

The resistor R_B can be adjusted to result in a t_a that equals t_b. In using Equation (A2.5), we must limit the value of R_B to lead to a finite voltage or less than $V_{CC}/3$ or the circuit will not oscillate. This imposes an upper limit on R_B of $R_A/2$.

Answers to Drill Problems

Sec. 1.2

1. 1010101011_2, 1253_8, $2AB_{16}$, 001 010 101 011 (BCO), 0010 1010 1011 (BCH).

2. 1001111.001001_2.

3. 101000101111_2, 5057_8, 2607_{10}.

4. 111001010_2, 0100 0101 1000 (BCD).

5. 86_{10}, 1010110_2.

6. 10101.0101110_2.

Sec. 1.4

1. The function table plots inputs and outputs in terms of high and low voltage levels. The truth table plots this information in terms of 1s and 0s.

2.

A_{in}	B_{in}	Out
0	1	0
1	0	1

3.

A	B	NOR	Out
0	0	1	0
0	1	0	1
1	0	0	1
1	1	0	1

The circuit performs the OR function.

● **Sec. 1.5**

1. $V_{OHmin} > V_{IHmin}$, $V_{OLmax} < V_{ILmax}$.

2. $I_{OH} = -200\,\mu A$, $I_{OL} = 9.6$ mA.

3. 0.2V

4. 24ns. Yes.

● **Secs. 2.1.1-2.1.3**

3. $X = \bar{A} + \bar{B} + C$

4. $F = ABC$

5. $AC + \bar{A}C = (A + \bar{A})C = 1C = C$

6. $AB + A\bar{B} + \bar{A}C = A(B + \bar{B}) + \bar{A}C = A + \bar{A}C = A + C$

● **Sec. 2.1.4**

1. a. $F = \bar{A}\bar{C} + \bar{A}\bar{D} + \bar{B}\bar{C} + \bar{B}\bar{D}$
   b. $F = AB\bar{C} + AB\bar{D}$
   c. $F = ABCD$
   d. $F = ACD + BCD$

2. $X = 0$

● **Sec. 2.1.6**

1. Change AND gate to NOR gate driven by A and B. Change NAND gate to OR gate driven by low asserted inputs. This gives

$$X = A + B + C + D$$

2. Change the NOR gate to an AND gate with low asserted inputs. This gives

$$X = \bar{A}\bar{B}\bar{C}\bar{D}$$

● **Sec. 2.2.1**

1. $X = AB + \bar{B}\bar{C}$
2. $X = ABC + AB\bar{C} + A\bar{B}\bar{C} + \bar{A}\bar{B}\bar{C}$
3. $F = \bar{A}\bar{B}\bar{C} + \bar{A}BC + A\bar{B}\bar{C} + AB\bar{C}$

● **Sec. 2.2.2**

1.

$\overset{\displaystyle A}{B}$	0	1
0	0	1
1	0	1

X

2.

$\overset{\displaystyle A}{B}$	0	1
0	1	1
1	0	1

Y

3.

$\overset{\displaystyle A}{B}$	0	1
0	1	0
1	1	1

Z

● **Sec. 2.2.3**

1. $X = AB + \bar{B}\bar{C}$
2. $X = ABC + \bar{B}\bar{C}D$
3. $Z = \bar{B}CD + BC\bar{D} + AB\bar{C}D$

● Sec. 2.2.4

1. $X = A\bar{C}D + AB\bar{D} + \bar{A}BD + \bar{A}\bar{C}\bar{D}$
2. $F = \bar{C}D + BC$
3. $\bar{F} = \bar{A}\bar{B}D + \bar{A}C\bar{D} + A\bar{B}\bar{D} + ACD$
4. $F = \bar{C}D + BC$

● Sec. 2.2.5

1. Hazard as A goes from 0 to 1

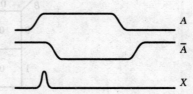

● Sec. 2.3.1-2.3.2

1. $X = \bar{A}B + A\bar{C}$
2. $Y = A\bar{B} + \bar{B}\bar{D}$
3. $F = \bar{C}D + BC$

● Sec. 2.3.3

1.

C	AB 00	01	11	10
0	$\bar{D}\bar{E} + \bar{D}E + DE$	0	DE	$DE + \bar{D}\bar{E}$
1	$\bar{D}\bar{E}$	DE	$D\bar{E}$	$\bar{D}E$

F

2.

B	A 0	1
0	$\bar{C}D$	$C\bar{D} + CD$
1	CD	1

$X = \bar{A}\bar{B}\bar{C}D + AC + AB + BCD$

● Sec. 2.4.2

1. $X = \bar{A}\bar{B} + AB$. Two AND gates plus an OR or one NEXOR gate.

2. $X = \bar{A}\bar{B} + AB + C$. Two AND gates plus an OR or 1 NEXOR plus an OR.

● Sec. 3.1.1

1. Connect: A to S_2, B to S_1, C to S_0; logic 0 to lines 3, 5, 6; logic 1 to lines 0, 1, 2, 4, 7.

2. Connect: A to S_1, B to S_0; logic 1 to line 0, $\bar{C}$ to lines 1 and 2, C to line 3.

3.

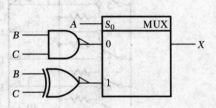

● Sec. 3.1.2

1. Inputs: Connect A to line 2, B to line 1, C to line 0. Outputs: Connect to inputs of NAND gate lines 0, 1, 2, 4, 7.

2. Inputs: Connect A to line 2, B to line 1, C to line 0. Outputs: Connect to inputs of NAND gate 1 lines 0, 1, 2, 3, 5, 6. Connect to inputs of NAND gate 2 lines 1, 2, 4. Connect to inputs of NAND gate 3 lines 0, 7.

3.

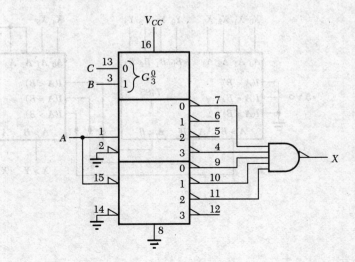

Sec. 3.2.1

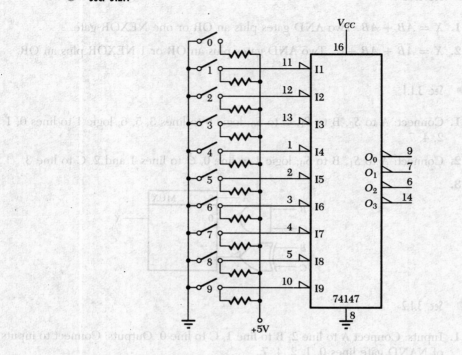

Sec. 3.2.2

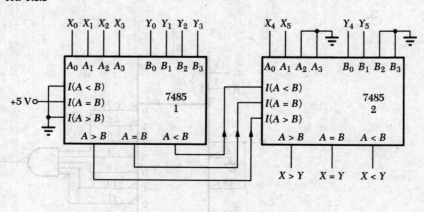

● Sec. 3.2.3

The equations for the segments are

$$a = A\bar{C} + \bar{A}\bar{B}C$$
$$b = A(B \text{ EXOR } C)$$
$$c = \bar{A}B\bar{C}$$
$$d = A\bar{B}\bar{C} + ABC + \bar{A}\bar{B}C$$
$$e = C + A\bar{B}$$
$$f = \bar{A}B + \bar{A}C + BC$$
$$g = \bar{A}\bar{B} + ABC$$

● Sec. 3.2.4

1.

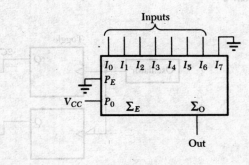

2.

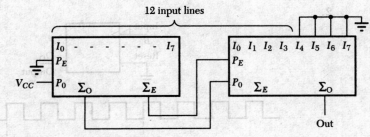

● Sec. 4.1.1

1.

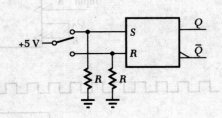

2. a. $Q = 0$
b. $Q = 1$
c. $Q = 0$

● **Sec. 4.1.2**

1. 0, 0, 1, 1, 1, 1.

● **Sec. 4.2.1-2**

1.

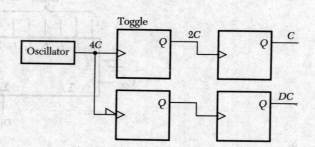

2.

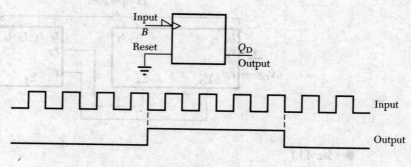

3.

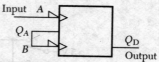

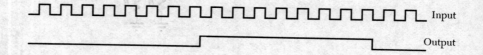

● Sec. 4.2.3

1.

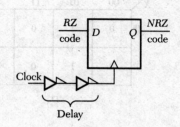

● Sec. 5.3.1

1. $D(t_0) = 1$, $D(t_1) = 1$, $D(t_2) = 0$, $D(t_3) = 1$.

2. $J(t_0) = \Theta$, $K(t_0) = 0$, $J(t_1) = \Theta$, $K(t_1) = 0$, $J(t_2) = \Theta$, $K(t_2) = 1$, $J(t_3) = 1$, $K(t_3) = \Theta$.

3.

Q_n	Q_{n+1}	T
0	0	0
0	1	1
1	0	1
1	1	0

● Sec. 5.3.2

1. This design is not possible due to conflicting requirements during different time slots.

● Sec. 5.3.4

1.

A	B	X	Next state A	B
0	0	0	0	0
0	0	1	0	1
0	1	0	1	0
0	1	1	1	0
1	0	0	0	0
1	0	1	0	1
1	1	0	0	0
1	1	1	0	0

2.

X\AB	00	01	11	10
0	0	1	0	0
1	0	1	0	0

D_A

X\AB	00	01	11	10
0	0	0	0	0
1	1	0	0	1

D_B

● **Sec. 5.3.5**

1.

● **Sec. 6.1**

1. $D_A = B$ $D_B = \bar{A} + \bar{B}$

2. $J_A = B$ $K_A = \bar{B}$
    $J_B = 1$ $K_B = A$

● **Sec. 6.3**

1. 3 flip-flops.

● **Sec. 6.4**

1. Assign states b, c adj. and d, e adj. from Prin. 1. Assign f, h adj. and g, h adj., and b, c, d, e adj. from Prin. 2.

2. Same assignment as Problem 1.

3. State assignment is arbitrary.

● **Sec. 6.5**

1.

2.

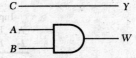

3.

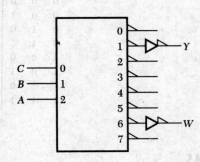

4.

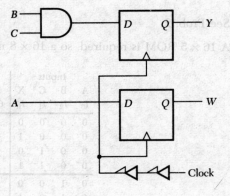

5.

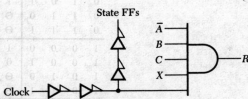

● **Sec. 7.1.1**

1. 111, 001.

2. Once state 001 is reached, the IFL is driven to cause a state change to either 101 or 011.

● **Sec. 8.2**

1.

Inputs			Outputs			
A	B	C		F3	F2	F1
I_2	I_1	I_0	O_4	O_3	O_2	O_1
0	0	0	Θ	1	0	0
0	0	1	Θ	0	1	0
0	1	0	Θ	0	1	0
0	1	1	Θ	0	0	1
1	0	0	Θ	0	1	0
1	0	1	Θ	0	0	1
1	1	0	Θ	0	0	1
1	1	1	Θ	1	0	0

● **Sec. 8.5**

1. See Prob. 2.

2. A 16×5 ROM is required, so a 16×8 might be used.

Inputs								
A	B	C	X	A	B	C	W	Y
I_3	I_2	I_1	I_0	O_5	O_4	O_3	O_2	O_1
0	0	0	0	1	0	0	0	0
0	0	0	1	0	1	0	0	0
0	0	1	0	0	0	0	0	1
0	0	1	1	0	0	0	0	1
0	1	0	0	0	0	1	0	0
0	1	0	1	1	1	0	0	0
0	1	1	0	Θ	Θ	Θ	Θ	Θ
0	1	1	1	Θ	Θ	Θ	Θ	Θ
1	0	0	0	1	1	0	0	0
1	0	0	1	0	0	1	0	0
1	0	1	0	Θ	Θ	Θ	Θ	Θ
1	0	1	1	Θ	Θ	Θ	Θ	Θ
1	1	0	0	0	0	0	1	0
1	1	0	1	0	0	0	1	0
1	1	1	0	Θ	Θ	Θ	Θ	Θ
1	1	1	1	Θ	Θ	Θ	Θ	Θ

● Sec. 9.2.1

1. 1's comp. = 00111110; 2's comp. = 00111111.

2. 1's comp. = 00101110; 2's comp. = 00101111.

3. 7360

4. 7361

● Sec. 9.2.2

1. 01111100_2

2. 01111100_2

● Sec. 9.2.3

1. 10010110_2

2. 10010110_2

● Sec. 9.3.1

1.

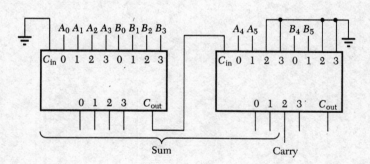

2.

● Sec. 9.3.4

1. $S_7 - S_0 = 11011100.$ $O_7 - O_0 = 00100011$, $G = 0$.
2. $S_7 - S_0 = 00100010.$ $O_7 - O_0 = 00100011$, $G = 1$.

● Sec. 9.5.2

1.
```
01111101
01000000
00011100
00011100
00011100
00011100
10011110
```

2. $MAR = T_1 + T_4(ADD + SUB + LOD + STO)$

● Sec. 9.5.3

1. 20 bits

Answers to Selected Problems

● Chapter I

1.2 Number of steps = 250.

1.4 a. $11101.1101_2 = 29.8125_{10}$
b. $10001.0001_2 = 17.0625_{10}$
c. $10.1101101_2 = 2.8515625_{10}$
d. $110010110101.01_2 = 3253.25_{10}$

1.7 $128.286_{10} = 1000000.0100100100_2$

1.10 $128.286_{10} = 10\,000\,000.010\,010\,010\,0_2 = 200.2220_8$
$= 010\,000\,000.010\,010\,010\,000$ (BCO)

1.14 $37.432_{10} = 25.6E8_{16}$

1.18 a. $748_{10} = 011101001000$ (BCD)
b. $5668_{10} = 0101011001101000$ (BCD)

1.23 The binary number is 110101. The binary bit (0 or 1) corresponding to a voltage of 1.7 V is not defined.

1.26

A	B	Out
0	0	0
0	1	0
1	0	1
1	1	0

1.29

A	B	C	D	X
0	0	0	0	0
0	0	0	1	0
0	0	1	0	0
0	0	1	1	0
0	1	0	0	0
0	1	0	1	0
0	1	1	0	0
0	1	1	1	0
1	0	0	0	0
1	0	0	1	0
1	0	1	0	0
1	0	1	1	0
1	1	0	0	0
1	1	0	1	1
1	1	1	0	1
1	1	1	1	1

1.34 Ten gates can be driven.

● **Chapter 2**

2.4 True

2.8 False

2.12 True

2.14 $X = \overline{\bar{A} + \bar{\bar{B}} + \bar{\bar{C}}}$

2.18 $F = \bar{A}BC\bar{D}\bar{E}$

2.22 $F = \overline{\overline{\bar{A}\bar{B}} + \overline{CD}} = \bar{A}\bar{B}CD$

2.25 $F = \bar{A}\bar{B}D + B\bar{C}D + ABC\bar{D}$

2.28 $F = ABC + \bar{A}B\bar{C}DE + \bar{A}BCD + AC\bar{D}\bar{E} + A\bar{B}\bar{C}DE$

2.32 $F = \bar{A}B + BC = B(\bar{A} + C)$

2.36 $F_R = AB + A\bar{C}$

2.39 $F_R = ABC + \bar{A}B\bar{C}DE + \bar{A}BCD + AC\bar{D}\bar{E} + A\bar{B}\bar{C}\bar{D}E$

2.43 $X = (A \oplus B) \oplus C = (A\bar{B} + \bar{A}B)\bar{C} + \overline{(A\bar{B} + \bar{A}B)}C = A\bar{B}\bar{C} + \bar{A}B\bar{C} + \bar{A}\bar{B}C + ABC$
No reduction.

2.46 $X = BC + \bar{A}C\bar{D} + A\bar{B}D$

● **Chapter 3**

3.1 Connect ABCD to select lines of a 16 : 1 MUX. Connect logic 1 to inputs 4, 6, 7, 8, 9, 12, 14, 15. Connect logic 0 to remaining input lines. Y is the output of the MUX.

3.4 Connect AB to the select lines of a 4 : 1 MUX. Input line 0 connects to logic 0, input line 2 connects to $\bar{C}$, and lines 1 and 3 connect to the output of a NAND gate driven by $\bar{C}$ and D.

3.8 Connect ABC to the input lines of a 3 line to 8 line decoder. The low-asserted output lines 0, 1, 2, and 4 should drive a low-asserted OR gate to form X.

3.12 Connect ABCD to decoder inputs with A as MSB. Follow decoder with 7 AND gates. These gates are equivalent to low asserted input OR gates with a low asserted output. Connect as follows:

Decoder outputs	to	Segment gate
0,2,3,5,7,8,9,13		a
0,1,2,3,4,7,8,9,12		b
0,1,3,4,5,6,7,8,9,11		c
0,2,3,5,6,8,10,11,13,14		d
0,2,6,8,10,14		e
0,4,5,6,8,9,12,13,14		f
2,3,4,5,6,8,9,10,11,12,13,14		g

3.15 Connect chips as shown in Fig. 3.29, but tie lines $\overline{10}$, $\overline{11}$, $\overline{12}$, $\overline{13}$, $\overline{14}$, and $\overline{15}$ high.

● **Chapter 4**

4.1

A	B	X
0	0	$\overline{Q1}_0$
0	1	1
1	0	0
1	1	Avoid

4.3 For TTL, the circuit would function if R_S were deleted, but would be more susceptible to noise.

4.7

	t_0	t_1	t_2	t_3
D	0	1	0	0
G	0	0	0	1
Q	1	1	1	0

4.11

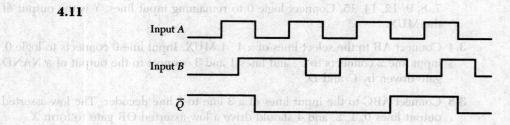

Input A

Input B

$\overline{Q}$

4.16

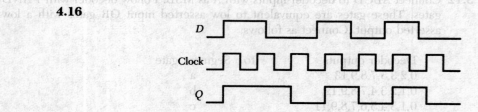

D

Clock

Q

4.20

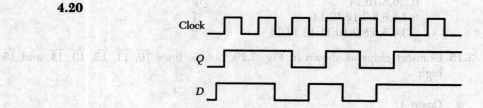

Clock

Q

D

4.25

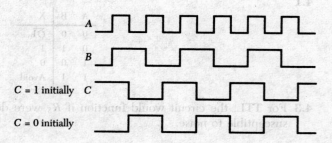

A

B

C = 1 initially   C

C = 0 initially

4.29 Connect as follows:

Pin	Connection
1	Out. En.
2	O1
3	D1
4	D2
5	O2
6	O3
7	D3
8	D4
9	O4
10	Gnd.
11	Data Strobe
12	O5
13	D5
14	D6
15	O6
16	O7
17	D7
18	D8
19	O8
20	+5 V

● **Chapter 5**

5.1 A synchronous state machine is a state machine driven by input signals that are referenced to the clock and having clock-driven state changes.

5.5

	t_1	t_2	t_3	t_4	t_5
Q1	-	0	1	1	0
Q2	-	1	0	1	1
D1	0	1	1	0	
D2	1	0	1	1	

5.9 Must first determine the signals required to drive D1 and D2.

	t_1	t_2	t_3	t_4	t_5
Q1	-	0	1	1	0
Q2	-	1	0	1	1
D1	0	1	1	0	
D2	1	0	1	1	

I1	I2	D1	D2
0	0	0	1
0	1	1	0
1	0	1	1
1	1	0	1

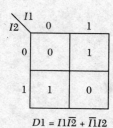

$D1 = I1\overline{I2} + \overline{I1}I2$

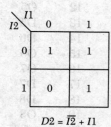

$D2 = \overline{I2} + I1$

5.12

	Prior to clock transition				After clock transition	
Clock	J1	K1	J2	K2	Q1	Q2
1	0	0	1	1	0	1
2	0	1	0	0	0	0
3	1	0	1	0	1	1
4	0	0	1	1	1	0
5	0	1	0	0	0	0
6	0	0	1	1	0	1
7	1	0	1	0	1	1

This is not a state machine since there is no feedback from output to input.

5.18

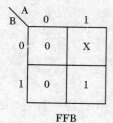

FFA   FFB

5.20

Prior to clock transition				After clock transition		
Clock	J_A	K_A	J_B	K_B	A	B	R
1	1	0	0	0	1	0	0
2	1	1	0	1	0	0	0
3	1	0	0	0	1	0	0
4	1	0	1	1	1	1	1
5	0	1	0	1	0	0	0
6	1	0	0	0	1	0	0

● **Chapter 6**

6.1

A	B	C	D_A	D_B	D_C
0	0	0	0	0	1
0	0	1	0	1	0
0	1	0	1	0	0
0	1	1	1	1	0
1	0	0	0	1	1
1	0	1	0	0	0
1	1	0	0	0	0
1	1	1	0	0	0

C \ AB	00	01	11	10
0	0	1	0	0
1	0	1	0	0

$$D_A = \bar{A}B$$

C \ AB	00	01	11	10
0	0	0	0	1
1	1	1	0	0

$$D_B = \bar{A}C + A\bar{B}\bar{C}$$

C \ AB	00	01	11	10
0	1	0	0	1
1	0	0	0	0

$$D_C = \bar{B}\bar{C}$$

6.4 From the next state maps of Prob. 7.3, the expressions for the flip-flop drive signals are found.

C \ AB	00	01	11	10
0	1	1	1	1
1	1	0	1	0

$$D_A = \bar{C} + \bar{A}\bar{B} + AB$$

C \ AB	00	01	11	10
0	1	1	0	1
1	1	0	0	1

$$D_B = \bar{B} + \bar{A}\bar{C}$$

C \ AB	00	01	11	10
0	1	1	0	1
1	0	1	1	1

$$D_C = \bar{A}B + \bar{B}\bar{C} + AC$$

The D flip-flops with proper IFL replace the JK flip-flops of Prob. 7.3.

6.10 There are 10 states, so $2^4 = 16$ gives $n = 3$ flip-flops.

6.13 There are several possibilities chosen by trial and error. The following is one.

C \ AB	00	01	11	10
0	a^S	Θ	c^R	e
1	Θ	b	Θ	d^Q

No transient crossing of states a, c, or d (output states).

6.17 Three 8 : 1 MUXs are used. Flip-flop outputs ABC connect to the select lines of all MUXs. The MUX that drives D_A has inputs 0 and 5 connected to ground, input 1 to $\bar{X}$, inputs 2, 6, and 7 to logic 1, and input 3 to X. The MUX that drives D_B has input 0 connected to X, inputs 1 and 3 connected to logic 1, inputs 2, 4, 5, 6, and 7 connected to ground. The MUX that drives D_C has inputs 0, 2, and 3 connected to X, inputs 1, 4, 5, and 6 connected to ground, and input 7 connected to $\bar{X}$.

6.24 Both outputs R and W are conditional. Clock suppression may be used to create these outputs and also eliminate transient glitches. The state assignment shown minimizes the OFL by placing output states near unused or "don't care" states.

Present state map

C \ AB	00	01	11	10
0	a	e	d	Θ
1	b	c	Θ	Θ

● **Chapter 7**

7.2 X is an asynchronous input, but Y is not. IFL can be set up to drive to a next state of 1101 then switch to 0111. It can also be set up to drive to 1110 then switch to this same value—no problems with this condition.

    1101 to 0111 has two possible paths: 1101 to 1111 to 0111 or 1101 to 0101 to 0111. The possible erroneous states are 0101 and 1111.

7.6 a. The only possible problem occurs as the system switches from state *a*. The two possible erroneous codes are 0010 and 0001. If the system reaches state *b*, it will move on to the next state before X is deasserted.

    b. If state *a* is changed to a code of 0010, erroneous state codes are avoided.

7.8 We note that the asynchronous inputs appear in the go-no go configuration. Thus, states a, b and b, c should be adjacent. To minimize IFL, states a, d should be adjacent (Principle 1). States b, c should be adjacent to minimize OFL. The following assignment is appropriate.

State assignment

B \ A	0	1
0	a	d
1	b	c

Chapter 8

8.1

Address				Contents			
0	0	0	0	0	0	0	1
0	0	0	1	0	0	1	0
0	0	1	0	0	0	1	1
0	0	1	1	0	1	0	0
0	1	0	0	0	1	0	1
0	1	0	1	0	1	1	0
0	1	1	0	0	1	1	1
0	1	1	1	1	0	0	0
1	0	0	0	1	0	0	1
1	0	0	1	1	0	1	0
1	0	1	0	1	0	1	1
1	0	1	1	1	1	0	0
1	1	0	0	1	1	0	1
1	1	0	1	1	1	1	0
1	1	1	0	1	1	1	1
1	1	1	1	0	0	0	0

8.4 A 6-bit code can have 64 unique combinations, thus 64 locations will be required.
  a. There will be 64 unused locations.
  b. Only six output lines are needed leaving 2 unused lines.
  c. Total bits in memory = $128 \times 8 = 1024$. The number required for 64 different 6-bit codes is $64 \times 6 = 384$. The total number of bits not requiring programming is $1024 - 384 = 640$.

8.12 With 5 states, 1 input, and 1 output, a 4 input, 4 output ROM (16×4) is needed. Clock suppression is used on the output as shown in Solution 6.20. Using the state assignment of Solution 6.20, the following ROM contents are necessary.

A I_3	B I_2	C I_1	Y I_0	D_A O_4	D_B O_3	D_C O_2	W O_1
0	0	0	0	0	1	0	1
0	0	0	1	0	1	0	1
0	0	1	0	1	1	0	1
0	0	1	1	1	1	0	1
0	1	0	0	0	1	1	0
0	1	0	1	0	0	1	0
0	1	1	0	1	1	0	0
0	1	1	1	1	1	0	0
1	1	0	0	0	0	0	0
1	1	0	1	1	1	0	0

● **Chapter 9**

9.1 $\bar{N}_4(10) = 10^4 - 2436 = 7564$

9.6 A = 10010010   B = 00110010

  1. Complement $B = 11001110$

  2. Add $A + B = 101100000$

  3. Two operands have different sign than result—out of range.

9.11 If result from first 3 bits is equal to or greater than 8, then $C_3 = 1$. By considering all combinations of $A_2A_1A_0$ and $B_2B_1B_0$, a K-map (6 variable) is plotted to give

$$C_3 = A_2B_2 + A_2A_1B_1 + A_1B_2B_1 + B_2B_1B_0A_0 + A_2A_1A_0B_0$$

$$+ B_1B_0A_2A_0 + B_2B_0A_1A_0$$

9.16 $S_7 - S_0 = 10101100, \, OF = 0.$

9.20 This memory has 2^{12} or 4096 locations that can be accessed. Each location has 8 bits or one byte. This is a 4 kbyte memory.

9.29 Twelve bits are required for the address part of the instruction word, leaving 4 bits for the op code. This leads to 16 instructions for the simple system.

9.31 Eight time slots are required.

AVG	
T_1	MAR ← PC
T_2	MBR ← M(MAR)
T_3	IR ← MBR
T_4	MAR ← ad(IR)
T_5	MBR ← M(MAR)
T_6	A ← A + MBR
T_7	A ← SR(A)
T_8	PC ← PC + 1

● **Chapter 10**

10.1 Noncritical races

Starting state	Ending state
10/00	11/10
10/00	00/01

Oscillation: Start in 00/01, oscillates between 01/01 and 01/00.

Critical race: Start in 00/01 can end in 01/00 or 01/11.

10.6 Change the entry in cell 11/01 to 11, change the entry under 00/10 to 11, and change the entry under 10/11 to 01. The entry under 10/10 can be changed to $\Theta\Theta$ for additional minimization.

The resulting map is

xy \ AB	00	01	11	10
00	00	00	11	01
01	00	$\Theta\Theta$	11	01
11	11	10	11	01
10	11	10	11	$\Theta\Theta$

Index